Benchmark Papers in Ecology

Series Editor: Frank B. Golley
University of Georgia

Related Titles in BENCHMARK PAPERS IN BEHAVIOR Series

 **Benchmark Papers
in Ecology / 9**

A BENCHMARK® Books Series

SYSTEMS ECOLOGY

Edited by

H. H. SHUGART

**Environmental Sciences Division
Oak Ridge National Laboratory**

and

R. V. O'NEILL

**Environmental Sciences Division
Oak Ridge National Laboratory**

 **Dowden, Hutchinson
& Ross, Inc.**

STROUDSBURG, PENNSYLVANIA

LIBRARY OF CONGRESS CATALOGING IN PUBLICATION DATA
Main entry under title:
Systems ecology.
    (Benchmark papers in ecology; v. 9)
    Includes indexes.
    1. Ecology—Mathematical models—Addresses, essays,
lectures.   I. Shugart, H. H.   II. O'Neill, Robert V., 1940–
QH541.15.M3S92   574.5'01'84   79-970
ISBN: 0-87933-347-2

Distributed world wide by Academic Press,
a subsidiary of Harcourt Brace Jovanovich,
Publishers.

SERIES EDITOR'S FOREWORD

Ecology—the study of interactions and relationships between living systems and environment—is an extremely active and dynamic field of science. The great variety of possible interactions in even the most simple ecological system makes the study of ecology compelling but difficult to discuss in simple terms. Further, living systems include individual organisms, populations, communities, and ultimately the entire biosphere; there are thus numerous subspecialties in ecology. Some ecologists are interested in wildlife and natural history, others are intrigued by the complexity and apparently intractable problems of ecological systems, and still others apply ecological principles to the problems of man and the environment. This means that a Benchmark Series in Ecology could be subdivided into innumerable volumes that represent these diverse interests. However, rather than take this approach, I have tried to focus on general patterns or concepts that are applicable to two particularly important levels of ecological understanding: the population and the community. I have taken the dichotomy between these two as the major organizing concept in the series.

In a field that is rapidly changing and evolving, it is often difficult to chart the transition of single ideas into cohesive theories and principles. In addition, it is not easy to make judgments as to the benchmarks of the subject when the theoretical features of a field are relatively young. These twin problems—the relationship between interweaving ideas and the elucidation of theory, and the youth of the subject itself—make development of a Benchmark series in the field of ecology difficult. Each of the volume editors has recognized this inherent problem, and each has acted to solve it in his or her unique way. Their collective efforts will, we anticipate, provide a survey of the most important concepts in the field.

The Benchmark series is especially designed for libraries of colleges, universities, and research organizations that cannot purchase the older literature of ecology because of cost, lack of staff to select from the hundreds of thousands of journals and volumes, or the unavailability of the reference materials. For example, in developing countries where a science library must be developed *de novo*, I have seen where the Benchmark series can provide the only background literature available to the students and staff. Thus, the intent of the series is to provide an authoritative selection of literature, which can be read in the original form, but that is cast in a matrix of thought provided by the editor. The volumes are designed to explore the historical development of a concept in ecology and point the way toward new developments, without being a historical study. We hope that even though the Benchmark Series in Ecology is a library-oriented series and bears an appropriate cost it will also be of sufficient utility so that many professionals will place it in their personal libraries. In a few cases the volumes have even been used as textbooks for advanced courses. Thus we expect that the Benchmark Series in Ecology will be useful not only to the student who seeks an authoritative selection of original literature but also to the professional who wants to quickly and efficiently expand his or her background in an area of ecology outside his or her special competence.

H. H. Shugart and R. V. O'Neill have developed in *Systems Ecology* a survey of one of the most rapidly developing and newest subdivisions of ecology. The concept of natural systems is old, of course, but the application of mathematics, cybernetic theory, and information theory to ecology is very new. The subject is important yet so recent that the authors must have struggled long for perspective. Fortunately, both authors have deep experience in the subject and both have contributed significantly to it. Shugart and O'Neill are staff members and colleagues at the ecology program of Oak Ridge National Laboratory, Oak Ridge, Tennessee. They were especially active in the modeling work of the United States International Biological Program, and contributed heavily to the development of systems studies at all levels of biological organization in that international effort. Both editors are active contributors to ecological literature.

FRANK B. GOLLEY

PREFACE

Systems ecology is a new and exciting subdiscipline of ecology. The field is characterized by its application of mathematical models to ecosystem dynamics. Although mathematical analysis has long been utilized in ecology, particularly population theory, the first papers in systems ecology appeared in the early 1960s. The field has seen phenomenal growth since then, and literally hundreds of models have been produced. An important factor in this growth was the International Biological Programme, particularly the U.S. Biome Programs. These programs provided substantial support and motivation for modeling activities.

The youth and rapid expansion of systems ecology makes the identification of benchmark papers problematic. It is still difficult to perceive which studies will have the greatest influence on later research. As a result, the selection of papers for this volume has been strongly influenced by the opinions and perceptions of the editors. Readers may well find themselves in disagreement with some of our choices. We have selected early papers that had a significant influence and more recent papers that represent innovative approaches. We have also included several papers that contain information or points of view that are important for understanding systems ecology.

One of the characteristics of systems ecology is the difficulty of communicating results in the traditional journal format. In some cases the models were developed as part of actively evolving IBP programs. The need to make the model available to researchers in the program led to publication in internal reports that are often difficult for nonparticipants to obtain. In other cases, the magnitude of the model made it impossible to condense into a single journal article. The most conspicuous example is the ELM model produced by G. S. Innis and his colleagues at Colorado State University as a part of the Grassland Biome, IBP project. This model is an important development in the field of systems ecology, yet it was not possible to include it in the present volume. The best access to the model is in book form: G. S. Innis, ed., *Grassland Simulation Model* (Springer-Verlag, N.Y., 1978).

For the reader who is interested in further explorations of the systems ecology literature, particularly models available only as internal reports, we recommend R. V. O'Neill, N. Ferguson, and J. A. Watts, eds., *A Bibli-*

ography of *Mathematical Modeling in Ecology*, EDFB/IBP-75/5 (U.S. Government Printing Office, Washington, D.C., 1977). This report contains over 900 references to the modeling literature.

H. H. SHUGART
R. V. O'NEILL

CONTENTS

Contents

CONTENTS BY AUTHOR

INTRODUCTION

Systems ecology is a robust hybrid of engineering, mathematics, operations research, cybernetics, and ecology. Depending on one's point of view, it has either borrowed from or given much to its sister subdisciplines: mathematical ecology and theoretical ecology. In this book we have collected papers that provide a sense of the history of systems ecology, that are important state-of-the-art contributions, or that represent directions the field may take in the future. Taken as a whole, these papers represent the content of systems ecology. Effort has been given to describe why we have chosen the various papers and to explain how the papers fit into the mainstream of the field. Our purpose in this Introduction is to convey something of the *gestalt* of systems ecology.

A decade ago, one of us and a fellow graduate student, George I. Child, developed a computer model of magnesium cycling in a Panamanian tropical moist forest. This model took almost a year to develop. Most of the time was spent debating the proper approach to use in applying systems ecology to the forest. George Child had spent two years in Panama collecting data on the standing crops and transport of several elements, including magnesium, in six different tropical ecosystems. The field work was intense; it included total harvesting of tropical trees, and the data were collected under the most primitive conditions. After two years of living in the wilds of Panama, George spoke Choco Indian dialect and Spanish fluently, had hiked across the Isthmus of Panama through the backcountry, had become an expert on the natural history of the area, and was dedicated to understanding tropical ecosystems. We argued for days.

The model had to fit on a small analog computer. We were limited to four to eight ordinary linear differential equations. We ended up with six. We argued whether one could express anything of interest about a tropical forest ecosystem using such a small number of equations. Since we worked with an analog computing device, analogies between voltage (the commodity of the analog computer), and magnesium rose and fell in importance. We were amazed to find that data that had taken George, Mike Duever, and about sixty Choco Indians almost six months to gather was represented as a single parameter in the model. On the other hand, analysis of the model revealed aspects of system stability that would not have been discovered otherwise (Child and Shugart 1972). We discussed how these model results might actually reflect the true nature of the Panamanian forest.

This anecdote points out two important characteristics of the systems ecologist's experience. First, there is the frustration in forcing the complexity of a real system into a few equations and parameters. As ecologists, we have been trained to deal with complexity and to appreciate the exceptions at least as much as we appreciate the generalities. Second, there is the excitement of hypothesizing new underlying behavior for the total ecosystem as a result of analyzing the model. For most, this excitement more than compensates for the initial frustration.

While the anecdotal account is useful for providing an impression of the working experience, it is also worth noting some of the unique attributes of systems ecology. Studies that make up the body of the field tend to differ from other ecological studies in the following attributes:

1. Consideration of ecological phenomena at large spatial, temporal, or organizational scales
2. Introduction of methodologies from other fields that are traditionally unallied with ecology
3. An emphasis on mathematical models
4. An orientation to computers (both digital and analog devices)
5. A willingness to develop hypotheses about the nature of ecosystems

We will elaborate on each of these five attributes.

Scales

Problems with temporal and spatial scale often plague data collection efforts. Frequently, the system changes more rapidly than one can measure it (e.g., insect populations). In other cases, changes occur so slowly that sampling intervals exceed the investigator's lifespan (e.g., tree populations, particularly in mature forests).

Systems ecologists frequently combine phenomena at glaringly different temporal or spatial scales. In some cases, the process is relatively simple. For example, if one is interested in the dynamics of forest canopy insects during the summer, it may be reasonable to assume that tree biomass is constant. Much of the "art" in modeling concerns which ecosystem attributes of an ecosystem must be considered explicitly and which attributes can be ignored. There is as much creativity in deciding what can be left out of a model as in designing the appropriate mathematical functions to mimic some ecological response. Needless to say, there is also ample opportunity for leaving out important phenomena that will embarrass the systems ecologist later.

Ecologists have only recently begun to work actively on the problem of scale in ecological models. Criteria for adding or leaving out a specific process has been considered part of the "tools of the trade" of the systems ecologist. Bits of information on scale have been passed among modelers largely as informal communications, in discussions at workshops, and, in the case of mistakes, as gossip. Unfortunately, there is no codification of what has been learned to date. Papers that treat the problem of modeling differing scales are Clymer (1969), Clymer and Bledsoe (1970), Zeigler and Weinberg (1970), Overton, (1972), and Overton et al. (1973). Allen and Koonce (1973) have used statistical transformations to determine scales in time and space for which various phenomena are important.

Use of Work From Other Sciences

Although most ecologists are willing to use proven techniques from other fields, systems ecologists are unique both in the breadth of other disciplines that they consider and in the fervor with which they search for these techniques. This exploration of other fields has been successful in the past and has become the *modus operandi* of many systems ecologists. Over the past decade, engineers, physicists, and mathematicians have colonized ecology, in part because the explorations of early systems ecologists created an interest in techniques from these disciplines. For example, the group with which we work (the Oak Ridge Systems Ecology Group) is presently composed of individuals with graduate training in astronomy, plasma physics, electrical engineering, zoology, botany, hydrology, and statistics. There is an excitement that is (or at least can be) attendant to crossing into other scientific fields for insights into ecosystem function.

Other sciences share characteristics with ecology that suggest

3

their methodologies might be transferable. For example, like ecologists, astronomers are forced to speculate about their system (the universe) using limited observational data and without recourse to experimentation. Therefore, techniques developed by astonomers, such as spectral analysis using the maximum entropy technique, may be applicable in ecology. But whether or not maximum entropy spectral analysis proves to be applicable, the point is that astronomers do not, on a regular basis, go to great lengths to assure that ecologists become aware of their findings. Systems ecologists have had to do the searching. Many systems ecologists feel this is their greatest contribution to ecology and take delight in the intellectual excitement produced in the melding of two otherwise disjunct aspects of science.

Models

If one were to pick the single most reliable fieldmark of the systems ecologist, it would be the use of mathematical models as a tool. However, there have been raging debates among systems ecologists on criteria for choosing types of models. These debates have, in several cases, amounted to divergent views on the fundamental nature of ecosystems. For example, the debate on whether ecosystems were linear or nonlinear has also identified possible dichotomies in management strategies. One might manage nonlinear systems in ways very different from linear systems. It is not clear at present that this debate is heading toward any resolution, nor is such resolution necessarily desirable. Diversity of opinion on appropriate mathematical formulations is healthy and should continue to be so in the future. The papers we have chosen in this book reflect part of the spectrum of the multifaceted argument, "How can one best model an ecosystem?"

Computers

Quantitative ecology was certainly an active field before the advent of the computer. However, the increased availability and reduction in cost (in the sense of cost per arithmetic operation) of computing devices has catalyzed the development of systems ecology over the past twenty years. In the 1950s, systems ecologists (notably H. T. Odum) actually built their own analog computing devices.

For a time, it was not clear whether analog or digital computers were more appropriate for ecological models. Analog computers had several advantages over digital computers until the mid-1960s. They were fast and relatively inexpensive. They allowed the ecologist to work "hands on" with a device that modeled flows directly (using electrical flow as an analog for energy or nutrient flow). Today, digital computers dominate the field because of their availability and the ease of programming. The ability to program an analog computer can be taken as the mark of an "old-timer" in the field.

Digital computers are readily available to ecologists at all major universities in the United States and Europe. Languages (such as IBM's CSMP–III Language) are being developed that greatly simplify the implementation of models and require little knowledge of numerical analysis. These languages promise to make simulation models available to students and working ecologists in the same way that "canned" statistical computer packages (e.g., SAS, SPSS, BMD, and so forth) have made elementary statistical analysis available in the past decade.

Hypothesis Development

Systems ecologists often break from "safe" descriptions of ecological systems to form interesting (and testable) *hypotheses* about system functions. We emphasize "interesting" because the hypothesis most frequently tested in ecological research is the uninteresting statistical null hypothesis: "the data indicate that there is no difference between these things." The hypotheses formulated by systems ecologists result from the juxtaposition of the assumptions and intuitions contained in a model. In fact, the ecosystem model itself can be viewed as a complex hypothesis about system functions.

Systems ecologists are anxious to formulate hypotheses that may someday lead to a theory of ecosystems. Probably most hypotheses developed today will be rejected. It is not the reward to success that stimulates hypothesis generation, it is the zeal for building a theoretical science that is the real motivation. Systems ecologists tend to feel that it is better to offer an hypothesis even though it may be rejected, than not to advance it in the first place. This seems an essential characteristic of systems ecology as it is of any young theoretical field.

In this Introduction, we have sought to provide a feeling for the spirit of a young field in the young science of ecology. This spirit is important to bear in mind while reading the papers that follow. In many respects, systems ecology is developing a theoretical tone to match the historical emphasis on accurate descriptions of ecological phenomena. If ecology develops in a manner analogous to older sciences such as physics, astronomy, and chemistry, it is possible that the spirit of systems ecology may be its most contagious aspect, with the potential to change ecological thought.

REFERENCES

Allen, T. F. H., and J. F. Koonce. 1973. Multivariate approaches to algae strategems and tactics in systems analysis of phytoplankton. *Ecology* **54**:1234–46.

Child, G. I., and H. H. Shugart. 1972. Frequency response analysis of magnesium cycling in a tropical forest ecosystem, pp. 103–35. In Patten, B. C. (ed.), *Systems Analysis and Simulation in Ecology*, vol. 2. Academic Press, New York, 592 pp.

Clymer, A. B. 1969. The modeling of hierarchical systems. Keynote Address, Conf. on Applications of Continuous System Simulation Languages, San Francisco.

Clymer, A. B., and L. J. Bledsoe. 1970. A guide to the mathematical modeling of an ecosystem, pp. 175–99. In Wright, R. G., and G. M. Van Dyne (eds.), *Simulation and Analysis of Dynamcis of a Semi-desert Grassland*, Range Sci. Ser. No. 6. Range Sci. Dep., Colorado State Univ., Fort Collins, 297 pp.

Overton, W. S. 1972. Toward a general model structure for a forest ecosystem, pp. 37–47. In Franklin, J. F., L. J. Dempster, and R. H. Waring (eds.), *Research on Coniferous Forest Ecosystems*, Proc. Symp. Coniferous Forest Ecosystems, Bellingham, Wash., Pacific Northwest Forest and Range Exper. Sta., U.S. Forest Service, Portland, Ore., 322 pp.

Overton, W. S., J. A. Colby, J. Gourley, and C. White. 1973. *FLEX 1 Users Manual*, Internal Rep. 126. Coniferous Forest Biome, Oregon State Univ., Corvallis.

Zeigler, B. P., and R. Weinberg. 1970. System theoretic analysis of models: Computer simulation of a living cell. *J. Theor. Biol.* **29**:35–56.

Part I

HISTORICAL ROOTS OF
SYSTEMS ECOLOGY

An excellent overall perspective on a field results from examining the historical roots from which it developed. This approach permits one to evaluate the predispositions, assumptions and aspirations that characterized the field at its inception. Frequently, these determine directions that are followed for many years.

Mathematical modeling of ecological phenomena is almost as old as the field of ecology itself. Early efforts focused on models for population growth (Gompertz 1825; Lotka 1931), predator-prey or competitive interactions (Volterra 1926, 1931; Lotka 1932) and other population phenomena (Volterra 1937). As pointed out by Caswell et al. (1972), compartment models have been used in related environmental fields, such as atmospheric science, since at least 1935 (Koztitzin 1935). Thus, ecologists were familiar with models long before the field of systems ecology developed.

In addition to these early applications of mathematical modeling, there was a general trend throughout this century away from "natural history" toward more quantitative approaches. Well before the development of systems ecology, statistics made a major impact on the ecological sciences. This trend toward "quantitative ecology" also predisposed ecologists to the introduction of the more complex analytical tools that characterize systems ecology.

Over the same period that digital computers were making exploration of large-scale models feasible, interest was developing in the study of total ecosystems. While the ecosystem concept had been pioneered much earlier, practical approaches to the study of whole systems only came with the introduction of the energy flow and trophic dynamic concepts (Lindeman 1942). The late 1950s saw the formulation of theoretical (Odum and Pinkerton 1955) and experimental (Odum 1957) methodologies. Thus, at the same time that adequate computational equipment was becoming available, a new research area was opening that encouraged the development of large-scale models.

The final ingredient in the genesis of the field was the rapid development of systems analysis during and following World War II. The general concept of a system was not new, but the techniques and approaches developed in systems analysis, operations research, and cybernetics only became available in the 1950s. Mathematical models and analysis applied to the control of large-scale systems showed a tremendous promise that strongly influenced many fields. The "systems approach" became identified with a new, innovative, and comprehensive viewpoint that was successful in addressing a broad range of problems.

All of these factors set the stage for systems ecology. The earliest report dealing specifically with this new field appears to be a brief mention by Olson in 1959. This was followed by pioneering studies of computer models of ecosystems (Olson 1960; Garfinkel 1962; Olson 1963). The success of these early studies encouraged a series of publications in the mid-1960s outlining the importance and potential of ecosystem modeling (Davidson and Clymer 1966; Holling 1966; Watt 1966). Thus, the parturition of systems ecology was firmly established by the end of the 1960s.

We have organized the presentation of the early formative papers in systems ecology by dividing the papers in Part I into four sections. The division illustrates defacto that systems ecology developed from a number of parallel and converging roots. The categories focus on four of the major emphases found in the pioneering papers and illustrate the differences in background of the researchers who formed the new field.

Physiological studies led to models for the kinetics of substances (e.g., calcium, isotopes, dyes) passing through the vertebrate body. These models were easily transferable to the movement of substances, particularly radioactive substances, between compartments of the ecosystem. Specific interest in the movement of fallout radionuclides in the environment, therefore, led to the development of some of the earliest ecosystem models.

The long development of modeling in *population theory* made this area particularly amenable to early applications of the new approach. Thus, a series of early papers discuss the potential of this new approach by the relatively simple juxtaposition of earlier population models with the capabilities of the digital computer.

Along a slightly different line, ecologists interested in total ecosystems saw the new *"systems approach"* mainly as a comprehensive and interdisciplinary methodology for problem solving. These papers show the first signs of a trend to incorporate models as an integral part of interdisciplinary research programs. In addition, the model is discussed as a means for integrating research and applying the results to environmental management.

The rapid expansion of *systems engineering* as a powerful new tool that was appplicable to a broad range of fields, made it natural for these techniques to find application in the developing field of systems ecology. Therefore, several papers (albeit slightly later in time) emphasize the transferal of mathematical and analytical tools developed in engineering to ecological problems.

REFERENCES

Caswell, H. H., H. E. Koenig, J. A. Resh, and Q. E. Ross. 1972. An introduction to systems science for ecologists, pp. 4–78. In Patten, B. C. (ed.), *Systems Analysis and Simulation in Ecology*, vol. 2. Academic Press, New York, 592 pp.

Davidson, R. S., and A. B. Clymer. 1966. Desirability and applicability of simulating ecosystems. *Ann. N.Y. Acad. Sci.* **128**:790–94.

Garfinkel, D. 1962. Digital computer simulation of ecological systems. *Nat.* **194**:856–57.

Gompertz, B. 1825. On the nature of the function expressive of the law of human mortality. *Phil. Trans.* **115**:513–85.

Holling, C. S. 1966. The strategy of building models of complex ecological systems, pp. 195–214. In Watt, K. E. F. (ed.), *Systems Analysis in Ecology*. Academic Press, New York, 276 pp.

Kostitzin, V. A. 1935. *Evolution de l'Atmosphere*. Herman, Paris.

Lindeman, R. L. 1942. The trophic dynamic aspect of ecology. *Ecology* **23**:399–418.

Lotka, A. J. 1931. The structure of a growing population. *Human Biol.* **3**:459–93.

Lotka, A. J. 1932. The growth of mixed populations, two species competing for a common food supply. *J. Wash. Acad. Sci.* **22**:461–69.

Odum, H. T. 1957. Trophic structure and productivity of Silver Spring, Florida. *Ecol. Monogr.* **27**:55–112.

Odum, H. T., and R. C. Pinkerton. 1955. Times speed regulator, the optimum efficiency for maximum output in physical and biological systems. *Am. Sci.* **43**:331–43.

Olson, J. S. 1963. Analog computer models for movement of isotopes through ecosystems. In Schultz, V., and A. W. Klements (eds.), *Proc. First Nat. Symp. Radioecol.*, Colorado State Univ. Published by Reinhold, New York and American Institute for the Biological Sciences, Washington, D.C.

Volterrra, V. 1926. Variazioni e fluttuazioni del numero d'individui in specie animali conviventi. *Mem Acad. Lincei.* **2**:31–113. (Translation in an appendix to Chapman, 1931, *Animal Ecology*. McGraw-Hill, New York.)

Volterra, V. 1931. *Lecons sur la theorie mathematique de la lutte pour la vie.* Paris, Gauthier-Villars.

Volterra, V. 1937. Principes de biologie mathematique. *Acta Biotheor.* **3**: 1–36.

Watt, K. E. F. 1966. The nature of systems analysis, pp. 1–14. In Watt, K. E. F. (ed.), *Systems Analysis in Ecology*. Academic Press, New York, 276 pp.

Editors' Comments
on Papers 1, 2, and 3

1   OLSON
Analog Computer Models for Movement of Nuclides Through Ecosystems

2   PATTEN and WITKAMP
Systems Analysis of ^{134}Cesium Kinetics in Terrestrial Microcosms

3   KAYE and BALL
Systems Analysis of a Coupled Compartment Model for Radionuclide Transfer in a Tropical Environment

DEVELOPMENTS IN PHYSIOLOGICAL MODELING AND RADIOISOTOPE THEORY

Physiological researchers in the 1930s became interested in the dynamics of various substances moving through the vertebrate body. The interest in kinetics led to experiments in which tracers (originally dyes) were injected and tracked as they moved through the body. For example, the time required for a dye, injected into an artery, to reappear in an adjacent vein yielded information on the rate of blood circulation. The dilution of the dye, when it returned, provided an estimate for the total volume of blood. This field of investigation was greatly stimulated by the availability of radioisotope tracers in the late 1940s.

Interest in the analysis of tracer experiments stimulated the development of mathematical theory (Teorell 1937; Zilversmit et al. 1943) to explain the dynamics. The theory was called tracer kinetics or compartment modeling (Hearon 1953; Solomon 1949; Sheppard and Householder 1951).

In tracer theory, the body is divided into a set of mutually exclusive compartments (e.g., blood, tissue spaces, bone, and so forth) that have distinctive turnover rates. Because the major interest was in kinetics, the rate of change in each compartment, x_i, was expressed by a differential equation. By applying the principle of conservation of mass, the rate of change in the compartment was expressed as the sum of the input rates minus the sum of the output rates:

$$\frac{dx_j}{dt} = \sum_j F_{ji} - \sum_i F_{ij},$$ (1)

where F_{ji} signifies the rate of transfer of material from some other compartment, x_i, to the compartment of interest, x_j.

Each of the flux rates, F_{ji}, was expressed as a constant fraction, a_{ji}, of the material in a "donor" compartment, x_i, which was transferred to the "recipient" compartment, x_j, over a unit of time:

$$F_{ji} = a_{ji} x_i.$$ (2)

The resulting model could then follow the movement of a tracer from some initial condition and predict the quantity to be found in each compartment at any point in time.

To simplify notation, the basic model was often written in matrix form. Since there is one equation (of the form of Eq. 1) for each compartment in the system, it was convenient to use the notation $x = (x_1, x_2, \ldots, x_n)$. We then designate a matrix A that contains the value a_{ji} in the j'th column and i'th row. The entire model can then be written in the simplified form:

$$\dot{x} = Ax.$$

The analogy between this approach and ecological models of energy or nutrient flow through an ecosystem is obvious. Ecological compartments (e.g., primary producers, herbivores, predators, decomposers, and so forth) can be specified, and changes in these compartments through time can be described by identical equations. Therefore, the transition from tracer kinetics models to linear, donor-controlled ecosystem models can be effected quite simply.

But the obvious application of these kinetics models might not have occurred without the stimulus of radioecologists. The availability of radioisotopes as research tools for ecological research and the incentive to study the movement of fallout radionuclides through the environment opened new areas for ecological research. The juxtaposition of the new research interest in radioactive tracers and the mathematical theory developed for tracer experiments made it natural for the linear "compartment" model to be introduced into ecology.

It is also understandable that the earliest developments in linear ecosystem models occurred at the Atomic Energy Commission National Laboratory at Oak Ridge (Olson 1959, 1960). Here the stimulus to develop ecological applications of radiotracers and the interest and expertise in tracer theory combined to initiate a major thrust in systems ecology.

All three of the papers that follow were generated by the ecological research group at the Oak Ridge National Laboratory. They illustrate the early efforts to develop and apply models. Paper 1 by Olson and Paper 2 by Patten and Witkamp both utilize the analog computer for implementing the models. Paper 3 by Kaye and Ball discusses the use of the digital computer and stresses the use of modeling to solve real environmental problems.

REFERENCES

Hearon, J. Z. 1953. The kinetics of linear systems with special reference to periodic reactions. *Bull. Math. Biophys.* **15**:121–41.

Sheppard, C. W., and A. S. Householder. 1951. The mathematical basis of the interpretation of tracer experiments in closed, steady-state systems. *J. Appl. Phys.* **22**:510–20.

Solomon, A. K. 1949. Equations for tracer experiments. *J. Clin. Invest.* **28**: 1297–307.

Teorell, T. 1937. Kinetics of distribution of substances administered to the body. *Arch. Int. Pharmacodyn.* **57**:205–40.

Zilversmit, D. B., C. Entenmann, and M. C. Fishler. 1943. On the calculation of turnover time and turnover rate from experiments involving the use of lasbeling agents. *J. Gen. Physiol.* **26**:325–31.

1

Reprinted from pages 121–125 of *Radioecology*, Proc. First Natl. Symp. Radioecol., Colorado State Univ., 1961, V. Schultz and A. W. Klement, Jr., eds., N.Y.: Reinhold Publ. Co., and Washington, D.C.: Inst. Biol. Sci., 1963, 746pp.

ANALOG COMPUTER MODELS FOR MOVEMENT OF NUCLIDES THROUGH ECOSYSTEMS

JERRY S. OLSON

Health Physics Division, Oak Ridge National Laboratory, Oak Ridge, Tennessee

INTRODUCTION

Analog computers should fulfill an increasing need in the interpretation of data in radioecology. Ecological tracer experiments and case studies of radioactive contamination of the environment often require us to interpret the net changes in radio-activity in some part or parts of our system in terms of simultaneous gains and losses. The basic operation of the analog computer is to keep a running balance of such gains and losses for all the major parts or "compartments" of a system, be it a reactor or an ecosystem.

Such computers are coming into frequent use for kinetic interpretation of physiological tracer experiments, concerned with the movement of nuclides between cells or organs of a single individual. They should eventually become even more helpful in more complicated ecological systems where other methods of unscrambling simultaneous transfers are not satisfactory. Ecological applications of electronic analog computers have been made to population problems (Wangersky and Cunningham, 1957a,b) and to the world-wide biogeochemical cycle of carbon (DeVries, 1958). H.T. Odum (1960) has used an electrical network analog for simulating certain features of an ecosystem. Neel and Olson (1962) have used the Oak Ridge National Laboratory Analog Computer Facility for similar problems which can be viewed in relation to the transfer of energy, carbon, biomass, nutrients, and radionuclides through ecosystems.

However, the analog computer is more than a handmaiden for deriving numbers from empirical research. It provides a physical simulation of the processes to which the model pertains. It is already serving as a valuable stimulus and aid in developing a body of quantitative ecological theory concerning the operation of ecosystems.

The present paper diagrammatically illustrates how the running balance, or integration, of gains and losses is accomplished. The Methods section points out the basic idea for a single compartment, namely photosynthetic vegetation. The main section, on Transfer of Carbon and Carbon-14, extends this example to a simple ecological model covering four compartments (vegetation tops, roots, litter, soil organic matter). Finally, a discussion of transfer of fission products and mineral nutrients through the ecosystem suggests how this basic methodology can eventually be extended further to models with more compartments and with other refinements.

METHODS

The basic unit of the analog computer is the integrator, which continually adds or subtracts voltages to the voltage already stored on a condenser contained within the integrator. If the photosynthetic rate P of producing vegetation were defined as a function of time and fed into the integrator (Fig. 1), then the output of the integrator would show the production which had occurred in the system since time zero. Schematic diagrams represent an integrator by a rectangular box next to a triangle which points in the "downstream" direction of transfer of material in the model. (Actually

INTEGRATOR

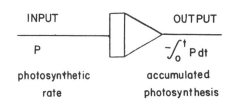

Figure 1. Simple integration of a single input (photosynthetic rate) to give output of cumulative photosynthesis on an integrator of an analog computer.

NEGATIVE FEEDBACK

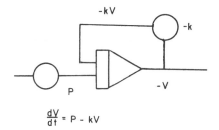

Figure 2. Integration of a positive input (photosynthesis) minus a negative feedback controlled by the feedback potentiometer setting k.

the sign of this voltage is inverted, so the negative voltage is shown on the output.) For example, if photosynthetic rate is constant, then accumulated photosynthesis would increase uniformly through time, just as would the level of water in a cylindrical bucket that was being filled by a steady inflow of water.

Now if we imagine instead a bucket with a hole that leaks at a rate proportional to the amount of water accumulated in it, the rate of rise of the water level increases rapidly at first, and then gradually more slowly until the rate of leakage at last nearly balances the rate of steady input. Similarly, in an ideal case in which the rate of loss of organic carbon was a constant fraction of the organic carbon accumulated by photosynthetic vegetation, the standing crop of vegetation would increase until losses equalled input and a steady state was attained. Either of these idealized situations can be simulated by the analog computer diagram of Figure 2. If we let the amount of accumulation (either vegetation biomass, or volume of water in the bucket) be V, then the rate of change of V is the sum of P and -kV, i.e.,

$$\frac{dV}{dt} = P - kV \qquad (1)$$

Since -V is simply the output of the integrator, a potentiometer is used to tap off a fraction k of this voltage to give the product -kV. This negative product, or negative feedback, is added to the positive part of the input represented by P, so the integrator actually works on the sum of these two terms. (A positive feedback has also been used to represent the buildup in photosynthetic rate as th

amount of photosynthetic tissue increase, but this
was omitted from Figure 2.)

The resulting integral is the familiar exponential equation which gradually approaches an upper asymptote in the way we had intuitively surmised. The level which is approached as a steady state is equal to the ratio of the income P to the "negative feedback" parameter k. For the initial condition V = 0 at t = 0, the model formula for the accumulation V in the vegetation is obtained by integrating with respect to the time up to the value t:

$$V = \int_0^t (P - kV)dt = \frac{P}{K}(1 - e^{-kt}) \qquad (2)$$

Chart records actually drawn by the analog computer are shown on Figure 3. The uppermost graph represents the constant input just assumed, which is here assumed to be 400 grams of carbon photosynthesized per square meter per year. Below this is the expected exponential equation for vegetation. The remaining graphs were obtained by using fractions of the output of the vegetation compartments as inputs for additional compartments as explained below.

A recent Oak Ridge National Laboratory report (Neel and Olson, 1962), based on a thesis by Robert Neel, provides further discussion and examples of the principles and techniques involved in this kind of electronic analog computer simulation. Earlier uses of electrical network analogs by DeVries (1958) and H.T. Odum (1960) are cited in it, along with some technical limitations of such networks and of the electronic computers discussed here.

TRANSFER OF CARBON AND CARBON-14

The first step in extending this basic methodology to models with several compartments is the preparation of a block design like that of Figure 4 showing the pathways of transfer. For carbon, this shows how one fraction of carbon in the above-ground portion of vegetation is delivered to the plant roots (R), and a small fraction is considered to be consumed by animals (which are not here shown as separate boxes, for the sake of simplicity of illustration). A third fraction contributes to dead organic litter (D). Both roots and litter in turn serve as inputs for the organic carbon or humus in the mineral soil (S). The k parameters here correspond with total loss rates, and the ϕ's with partial transfer coefficients, indicating the redistribution just mentioned. The numbers cited here were somewhat arbitrary, for illustrative purposes.

The second step which can follow directly from this block diagram is the construction of an analog computer circuit diagram (Figure 5). Corresponding with each compartment is a little sub-circuit like that of Figure 2. Additional potentiometers are used to indicate what fraction of the output from one compartment serves as input for the next compartment. An initial condition other than zero volts could be provided for on any compartment; as if, for example, we wished to start counting t = 0 at a time when there was already a cover of vegetation V_0 present on the ground.

Instead of the rather unrealistic assumptions, such as of constant rate of photosynthesis, various elaborations have been introduced to this model one at a time. For example, in Figure 6, we have an

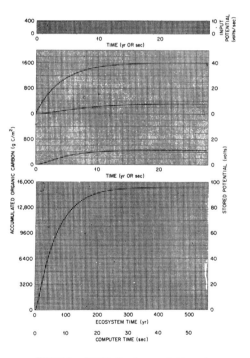

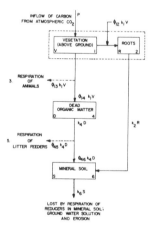

Figure 3. Hypothetical accumulation of carbon in above-ground vegetation, roots, organic litter, and inorganic matter in mineral soil, from steady photosynthetic input into model system described in Figure 4.

Figure 4. Box diagram indicating pathways of movement of carbon in the larger compartments of a model ecosystem, and arbitrary selection of parameters for illustrations in Figures 3 and 6.

15

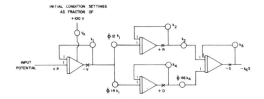

INITIAL CONDITION SETTINGS
AS FRACTION OF

X - POINTS OF ANALOG COMPUTER CIRCUIT CONTINUOUSLY MONITORED BY
VOLTAGE CHART RECORDERS

Figure 5. Simple analog computer diagram showing transmission of changes in voltage from input on left, through integrators corresponding to above-ground vegetation (V), roots (R), dead organic litter (D), and soil organic matter (S).

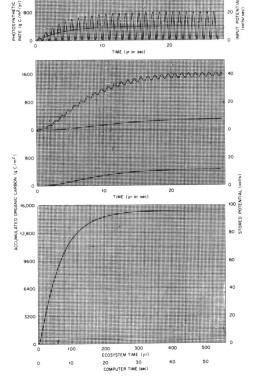

Figure 6. Hypothetical accumulation of carbon in a system with an annual cycle of high and low photosynthetic rate, initially averaging 0 grams of carbon per square meter per year, gradually rising to average of 400 grams of carbon per square meter per year. Sinusoidal oscillations attenuated in vegetation and almost eliminated in roots and litter under the assumed conditions. Note scale shift for soil organic matter.

input which oscillates to simulate the change in production between summer and winter. Furthermore, the mean annual rate of photosynthesis is allowed to vary instead of being kept as a constant over a period of years. Sinusoidal terms are evident in the

graph for the vegetation compartment, though somewhat damped in amplitude compared with the oscillations in photosynthetic rate. Each additional stage of transfer, such as that to roots or litter, further attenuates the amplitude of the oscillatory term, so that graphs for these compartments are virtually identical with those that would be expected for a model with an input having no annual oscillation of photosynthesis. Because of the very slow turnover assumed for soil organic carbon, the use of a compressed scale for both axes was necessary for showing this compartment on the same size graph paper.

While elementary differential equations can be used to solve these equations analytically, the solutions become awkward even for the four-compartment model considered here. They become more difficult to handle when additional compartments are added, and quite intractable when certain "nonlinear" features are introduced into the ecological model.

Fortunately, the electronic analog computer circuit has a variety of special devices such as function generators, servomultipliers and diode-limiters, which make the computer about as satisfactory for working with nonlinear models as with the highly simplified linear model considered previously.

One of the special devices, the function generator, was used by Neel and Olson (1962) as a first approach to extending the present model to see how a sudden increase of carbon-14 in the atmosphere as a result of thermonuclear tests would modify the carbon-14 content of the other parts of the same model ecosystem considered above (Figure 7). It was assumed that the "normal" amount of carbon-14

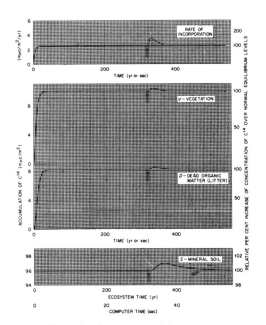

Figure 7. Response of model ecosystem described in Figure 4 to initial accumulation of "normal" carbon-14 in proportion to total carbon, and subsequent increase in carbon-14 fixation due to thermonuclear tests. Note lag in response of soil compartment.

produced in cosmic rays had previously been incorporated in the ecosystem in direct proportion to the content of total carbon, during the development of vegetation and litter. The soil humus was assumed to have had its equilibrium content of organic carbon and carbon-14 already. At the time indicated by the arrow, the analog computer was switched to the "hold" position and the function generator was switched into the circuit. The function generator then fed in a somewhat higher rate of incorporation of carbon-14 (Figure 7, top) which had previously been set up to correspond to the increase projected by Broecker and Olson (1960).

Because of the relatively fast time constants assumed for vegetation and litter, these compartments responded fairly promptly in showing an increase in carbon-14. The slow time constant which was assumed for the turnover of humus incorporated in the mineral soil resulted in a slower response to the "pulse" of carbon-14, and a markedly slower return to the "pre-nuclear test" levels (assuming that atmospheric nuclear testing and nuclear warfare were not resumed for centuries following the 1958 moratorium on testing). The absolute quantity of carbon-14 projected for each square meter of ground surface was somewhat higher for soil humus than for vegetation and litter, but because of the larger storage of carbon, the percentage increase was much less.

Examples of the kinds of ecological conclusions which might follow from more extended analyses of this type include the following: (1) The substantial increase in specific activity of atmospheric carbon-14 from weapons tests will have a diluted influence on the specific activity of carbon-14 with each transfer through the ecosystem, but the amounts of dilution and promptness of response will vary markedly depending on the parameters of different kinds of ecosystems. (2) If the increase in radiation hazard relative to existing background can be considered small (Totter, Zelle and Hollister, 1958) for a comparatively radiosensitive organism like man, who derives food fairly directly from photosynthetic producers, then the radiation influence on other parts of the ecosystem from this source is even less likely to be important in comparison with such sources of radioactivity as potassium-40 and fission products. (3) Much more widespread atmospheric contamination which might arise in the event of nuclear war would similarly be less spectacular at first than that from fission products, but the long half-life and the capacity for storage of carbon-14 in humus as well as in the oceans would mean that any environmental problem involving carbon-14 might persist for many generations to follow. (4) The increase in specific activity of carbon already produced from nuclear tests provides an unprecedented opportunity for gathering information on the probable holdup of carbon-14 in different compartments of the biosphere and for evaluating the consequences of further contamination of this type. (5) Information on the turnover of carbon is valuable for basic research on the time parameters of carbon transfer in ecosystems, since these parameters influence the rates of many other ecological processes.

The interpretation of such measurements of response to changes in atmospheric carbon-14 would have to go hand in hand with definite hypotheses concerning the several pathways of income and loss of carbon. Such interpretations would presumably be expedited by analog computer models appropriate for the systems where measurements were being taken.

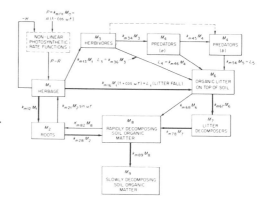

Figure 8. Flow of biomass and organic matter (M) through an herbaceous terrestrial ecosystem. Main units of M are grams per square meter. Heavy lines show main pathway of flow. Losses to atmosphere as carbon dioxide occur at many stages. The k_m's express instantaneous rates of transfer, as a fraction of source compartment, moving between the compartments numbered.

TRANSFER OF FISSION PRODUCTS THROUGH ECOSYSTEMS

In 1961, several obvious limitations of the preceding small-scale models were overcome. (1) Additional special computer features such as servomultipliers and diode limiters were used to make the behavior of the models correspond more closely to natural processes. (2) The large capacity of the National Laboratory Analog Computer Facility at Oak Ridge was more fully utilized for a larger number of ecological compartments (Figure 8), including the various trophic levels of the animal food chain, litter decomposers, and separate compartments for rapidly decomposing organic matter and slowly decomposing organic matter. (3) Finally, a second console of the computer was operated in tandem with the first in order to represent the transfer of mineral nutrient elements (or their radioactive nuclides) which follow some pathways of movement different from carbon.

With slight changes in definitions of terms, the first console can be used for models of transfer of energy, carbon, or of biomass. The transfer of free energy of course involves loss of availability at every stage of transfer, so there is no recycling. Each transfer of carbon involves a release of carbon dioxide which (for terrestrial systems) mostly becomes mixed with the general atmospheric pool before being reincorporated into the biomass as a result of new photosynthesis. Most of the biomass of organisms and soil organic matter follows the same pathway as carbon, from photosynthetic parts of vegetation, down through roots, or through litter and microorganisms, to soil.

By contrast, on the second console whose block diagram is shown in Figure 9, there is a major pathway of uptake from some exchangeable pool of nutrient material in the soil, to roots, up into plant tops, and only then along the above-mentioned pathway to litter and back to soil. For either mineral nutrient elements, or the majority of fission products or other radioactive nuclides, this kind of

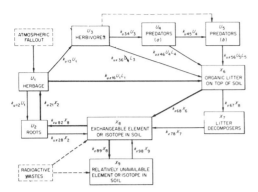

Figure 9. Flow of inorganic nutrient elements or radioactive nuclides through the system whose biomass transfers were shown in Figure 8. In U, units are expressed per gram of biomass or organic matter, while in X, units are per square meter. (The k_{UX} and k_{XU} are factors for appropriate conversion between these units; L represents rates of litter fall in grams per square meter.) Pathways of movement of nuclides from radioactive fallout or waste disposal can be traced.

pathway would be followed, and would involve the possibility of repeated cycling of materials from soil to organisms and back to soil. For some elements like cesium, there is a strong tendency toward fixation into a relatively unavailable form within the soil On the other hand, even such a readily fixable element can be recycled by root uptake before it happens to come into contact with sites of fixation, or may even be taken up directly by roots from decomposing litter (Witherspoon, 1963).

Further explanation of the terminology involved in Figures 8 and 9 can be consulted in the forest ecology section of the 1961 Oak Ridge National Laboratory Health Physics Progress Report (Olson et al., 1961). Time does not permit coverage of the analog computer circuit diagrams which have been developed on the basis of these block diagrams. The preceding discussion may indicate how electronic analog computers can be used for a variety of models for the movement of radionuclides through ecosystems.

SUMMARY

The interpretation of information on the rates of income and loss of radionuclides (or of nutrients, carbon, biomass, or free energy) between compartments of an ecosytem involves assumptions about the

system which can be formalized in mathematical models. Electronic analog computer techniques can be viewed either as a means of solving the differential equations describing the income and losses for each compartment of the system, or as a device for stimulating the natural processes operating in the system.

An example of a small analog computer model for the transfer of organic carbon or carbon-14 from photosynthetic vegetation, to roots or organic litter, to humus in mineral soil was discussed and further details are given in an Oak Ridge National Laboratory Report (Neel and Olson, 1962). Newer work in 1961 involved the use of two consoles, one to simulate the transfer of energy, of carbon, or of biomass, and a second one to simulate the cycling movement of inorganic nutrient elements or of radioactive nuclides.

ACKNOWLEDGMENTS

This work was performed under contract number W-7405-eng-26 between the U. S. Atomic Energy Commission and the Union Carbide Corporation.

REFERENCES

Broecker, W.S., and E.A Olson. 1960. Radiocarbon from nuclear tests, II. Science 132(3429): 712-721.

DeVries, H. 1958. Variation of concentration of radiocarbon with time and location on earth. Proc. Koninkl. Ned. Akad. Vetenschap. B-61: 94-102.

Neel, R.B., and J.S. Olson. 1962. Use of analog computer models for simulating the movement of isotopes in ecosystems. Oak Ridge Natl. Lab., U. S. AEC report ORNL-3172. xii, 108 pp.

Odum, H.T. 1960. Ecological potential and analogue circuits for the ecosystem. Am. Scientist 48(1): 1-8.

Olson, J.S., et al. 1961. Forest Ecology. In Health Physics Div. Ann. Prog. rep. for period ending July 31, 1961. Oak Ridge Natl. Lab., U. S. AEC report ORNL-3189. pp. 105-128.

Totter, J.R., M.R. Zelle, and H. Hollister. 1958. Hazard to man of carbon-14. Science 128(3337): 1490-1495. (Also: U. S. AEC report WASH-1008).

Wangersky, P.J., and W.J. Cunningham. 1957a. Timelag in prey-predator population models. Ecology 38 (1): 136-139.

- - - 1957b. Timelag in population models. Cold Cold Spring Harbor Symposia on Quantitative Biology 22: 329-338.

Witherspoon, J.P. 1963. Cycling of cesium-134 in white oak trees on sites of contrasting soil type and moisture. I. 1960 Growing Season. In this volume, pp. 127-132.

2

SYSTEMS ANALYSIS OF [134]CESIUM KINETICS IN TERRESTRIAL MICROCOSMS[1]

BERNARD C. PATTEN AND MARTIN WITKAMP

Radiation Ecology Section, Health Physics Division,
Oak Ridge National Laboratory, Oak Ridge, Tennessee

(Accepted for publication September 12, 1966)

Abstract. Laboratory experiments were conducted to determine patterns and rates of [134]Cs exchange in microecosystems composed of different combinations of radioactive leaf litter, soil, microflora, millipedes, and aqueous leachate. Rate constants were determined by fitting models to data with an analog computer. Simulations with the models permitted comparisons of different microcosms in terms of time to radiocesium equilibrium, steady state concentrations, concentration factors, input and output fluxes, turnover rates, and stability.

Rate constants varied with different compartment combinations, indicating both qualitative and quantitative differences in the cesium exchange patterns within different systems. This result is generalized: transfers of energy and matter in ecosystems are functions of networks which define intercompartmental interactions; internal coupling should be considered a significant variable in investigations of ecosystem processes.

Points of particular interest derived from the computer simulations are (i) organism concentration factors for a material may vary in different systems, depending upon how the organism is coupled to other compartments; (ii) total flux of a material in a steady state system may vary considerably from that in another system which receives identical input; (iii) material turnover within compartments and in the system as a whole tends to increase as more compartments are added; and (iv) stability of material concentration does not appear to increase with system complexity.

[1] Research sponsored by the U. S. Atomic Energy Commission under contract with the Union Carbide Corporation.

INTRODUCTION

Investigations of mineral cycling in terrestrial ecosystems are hindered by difficulties of separating soil organisms, plant roots, dead organic matter, and the mineral soil. Consequently, most studies consider only two compartments, such as soil-litter or soil-plant, and usually only at some experimental endpoint (harvest). Investigations of mineral flows between multiple compartments have been few (Jansson 1958, Remezov 1959, Witherspoon 1964, Witkamp and Frank 1964, Neel and Olson 1962, Olson 1965), and only the last two of these have given explicit attention to exchange kinetics. Whittaker (1961) has attempted to determine the kinetics of ^{32}P in aquatic microcosms. The present study employs two innovations to develop fuller understanding of mineral cycling in a series of experimental microecosystems, one involving laboratory syntheses of progressively more complex systems, and the other involving use of an analog computer to determine the kinetics of intercompartmental transfers.

Microcosms of increasing complexity were synthesized by adding one compartment at a time. Five compartments were used: ^{134}Cs-labeled oak leaves, mineral soil, microflora, millipedes and an aqueous leachate. Radioactivity in each compartment was determined as a function of time, with ^{134}Cs in the microflora computed as the difference between activity leached from sterile and nonsterile leaves. This stepwise approach to construction of multicompartmental systems permits evaluation of the effects of each added compartment on the mineral's dynamics.

Empirical data take the form of curves of radiocesium activity in each compartment graphed against time. Such information by itself does not specify transfer pathways, rates and fluxes along these pathways, or transfer mechanisms. It is possible, however, to determine these system characteristics by adjusting an analog computer model to fit the experimental data. This procedure results in a description of radiocesium kinetics for each microcosm in terms of a system of differential equations. The equations specify and quantify transfer pathways, and permit additional information to be developed by computer simulations of experiments which would be difficult or impossible to perform in the laboratory.

Miss Bonnie McGurn assisted in conducting the laboratory experiments, and Drs. D. A. Crossley and D. E. Reichle provided helpful criticism and discussion.

MATERIALS AND METHODS

Experimental

Microcosms were established in 60 ml glass funnels, inside diameter 4 cm, with fritted, medium porosity filters. The funnels were seated on 125 ml suction flasks each containing a counting tube for collection of leachate. Triplicates of the following systems were established: (i) sterile white oak (*Quercus alba* L.) leaves, (ii) leaves and microflora, (iii) leaves, microflora and millipedes, and (iv-vi) same as i-iii but placed on the surface of a 1 cm layer of silt-loam (pH 6.4, 12.6% organic matter, 27% moisture). The microcosms were maintained under identical conditions at room temperature (about 25°C).

The radioactive leaves were collected in July 1961 from small trees trunk-labeled (Witherspoon 1964) with ^{134}Cs a year before. Each funnel received 1 g of air-dried leaf material cut slightly smaller than 4 cm diameter. Leaves and soil were sterilized with 12% ethylene oxide (16 hr, 8 psi, 40°C) after which microflora and millipedes were added. Five millipedes (*Dixidesmus erasus* Loomis), averaging 392 mg total fresh weight and newly-collected from neighboring forests, were added to each funnel requiring millipedes. Dead millipedes were replaced daily if necessary and assayed for ^{134}Cs activity. Microflora was introduced in 1 ml of supernatant from 10 g wet weight of oak leaves shaken for 15 min with 100 ml of sterile water. This inoculum was administered to funnels requiring microflora as part of the first leaching at the start of the experiments. For each leaching, 8 ml of sterile water was dripped from a pipette as uniformly as possible over the contents of each funnel. This procedure was repeated every 2–3 days to approximate mean precipitation in Oak Ridge (2.5 cm/wk). During leaching, 10 cm Hg vacuum was applied. Leachate volumes were recorded and radiocesium activity of 5 ml aliquots was determined by NaI(Tl) scintillation counting. Between leachings, positive pressure of about 1 cm Hg was maintained in the flasks to prolong millipede survival by providing aeration. After 18 days and 9 leachings, the experiments were terminated and ^{134}Cs activity of remaining litter, soil and millipedes measured.

Computational

A block diagram of the litter-soil-microflora-millipede-leachate microecosystem, indicating all possible transfer routes, is shown in Figure 1. This system is reducible to fewer compartments by letting ^{134}Cs transfers be zero along appropriate pathways. All the microcosms thus implied in Figure 1 can be regarded as closed with respect

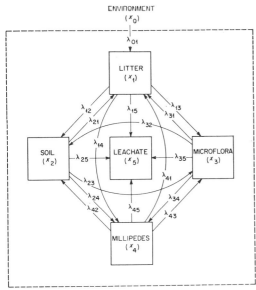

FIG. 1. Block diagram of litter-soil-microflora-millipede-leachate microecosystem showing all possible routes of cesium transfer and associated rate constants, $\lambda_{ij}(i = 0,1,\ldots,4;\ j = 1,2,\ldots,5)$.

to the radioisotope, even though radioactivity was removed in successive leachings, by considering the cumulative leachate as a sink (no back transfers to other compartments) within the system of definition. Thus, the only loss of activity for computational purposes is by radioactive decay, and this is negligible in an 18-day experiment ([134]Cs half-life = 2.07 y).

Let X (a constant) be the initial radioactivity introduced in oak litter, and hence the total activity of the microcosm throughout an experiment. This activity will move from one compartment to another with time. If $X_j(t)$ is the total amount of radioactivity in compartment j at time t, then, since no activity is lost in an experiment, $\Sigma_j X_j = X =$ constant. If the mass of compartment j at time t is $m_j(t)$, then the concentration or activity-density of radiocesium in the compartment at time t is $x_j = X_j/m_j$. Letting $\lambda_{ij}(t)$ be a function expressing the rate of [134]Cs transfer from compartment i to compartment j, the rate of change of radioisotope concentration in the j'th compartment is the balance between incomes and losses:

$$\dot{x}_j = \Sigma_i \lambda_{ij} x_i - \Sigma_i \lambda_{ji} x_j, \qquad (1)$$
$$(i,j = 1,2,\ldots,5;\ i \neq j;\ 0 \leq \lambda_{ij}, \lambda_{ji} < \infty;$$
$$0 \leq x_i, x_j \leq 1)$$

where $\dot{x}_j$ is the first derivative of x_j with respect to time. Since the experimental systems were all

non-stationary (at least one $\dot{x}_j \neq 0$ during the period of observation) and since a transient solution of a differential equation corresponds to a unique transient behavior, equations (1) can be made to describe uniquely [134]Cs exchanges in the microcosms by finding appropriate functions, $\lambda_{ij}(t)$. To do this a simplest possible model was programmed for study on an Electronic Associates TR–10 analog computer. This proved adequate for description of the experimental results, and further refinements were unnecessary.

The model assumed first order kinetics: that transfer of [134]Cs from compartment i to compartment j is directly proportional to the radioisotope concentration in compartment i. With this assumption, the λ_{ij} in equations (1) become rate constants, having units $\{t^{-1}\}$.

To normalize the system equations so that different microcosms with different values of X could be compared, the fraction of radioactivity in each compartment relative to total system activity, $y_j = x_j/X$, was used as a variable for work on the computer. This has no effect on the rate constants to be determined, as the normalized system equations show:

$$\dot{y}_j = \frac{\dot{x}_j}{X} = \Sigma_i \lambda_{ij} \frac{x_i}{X} - \Sigma_i \lambda_{ji} \frac{x_j}{X}$$
$$= \Sigma_i \lambda_{ij} y_i - \Sigma_i \lambda_{ji} y_j. \qquad (2)$$
$$(0 \leq y_i,\ y_j \leq 1)$$

The computing procedure was to fit program-generated curves of $y_j(t)$ to empirical curves by adjusting potentiometers representing the λ_{ij} and λ_{ji}, thereby obtaining numerical values for these parameters and simultaneous solutions of the system equations (1 and 2) for each experimental microcosm.

RESULTS

The five compartments can be labeled as follows: $i,\ j = 1 \equiv$ oak litter, $2 \equiv$ soil, $3 \equiv$ microflora, $4 \equiv$ millipedes and $5 \equiv$ leachate (Fig. 1). The environment is compartment 0. Possible combinations of litter-leachate and other compartments are 1–5, 1–2–5, 1–3–5, 1–4–5, 1–2–3–5, 1–2–4–5, 1–3–4–5, and 1–2–3–4–5. Of these, 1–4–5 and 1–2–4–5 are not experimentally feasible because of difficulties of sterilizing millipedes. The remaining combinations, labeled I-VI, respectively, were investigated in six experiments whose results are summarized in Figure 2 and Table 1.

Combination I (litter-leachate).—As shown in Figure 2a, [134]Cs was transferred from litter to leachate at a rate of $\lambda_{15} = 0.037$ day^{-1}, 3.7% per

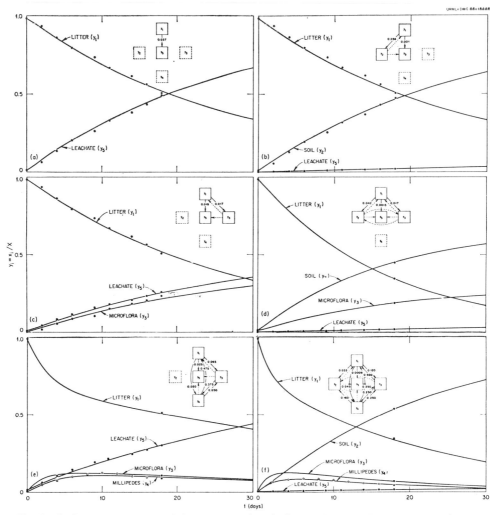

FIG. 2. Analog computer curves of changes in relative distribution (y_j) of radiocesium in the microcosms of Combs. I–VI (Figs. a–f, respectively). Means of three replicates are shown as solid circles. Open circles represent estimates of relative cesium concentration in millipedes based on radioassays of animals which died. Insets show compartment combinations and rate constants (day^{-1}) for each indicated transfer route. Pathways over which no cesium was transferred are illustrated by broken arrows.

day. System equations will not be written explicitly hereafter, but for the purpose of illustrating how the microcosm behaviors depicted in Figure 2 are described by the model (equations 1 and 2), the system equations for the present experiment are

$$\begin{cases} \dot{x}_1 = -0.037\,x_1 \\ \dot{x}_5 = 0.037\,x_1 \end{cases} \text{ or } \begin{cases} \dot{y}_1 = -0.037\,y_1 \\ \dot{y}_5 = 0.037\,y_1 \end{cases}$$

Combination II (litter-soil-leachate).—Addition of soil sharply reduced transfer of radiocesium from litter to leachate (Fig. 2b). The total loss rate from litter was the same as in Combination I

($\lambda_{12} + \lambda_{15} = 0.037$ day^{-1}, Table 1), but now 0.036 day^{-1} went to soil and only 0.001 day^{-1} to the leachate. Once in the soil radiocesium was held tightly, as indicated by non-leachability ($\lambda_{25} = 0$) and absence of back transfer to litter ($\lambda_{21} = 0$).

Combination III (litter-microflora-leachate).—Loss of ^{134}Cs from litter was essentially the same as in Combination I, 0.036 day^{-1} (Table 1). Of the total litter loss, 0.017 day^{-1} was transferred to microflora and 0.019 day^{-1} to the leachate (Fig. 2c, Table 1). In agreement with *in vitro* leaching experiments with microflora, radiocesium in

TABLE 1. Rate constants, day^{-1}

Parameters	I (1-5)	II (1-2-5)	III (1-3-5)	IV (1-2-3-5)	V (1-3-4-5)	VI (1-2-3-4-5)
			Combinations			
λ_{12}	—	0.036	—	0.042	—	0.033
λ_{13}	—	—	0.017	0.017	0.065	0.120
λ_{14}	—	—	—	—	0.020	0.045
λ_{15}	0.037	0.001	0.019	0.0013	0.025	0.0009
$\Sigma\lambda_{1j}$	0.037	0.037	0.036	0.060	0.110	0.199
λ_{21}	—	0	—	0	—	0
λ_{23}	—	—	—	0	—	0
λ_{24}	—	—	—	—	—	0
λ_{25}	—	0	—	0	—	0
$\Sigma\lambda_{2j}$	—	0	—	0	—	0
λ_{31}	—	—	0	0	0	0.500
λ_{32}	—	—	—	0	—	0
λ_{34}	—	—	—	—	0.375	0.250
λ_{35}	—	—	0	0	0	0
$\Sigma\lambda_{3j}$	—	—	0	0	0.375	0.750
λ_{41}	—	—	—	—	0.475	0.250
λ_{42}	—	—	—	—	—	0.160
λ_{43}	—	—	—	—	0.050	0.250
λ_{45}	—	—	—	—	0	0
$\Sigma\lambda_{4j}$	—	—	—	—	0.525	0.660

the microflora was non-leachable ($\lambda_{35} = 0$). No reverse transfer from microflora to litter was indicated ($\lambda_{31} = 0$). The rate of leachate gain was reduced compared to when microflora were absent (Comb. III vs. I), but not as much as when soil was present (Comb. III vs. II).

Combination IV (litter-soil-microflora-leachate)—Soil and microflora together had a synergistic effect which markedly increased the rate of cesium loss from litter ($\lambda_{12} + \lambda_{13} + \lambda_{15} = 0.060$ day^{-1}, Table 1). As shown in Figure 2d and Table 1, transfer from litter to soil was increased to $\lambda_{12} = 0.042$ day^{-1} compared to 0.036 in Combination II. Accumulation by the microflora, however, was unaffected by soil ($\lambda_{13} = 0.017$ as in Comb. III). Leachability of the litter ($\lambda_{15} = 0.0013$) was essentially as in Combination II. Hence, the synergism appears to be an effect mainly on the ability of soil to acquire cesium from the litter, which is difficult to understand in view of previous results indicating that microflora do not alter the rate of loss from the litter (Comb. III vs. I). Furthermore, no direct exchanges between soil and microflora are indicated ($\lambda_{23} = \lambda_{32} = 0$), and consequently the character of the microflora effect on litter-soil exchange is not ap-

parent. As in Combinations II and III, soil and microflora radioactivity was non-leachable ($\lambda_{25} = \lambda_{35} = 0$), and also there was no cesium feedback from these compartments to the litter ($\lambda_{21} = \lambda_{31} = 0$).

Combination V (litter-microflora-millipedes-leachate).—The open circles in Figure 2e represent estimates of y_4 for total millipedes based on radioassays of animals which died and were removed. The fit of the model to these data is not too satisfactory: the millipede curve does not rise high enough between 6 to 14 days and does not fall rapidly enough toward the end of the observation period. Whether this difficulty is primarily with the model or with the extrapolated data is uncertain, but the fact that a better fit was obtained in the next, more complex combination which included soil (Fig. 2f) is good reason for confidence in the model.

The Combination V results (Fig. 2e, Table 1) indicate that when both microflora and millipedes are present the total rate of cesium loss from litter ($\lambda_{13} + \lambda_{14} + \lambda_{15} = 0.110$, Table 1) was tripled compared to Combinations I–III and almost doubled compared to Combination IV. The rate of transfer from litter to microflora was four times

23

greater than in Combination III ($\lambda_{13} = 0.065$ vs. 0.017), presumably due to action of millipedes upon the litter. This action is also reflected in increased leachability of litter cesium ($\lambda_{15} = 0.025$ vs. 0.019 in Comb. III).

Despite these indications of millipede effects on litter, the rate of radiocesium transfer from litter to millipedes was fairly low, $\lambda_{15} = 0.020$ day^{-1}. In contrast, uptake rate from the microflora was about 19 times greater ($\lambda_{34} = 0.375$). Based on these results, and prompted by laboratory observations of no visible fungal mycelia in microcosms containing *Dixidesmus* but abundant growths otherwise, it was originally thought that the model denoted a 19-fold perference of *Dixidesmus* for microflora over litter. This is not necessarily true, as illustrated by the following examples of two-compartment exchanges.

Let compartment i have biomass m_i and cesium concentration x_i, compartment j have biomass m_j and concentration x_j, and suppose $x_i > x_j$. (i) If biomass is transferred from i to j ($\lambda_{ij} > \lambda_{ji} = 0$), then compartment i loses mass but its cesium concentration does not change (homogeneous distribution of the mineral within the mass is assumed for purposes of the argument). Compartment j gains biomass which is of higher radioisotope concentration than that within the compartment, and therefore x_j increases. As the process continues, both activity-densities approach equality, $x_j \to x_i$. (ii) If biomass is transferred from j to i ($0 = \lambda_{ij} < \lambda_{ji}$), compartment i gains biomass with cesium concentration x_j, and therefore x_i decreases. Compartment j loses mass but its concentration of radiocesium is unchanged. Continued exchange leads in the direction $x_i \to x_j$. (iii) If both forward and reverse exchanges occur simultaneously ($\lambda_{ij}, \lambda_{ji} > 0$), the cesium concentrations will move toward an equilibrium defined by $x_{j(eq)} = \dfrac{\lambda_{ij}}{\lambda_{ji}} x_{i(eq)}$. The second example, where compartment i gains biomass and consequently has its cesium concentration reduced, illustrates why biomass and tracer transfers cannot both be represented by the same rate constants. A 19-fold difference in rates of cesium input from two compartments cannot be interpreted as a 19-fold food preference.

As with the microflora in this and preceding combinations, millipede radioactivity was non-leachable ($\lambda_{35} = \lambda_{45} = 0$). However, unlike the microflora which did not transfer ^{134}Cs back to the litter ($\lambda_{31} = 0$), the millipedes did so at a high rate ($\lambda_{41} = 0.475$). In addition, the animals also transferred radiocesium to the microflora at a rate of $\lambda_{43} = 0.050$ day^{-1}.

Combination VI (litter-soil-microflora-millipedes-leachate).—Results with this most advanced microcosm are given in Figure 2f and Table 1. Numerous points of comparison with previous combinations are possible, and the most systematic way to proceed is down the last column of Table 1.

The rate of cesium transfer from litter to soil ($\lambda_{12} = 0.033$) was less than in other combinations in which soil was present. In Combination IV, compared to II, microflora increased the rate of movement from litter to soil. With millipedes present (Comb. VI), this effect is apparently nullified. One reason is that the rate of movement from litter to microflora was about 7 times greater ($\lambda_{13} = 0.120$ in Comb. VI vs. 0.017 in Comb. IV). Another is that the rate of transfer from litter to millipedes was more than doubled in the presence of soil ($\lambda_{14} = 0.045$ vs. 0.020 in Comb. V). The basis for this soil-millipede interaction is not apparent except, perhaps, that the normal habitat of these animals was better approximated here than in Combination V, resulting in more normal rates of food consumption. Leachability of litter cesium, $\lambda_{15} = 0.0009$ day^{-1}, was low as in Combinations II and IV, which also had soil present. That microflora and millipedes substantially increased the mobility of litter cesium is clear from the total litter loss rate ($\lambda_{12} + \lambda_{13} + \lambda_{14} + \lambda_{15} = 0.199$), which is almost 6 times greater than in Combinations I–III, more than 3 times greater than in Combination IV, and almost twice as great as in Combination V.

Soil is clearly a cesium sink at the low concentrations in these experiments, since none of the microcosms studied contained a compartment capable of acquiring the mineral from soil ($\lambda_{2j} = 0$ for all $j \neq 2$; Table 1).

Unlike previous combinations with microflora, radiocesium in this combination was transferred at a high rate from microflora to the litter ($\lambda_{31} = 0.500$). This may be due to millipede feeding since Combination V indicated that these animals obtain cesium from the microflora ($\lambda_{34} = 0.375$ in Comb. V vs. zero in Combs. III and IV without millipedes). The transfer to litter rather than soil ($\lambda_{32} = 0$) verifies the experimentally observable fact that mycelia and other microbial components are associated with litter more than soil. Lack of transfer from microflora to litter in Combination V is unexplained.

The transfer rate from microflora to millipedes, while still high ($\lambda_{34} = 0.250$), is nevertheless reduced compared to Combination V. Soil, the only variable between these two combinations, must be responsible. A physiological difference between microflora in this combination and Combination V is indicated by doubling of the radiocesium loss rate in the presence of soil ($\lambda_{31} + \lambda_{32} + \lambda_{34} + \lambda_{35} = 0.750$ day^{-1}). Nothing is known of the microflora species composition, but it must have been different from Combination V, and this could account for some difference in transfer to millipedes.

Cesium transfer from millipedes to litter was reduced almost half compared to Combination IV ($\lambda_{41} = 0.250$). This is due partly to soil affinity for excreted cesium ($\lambda_{42} = 0.160$), but the combined loss to soil and litter ($\lambda_{41} + \lambda_{42} = 0.410$) is still somewhat less than the lost to litter in Combination V ($\lambda_{41} = 0.475$). The high transfer from millipedes to microflora ($\lambda_{43} = 0.250$ vs. 0.050 in Comb. IV) may be due to improved nutritional and moisture conditions with soil present. This contributed to a higher observed elimination rate for *Dixidesmus* than in Combination V ($\lambda_{41} + \lambda_{42} + \lambda_{43} + \lambda_{45} = 0.660$ vs 0.525), although this difference appears to be due largely to the method of obtaining millipede data: after each leaching, dead millipedes were removed for radioassay and replaced by new animals. Data from the dead specimens were used to represent live animals. No difficulties apparently were encountered through cesium leaching from the dead material ($\lambda_{45} = 0$), but microbial decomposition ($\lambda_{43} = 0.250$) had begun. Assuming no direct transfer to microflora from live millipedes, corrected elimination rates for *Dixidesmus* in Combs. V and VI, respectively, would be 0.475 and 0.410 day^{-1}.[2] These values compare favorably with 0.29 day^{-1} (biological half life = 2.4 days at 20°C), reported for assimilated cesium by Reichle and Crossley (1965), considering the higher temperature of the present experiments and the fact that both assimilated and unassimilated components are taken into account.

In general, the Table 1 results show that, under otherwise similar conditions, rate constants change in microcosms composed of different combinations of the same compartments. From this it can be concluded that patterns of intercompartmental coupling are prime variables to consider in studies of energy and material flows in ecosystems.

DISCUSSION

Pathways of mineral transfer in coupled microecosystems are both identifiable and quantifiable by the procedures employed, namely the simul-

[2] The values 0.475 and 0.410 may not be significantly different because of subjectivity in determining best fit in the curve-fitting procedure. Numerical approaches for use with digital computers are now being explored. A preliminary value of λ_{41} is 0.414 day^{-1} (G. M. Van Dyne, personal communication), and that by another 0.470 day^{-1}.

tancous solution of systems of differential equations with an analog computer. (i) Qualitatively, actual transfer routes of those which are possible were determined. These are illustrated in the insets of Figure 2, the pathways over which no transfers occurred being shown as broken arrows. The numbers of actual routes in the systems of Combinations I–VI were, respectively, 1, 2, 2, 3, 6 and 9. The ratios of actual pathways to possible ones were 1.00, 0.50, 0.50, 0.33, 0.67 and 0.56 (ii) Quantitatively, rates of cesium transfer along the actual routes are given by the values of the parameters, λ_{ij}.

In general, two systems are identical if their parts are the same and the connections between the parts are the same. In Figure 3, systems 1

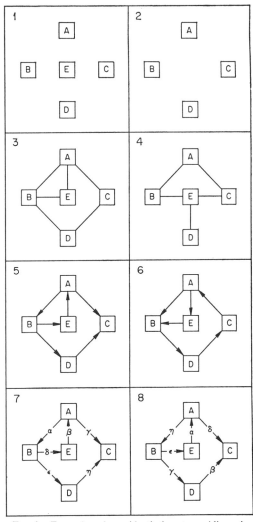

Fig. 3. Examples of non-identical systems (discussion in text).

and 2 are different because their components are different; systems 3 and 4 differ because the connections are not the same in the "qualitative" sense given above; in systems 5 and 6 the connections are different in a "relational" or "directional" sense; and systems 7 and 8 differ in the "quantitative" sense that their rate functions are different. None of the systems of Figure 3 are identical, and consequently none will behave precisely the same as another whether it be in reaction to some environmental stimulus, in utilization of energy, or in the cycling of minerals, if these be appropriate system activities. The "structure and function" of these systems is said to be different. Similarly, the six microecosystems of this study have been shown to differ in radiocesium kinetics, with examples (Fig. 2, insets) of each of the reasons for system differences, illustrated in Figure 3, being represented.

Analog Computer Simulations

The microcosm models which have been obtained are operational, and it now becomes possible to study comparative transient and stationary behavior of these systems on the computer, in effect performing experiments which would be difficult or impossible to conduct in the experi-

mental laboratory. The number of experiments possible is infinite, and the ones actually performed were selected to bring out aspects of microecosystem behavior of general ecological interest.

To generate system behavior, a driving or forcing function, $\lambda_{01} x_0$, was introduced via the litter compartment (Fig. 1). In the six experiments preceding, there was only an initial cesium concentration in the litter, but no subsequent input from the environment ($\lambda_{01} = 0$). Driving the systems through the litter is somewhat artificial, just as establishing the microcosms with tagged litter is, because in nature each compartment is exposed directly to cesium input as environmental fallout. The experiments might have been more realistic, therefore, if an initial cesium dose had been rained on as part of the first leaching rather than introduced as tagged litter. This would not have changed any system parameters (Table 1), but it would have yielded additional parameters of interest (λ_{0j}; $j = 1, 2, \ldots, 5$). The simulations could then have been driven by multiple inputs. The present model is still very useful for relative comparisons of cesium behavior in the six microcosms. Relationships between radioactivity and voltage, and real time and computer time are clarified in the appendix.

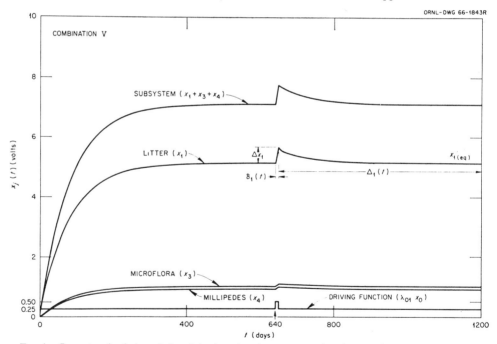

Fig. 4. Computer simulation of time behavior of cesium concentration (x_j) in the Combination V system, with constant input and perturbation after equilibrium. Soil and leachate did not equilibrate and are omitted. Shown is the driving function of 0.25 v, increased to 0.50 v between 640 and 648 days; also litter, microflora and millipede compartments, and the subsystem of these three compartments. The variables $\delta_j(t)$, $\Delta_j(t)$, $x_{j(eq)}$ and Δx_j are illustrated for the litter compartment.

Experiments of 1200 days (3.3 years) duration were run (Fig. 4). The driving function arbitrarily selected was a constant $\beta^{-1}\lambda_{01} x_0 = 0.25$ v sec^{-1}, obtained by letting $x_0 = 10$ v, $\lambda_{01} = 0.0125$ day^{-1}, and β (time-scale factor, see appendix) $= \frac{1}{2}$ sec day^{-1}. All compartments were radiocesium-free initially ($x_j(0) = 0$, all $j > 0$). Figure 4 shows results for the Combination V system. A similar graph was obtained for each of the other microcosms from which the data of Table 2 were then read off and computed.

Figure 4 illustrates general aspects of system behavior in these simulations under the influence of the constant "radiocesium" input. Three com-

TABLE 2. Equilibrium characteristics of compartments, and subsystems composed of equilibrated compartments.

Combinations	Litter	Soil	Microflora	Millipedes	Leachate	Equilibrium Subsystems
Time to Equilibrium (days)						
I	264	—	—	—	∞	264
II	328	∞	—	—	∞	328
III	264	—	∞	—	∞	264
IV	168	∞	∞	—	∞	168
V	616	—	448	432	∞	616
VI	232	∞	186	168	∞	232
[134]Cs Concentration (v)						
I	3.54	—	—	—	∞	3.54
II	3.52	∞	—	—	∞	3.52
III	3.52	—	∞	—	∞	3.52
IV	2.20	∞	∞	—	∞	2.20
V	5.10	—	1.01	0.93	∞	7.04
VI	2.28	∞	0.47	0.33	∞	3.08
Concentration Factors						
I	14.16	—	—	—	∞	14.16
II	14.08	∞	—	—	∞	14.08
III	14.08	—	∞	—	∞	14.08
IV	8.80	∞	∞	—	∞	8.80
V	20.40	—	4.04	3.72	∞	12.27
VI	9.12	∞	1.88	1.32	∞	5.76
Input (v day⁻¹)						
I	0.13	—	—	—	0.13	0.26
II	0.13	0.13	—	—	0.003	0.26
III	0.13	—	0.06	—	0.07	0.26
IV	0.13	0.09	0.04	—	0.002	0.26
V	0.57	—	0.38	0.48	0.13	1.56
VI	0.44	0.13	0.36	0.22	0.02	1.17
Output (v day⁻¹)						
I	0.13	—	—	—	0.00	0.13
II	0.13	0.00	—	—	0.00	0.13
III	0.13	—	0.00	—	0.00	0.13
IV	0.13	0.00	0.00	—	0.00	0.13
V	0.56	—	0.38	0.49	0.00	1.43
VI	0.45	0.00	0.35	0.22	0.00	1.02
Turnover (% day⁻¹)						
I	3.70	—	—	—	0.00	3.70
II	3.70	0.00	—	—	0.00	3.70
III	3.60	—	0.00	—	0.00	3.60
IV	6.03	0.00	0.00	—	0.00	6.03
V	11.00	—	37.50	52.50	0.00	101.00
VI	19.89	0.00	75.00	66.00	0.00	160.89
Stability (%)						
I	21.60	—	—	—	—	21.60
II	18.08	—	—	—	—	18.08
III	23.44	—	—	—	—	23.44
IV	18.08	—	—	—	—	18.08
V	20.40	—	24.96	29.04	—	21.61
VI	10.43	—	12.23	10.64	—	10.02

partments—litter, microflora and millipedes—went through a transient response followed by a steady state. Soil and leachate never equilibrated because they are cesium sinks at the concentrations considered. Equilibrial characteristics of interest include (i) *time to equilibrium,* {days}; (ii) radiocesium *concentration,* {v}; (iii) *concentration factor,* a dimensionless property defined for present purposes as the ratio of compartment concentration to the voltage of the forcing function, 0.25 v; (iv) cesium *input flux* (input to compartment $j = \Sigma_i \lambda_{ij} x_i$), {v day^{-1}}; *output flux* (output of $j = \Sigma_i \lambda_{ji} x_j$), {v day^{-1}}; (vi) *turnover rate,*

calculated for compartment j either as $\Sigma_i \dfrac{\lambda_{ij} x_i}{x_j}$ or as

$\Sigma_i \dfrac{\lambda_{ji} x_j}{x_j} = \Sigma_i \lambda_{ji}$, {day^{-1}}; and (vii) *stability.*

Stability is the capacity of a system, subsystem, or compartment to minimize effects of a disturbance. To assess this for the six microcosms, the systems were disturbed after compartments which would equilibrate had reached equilibrium. This was accomplished by doubling the input to $\beta^{-1} \lambda_{01} x_0 = 0.5$ v sec^{-1} for 8 days and then returning it to its former level (Fig. 4: the driving function is doubled between 640 and 648 days). The disturbance caused a perturbation in equilibrated compartments (the nonsteady state ones were ignored; as cesium sinks they do not feed back to the remaining system and hence cannot affect it). Removal of the disturbance was followed by return to former equilibrium levels (Fig. 4: the graph shows deflections between 640 and 648 days, followed by recovery of equilibrated compartments—litter, microflora and millipedes— and of the subsystem comprised of these compartments). If $x_{j(eq)}$ is the equilibrium cesium concentration of compartment or subsystem j, Δx_j the perturbation of the same unit, $\delta_j t$ the duration of the perturbation, and $\Delta_j t$ the time for j to return to the original equilibrium (see Fig. 4), then a good measure of the stability of j is provided by $(x_{j(eq)}) (\delta_j t) / (\Delta x_j) (\Delta_j t)$. This is because the less the disturbance, (Δx_j), relative to the equilibrium concentration, $(x_{j(eq)})$, and the more rapid the recovery time, $(\Delta_j t)$, relative to the duration of the disturbance, $\delta_j t$, the greater is the stability of j. The measure is dimensionless.

The following discussion will be limited to a general comparison of principal system differences which appear in the table.

Equilibration time.—Given a nonzero input, in order for a system, subsystem, or compartment to achieve a steady state it must also have an output which is nonzero. Thus, while soil and leachate

in all of the microcosms studied had cesium inputs, they lacked outputs (Fig. 2, insets), and consequently never equilibrated (Table 2). Similarly, the microflora of Combinations III and IV accumulated radiocesium, but did not lose it. All other compartments eventually equilibrated at cesium levels defined by the equilibrium solutions of equations (1).

As shown in Table 2, the time to achieve this varied almost 4-fold, from 168 days for the litter of Combination IV and the millipedes of Combination VI to 616 days for the litter of Combination V (values to the nearest 8 days). Two general points illustrated by the tabulated data are that different compartments of the same microcosm and the same compartments in different microcosms take different periods of time to reach equilibrium.

Equilibrium Concentration of ^{134}Cs.—Compartments which do not equilibrate are indicated in Table 2 by "∞." Equilibrium concentrations, $x_{j(eq)}$, varied from 0.33 to 5.10 v in individual compartments, and from 2.20 to 7.04 v in subsystems composed of the steady state compartments. The litter value of 5.10 for Combination V is unusually high and more than double the equilibrium level of this compartment in Combination VI. The only difference between these two microcosms is that one contained soil. Since the environmental input to all of these systems was identical ($\lambda_{01} x_0 = 0.125$ v day^{-1}), the data of Table 2 indicate considerable variability in steady state cesium levels. This is a function of the components present and how they are coupled together.

Concentration Factors.—The observed range was 1.32 for millipedes in Combination VI to 20.40 for litter in Combination V. Since the driving function was constant, these figures are multiples (4×) of the equilibrium cesium levels. Expressed as concentration factors, though, they give some additional insights. The change in relative cesium-concentrating ability of *Dixidesmus* in Combinations V and VI can be interpreted in terms of the defining equations. The equilibrium solutions of equations (1) are

$$0 = \Sigma_i \lambda_{ij} x_{i(eq)} - \Sigma_i \lambda_{ji} x_{j(eq)},$$

signifying input = output,

$$\Sigma_i \lambda_{ij} x_{i(eq)} = \Sigma_i \lambda_{ji} x_{j(eq)},$$

and giving for the equilibrium concentration of compartment j

$$x_{j(eq)} = \left(\Sigma_i \lambda_{ij} x_{i(eq)} \right) \Big/ \Sigma_i \lambda_{ji}. \qquad (3)$$

Note (definitions p. 822) that the numerator

is the compartment's input flux and the denominator its turnover rate.

$$(V) \quad x_{4(eq)} = \frac{\lambda_{14}x_{1(eq)} + \lambda_{34}x_{3(eq)}}{\lambda_{41} + \lambda_{43}} = \frac{(0.020)\,(5.10) + (0.375)\,(1.01)}{0.475 + 0.050} = 0.92 \text{ v}$$

$$(VI) \quad x_{4(eq)} = \frac{\lambda_{14}x_{1(eq)} + \lambda_{34}x_{3(eq)}}{\lambda_{41} + \lambda_{42} + \lambda_{43}} = \frac{(0.045)\,(2.28) + (0.250)\,(0.47)}{0.250 + 0.160 + 0.250} = 0.33 \text{ v}$$

Note that these values are almost identical to those given in Table 2, which were read off the computer output graphs. These calculations indicate, subject to the limitations of counting dead millipedes for radioactivity, as discussed earlier, that the equilibrium cesium concentrations—and hence the concentration factors for *Dixidesmus*—differ between Combinations V and VI. Correcting for loss to microflora through decomposition of the dead animals (i.e., assuming $\lambda_{43} = 0$), the concentration factors would still be different because $x_{4(eq)} = 0.29$ v in Combination V and 0.54 v in VI. In general, concentration factors for given compartments, j, would be expected to change in communities with different compartment compositions (different i's) as the input flux (numerator of equation 3), and possibly also the turnover (denominator), changed due to differences in the i's. This is borne out in Table 2 for all equilibrated compartments and subsystems.

Input and Output Fluxes.—Compartments, subsystems or systems in steady state with respect to a material have input = output. This is illustrated in Table 2. For equilibrated compartments in the six microcosms, inputs and outputs ranged from 0.13 to 0.56 v day^{-1}. For subsystems of equilibrated compartments, the input range was 0.26 to 1.56 v day^{-1}, and the output 0.13 to 1.43 v day^{-1}. These data illustrate the following points about material equilibria in coupled systems, all a consequence of network organization. Within a system, different compartments will tend to have different stationary input/output fluxes. Between systems driven by identical forcing functions, the same compartments may have quite different inputs and outputs. Therefore, the total flux of a material in a steady state system may vary considerably from that in another system which receives identical input from the environment.

Turnover.—The turnover rate of a compartment, j, is most easily computed as the sum of output rates, $\Sigma_i \lambda_{ji}$, rather than as the ratio of input flux to concentration, $\Sigma_i \frac{\lambda_{ij}x_i}{x_j}$. Turnover is thus equivalent to elimination rate. Calculations of turnover as $\Sigma_i \lambda_{ji}$ can be made directly from Table 1, and consequently are independent of the hypothetical simulations. The turnovers listed in

Table 2, therefore, are real values for the experimental microcosms.

Table 2 indicates that each compartment tends to have higher turnover or elimination rates in systems with more compartments. This is because there are more compartments, i, from which to receive inputs in $\Sigma(\lambda_{ij}x_i)/x_j$, and to which to transfer outputs in $\Sigma\lambda_{ji}$. The result does not apply necessarily to compartments whose elimination is controlled by internal physiology, and the millipede data have already been qualified (p. 819, and footnote 2). The corrected daily millipede turnover rates are 47.5% in Combination V, and 41.0% in Combination VI. As more compartments are accrued in a system, total system turnover tends to increase. This is illustrated in the right-hand column of Table 2, and is consistent with general notions about energy and material flows in relation to ecosystem complexity.

Stability.—Data tabulated in Table 2 range from 10.43×10^{-2} to 29.04×10^{-2} for individual compartments, and 10.02×10^{-2} to 23.44×10^{-2} for the subsystems of equilibrated compartments. It is indicated that different compartments of a system can have different, although here generally similar, stabilities, and that the stability of cesium concentration in a given compartment varies from system to system. There appears to be no correlation between system complexity and ability to minimize and damp perturbations of radiocesium concentrations, which is contrary to inference from the stability theory of MacArthur (1955).

CONCLUSIONS

This investigation demonstrates that mineral kinetics in small laboratory microecosystems of limited complexity can be successfully modeled, and then their transient and stationary behavior studied comparatively with an analog computer. Environmental conditions and the compartments employed were the same for all systems, but the compartment combinations were different, and this was the principal experimental variable. It resulted in different patterns and degrees of intercompartmental coupling, producing radiocesium kinetics which were unique for each microcosm. The variable kinetics defined in turn variable system behavior. System and compartment attributes which changed with the nature of the coupling networks were duration of transient response, equilibrial cesium concentrations, concentration factors, input and output fluxes, turnover rates, and stability.

These results focus attention on the exceeding importance in natural, complex ecosystems of the organizational networks which define compartment interactions. Only one of numerous substances simultaneously transferred in the present microcosms, probably each with unique kinetics, was actually studied. Changes in the concentration or dynamics of any of these, especially major nutrients, would probably have altered the cesium kinetics. The multiplicity of material transfers and interactions conceivable in macroecosystems, together with the effects of intrasystem coupling as revealed in this investigation, make it apparent that to understand ecosystems ultimately will be to understand networks.

Appendix

Relationships between the microcosms and their models are as follows. Let

$x_j \equiv$ radionuclide activity-density of compartment j, in appropriate units, $\{a\}$.

$t \equiv$ real time, $\{days\}$.

$\lambda_{ij} \equiv$ rate constant for transfer of radioisotope from compartment i to j, $\{t^{-1}\}$.

$\lambda_p \equiv$ rate constant for physical radioactive decay $\{t^{-1}\}$; $\lambda_p = -0.0009$ day^{-1} for ^{134}Cs.

$v \equiv \{volts\}$.

$\alpha_j \equiv$ scale factor relating radioactivity to computer voltage for compartment j, $\{va^{-1}\}$.

$\tau \equiv$ computer time, $\{sec\}$.

$\beta = \tau t^{-1} \equiv$ scale factor relating real time to computer time, $\{sec\ day^{-1}\}$.

The system equations (1), prior to being corrected for radioactive decay, would be

$$\dot{x}_j = \sum_i \lambda_{ij} x_i - \sum_i \lambda_{ji} x_j - \lambda_p x_j .$$

Correction of the primary data before curve-fitting eliminates physical decay from further consideration, both in determination of rate constants, and in simulations with the determined coefficients. In (1), both sides of each equation have the units $\{at^{-1}\}$. With voltage scaling, (1) becomes

$$[\alpha_j \dot{x}_j] = \sum_i \frac{\alpha_j(\lambda_{ij})[\alpha_i x_i]}{\alpha_i} - \sum_i \frac{\alpha_j(\lambda_{ji})[\alpha_j x_j]}{\alpha_j}$$

$$= \sum_i \left(\frac{\alpha_j}{\alpha_i}\right)(\lambda_{ij})[\alpha_i x_i] - \sum_i (\lambda_{ji})[\alpha_j x_j] ,$$

and the new units are $\{vt^{-1}\}$. The ratios (α_j/α_i) are input gains on integrators, and terms in brackets are scaled computer variables. Substituting $\beta\tau^{-1}$ for t^{-1} to achieve time scaling,

$$\beta[\alpha_j \dot{x}_j] = \sum_i \left(\frac{\alpha_j}{\alpha_i}\right)(\lambda_{ij})[\alpha_i x_i] - \sum_i (\lambda_{ji})[\alpha_j x_j] ,$$

or

$$[\alpha_j x_j] = \sum_i \left(\frac{\alpha_j}{\alpha_i}\right)\left(\frac{\lambda_{ij}}{\beta}\right)[\alpha_i x_i] - \sum_i \left(\frac{\lambda_{ji}}{\beta}\right)[\alpha_j x_j] ,$$

with units $\{v\tau^{-1}\}$. Thus, the real systems, with units $\{at^{-1}\}$, are converted to computer systems with units $\{v\tau^{-1}\}$.

When determining values of λ_{ij} in the Results section, voltage scale factors were $\alpha_i = \alpha_j = 10v/100\%$ radioactivity $= 0.1\{va^{-1}\}$. The time-scale factor was $\beta = 15$ sec/30 days $= \frac{1}{2}\{\tau t^{-1}\}$. The equations actually used instead of equation (2) were then

$$[0.1 \dot{y}_j] = \sum_i 2(\lambda_{ij})[0.1 y_i] - \sum_i 2(\lambda_{ji})[0.1 y_j] ,$$

where the y's are dimensionless.

In the simulations (Discussion), x_j was in arbitrary units, requiring no explicit scale factors, α_j. Computing time was $\tau = 600$ sec, representing a real time of 1200 days ($\beta = \frac{1}{2}\{\tau t^{-1}\}$). Scaled equations for all compartments except litter were

$$[\dot{x}_{j \neq 1}] = \sum_{i=1}^{5} 2(\lambda_{ij})[x_i] - \sum_{i=1}^{5} 2(\lambda_{ji})[x_j] .$$

For the litter, driven by a forcing function as input,

$$[\dot{x}_1] = \sum_{i=0}^{5} 2(\lambda_{i1})[x_i] - \sum_{i=1}^{5} 2(\lambda_{1i})[x_1] .$$

Literature Cited

Jansson, S. L. 1958. Tracer studies on nitrogen transformations in soil with special attention to mineralisation-immobilization relationships. Ann. Royal Agric. Coll. Sweden 24: 101–361.

MacArthur, R. 1955. Fluctuations of animal populations, and a measure of community stability. Ecology 36: 533–536.

Neel, R. B. and J. S. Olson. 1962. Use of analog computers for simulating the movement of isotopes in ecological systems. Oak Ridge Nat. Lab. Rep. 3172: 1–111.

Olson, J. S. 1965. Equations for cesium transfer in a *Liriodendron* forest. Health Phys. 11: 1385–1392.

Reichle, D. E. and D. A. Crossley. 1965. Radiocesium dispersion in a cryptozoan food web. Health Phys. 11: 1375–1384.

Remezov, N. P. 1959. Methods of studying the biological cycle of elements in forests. Sov. Soil Sci.: 59–67.

Whittaker, R. H. 1961. Experiments with radiophosphorus tracer in aquarium microcosms. Ecol. Monogr. 31: 157–188.

Witherspoon, J. P. 1964. Cycling of Cs^{134} in white oak trees. Ecol. Monogr. 34: 403–420.

Witkamp, M. and M. L. Frank. 1964. First year of movement, distribution and availability of Cs^{137} in the forest floor under tagged tulip poplars. Rad. Bot. 4: 485–495.

3

Reprinted from pages 731–739 of *Symposium on Radioecology*, Proc. Second Natl. *Symp.*, Ann Arbor, Mich., 1967, D. J. Nelson and F. C. Evans, eds., Washington, D.C.: U. S. At. Energy Comm., 1969, 744pp.

SYSTEMS ANALYSIS OF A COUPLED COMPARTMENT MODEL FOR
RADIONUCLIDE TRANSFER IN A TROPICAL ENVIRONMENT[1]

Stephen V. Kaye and Sydney J. Ball

*Health Physics Division and Instrumentation and Controls Division,
Oak Ridge National Laboratory, Oak Ridge, Tennessee*

Introduction

One of the principal objectives of the on-site investigations for the sea-level canal feasibility study is to gather and analyze environmental data that will be used to identify the pathways of radionuclide transfer to man, in order to estimate the radiation doses to man from radionuclides released by nuclear devices which might be used for excavation. Fulfillment of this objective will require a large, well-coordinated effort for data acquisition, advanced methods for machine handling of field data, and proper use of these data in estimating potential radiation doses for the feasibility study. The problems and questions are many. For instance, which of the over 100 radionuclides to be produced by the detonations may represent a negligible hazard, and which may pose a significant hazard that will in turn require further detailed investigations? How will each of these radionuclides move through the complex ecosystems of the tropical environment?

The major objective of this paper is to present a preliminary scheme for analyzing the pathways of radionuclide transfer in the environment. This preliminary scheme for environmental pathway analysis is based partly on the use of systems analysis techniques. This approach uses an income and loss model which allows one to calculate the radionuclide concentration in any environmental compartment. Two of the most important kinds of information required are: (1) an accurate compartmentalized representation of the tropical environment (including the Atlantic Ocean, Pacific Ocean, terrestrial habitats, freshwater habitats, and estuaries); and (2) the rate constants which quantify the intercompartmental transfer of radionuclides. Obviously, there will have to be numerous compromises because of the complexity of the system. An extensive field program is in progress now, and on-site data are being collected which will be useful for testing and refining the models.

[1]Research sponsored by the U.S. Atomic Energy Commission under contract with the Union Carbide Corporation. Parts of this effort were also supported under Purchase Contract No. S6230 from Battelle Memorial Institute, Columbus, Ohio, under the auspices of the AEC Nevada Operations Office for the Atlantic-Pacific Interoceanic Canal Commission.

Coupled Compartment System

A convenient way of dealing with environmental pathways for analytical purposes is the diagrammatic representation of the environment by a network of coupled compartments. Each compartment represents an environmental unit (e.g., grasses and herbs, surface water, bananas, sea turtles, etc.). Each has income and loss terms (via environmental pathways) which alter the inventory of material within the compartment. The types of inventory of interest include stable element, radionuclide, and in some cases biomass. The aspect of biomass of most interest is the flux of food and water to man, since they are the vehicles for transfer of radionuclides and stable elements. Amounts of food must be considered, along with activity per gram of food or per gram of element, to estimate nuclide intake in the terms needed for internal-dose calculations.

Income and Loss Model

Net flux to a compartment is the difference of income and loss rates. This simple principle may be represented in various ways such as the equation

$$m_j \frac{dX_j}{dt} = \sum_{\substack{i=0 \\ i \neq j}}^{n} \lambda_{ij} m_i K_{ij} X_i - \sum_{\substack{i=0 \\ i \neq j}}^{n} \lambda_{ji} m_j K_{ji} X_j - \lambda_r m_j X_j \ ,$$

where

j = a subscript designating the compartment of reference,
i = a subscript designating any donor or receptor compartment other than compartment j,
X = radionuclide concentration (μCi/g),
λ = environmental transfer coefficient $\lceil$(g transferred)/(total g present) (day) = (days^{-1})$\rceil$,
λ_r = radioactive decay constant (days^{-1}),
m = biomass per unit area (g/m^2), the subscript designation depending upon how the biomass transfer is controlled,
K = selectivity factor which adjusts the environmental transfer coefficient when it differs quantitatively from the radionuclide transfer coefficient.

This equation is easily adapted to uses of specific activity measurements from the relationship $X = C \cdot S$, where X is as defined above, C is the concentration of element in compartment material (g element/g material), and S is the specific activity in the compartment (μCi/g element).

Coupled Compartment Diagram of the Tropical Environment

A general diagram of a compartmental model of coupled pathways is shown in Fig. 1. One purpose of the entire field program of the feasibility study is to define the compartments and pathways, and to estimate the transfer coefficients involved in the food chains leading to indigenous man. The model diagrammed in Fig. 1 is based only on preliminary information, and it is presented to illustrate the method of tracing radioactivity through the food web from point of entry to man. As other data become available the model can be modified to provide a more detailed, more realistic representation of the important compartments and critical food-chain pathways. Proper implementation of this model may result in a step toward estimating the fraction of the radioactivity entering the environment which contributes to the internal radiation dose received by man.

The left side of each compartment is the income side, and the right side is the loss side. Compartments which have direct food-chain inputs to man have an arrow leaving the bottom side of the compartment box.

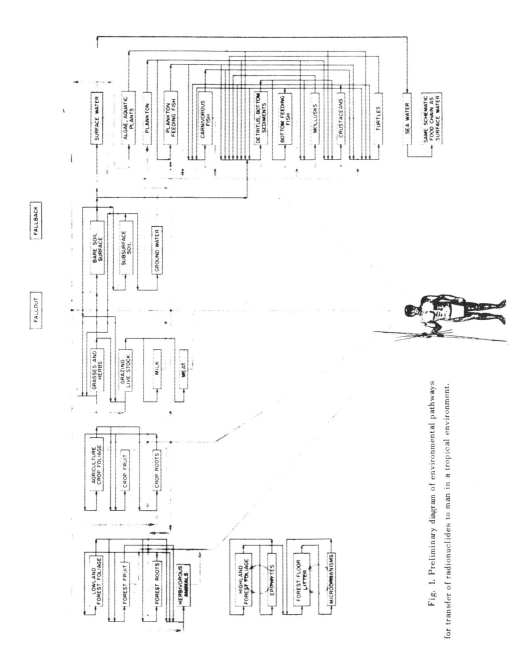

Fig. 1. Preliminary diagram of environmental pathways
for transfer of radionuclides to man in a tropical environment.

33

Fallout and fallback comprise two source compartments for the model. The loss arrow from the fallout compartment can be followed to see where it becomes income from the primary interceptor compartments. Visualize the fallout descending on the environment and falling on forests, agricultural crops, grasses and herbs, bare soil, surface water, aquatic plants, and seawater. There may even be inhalation by livestock and man. The other source compartment, fallback, has inputs to subsurface soil, ground water, and surface water. The income to any primary receptor compartment from the two source compartments is proportioned out as loss from the primary receptor compartments to arrive as income to other compartments in the coupled system.

Forests have been arbitrarily divided into a highland ecosystem ($\sim$ 300 m elevation, or ecologically similar to the El Verde forest in Puerto Rico) and a lowland ecosystem ($<$300 m elevation). The highland forest may be subject to runoff and possible erosion and the lowland ecosystem may accumulate water or sediments from the uplands. Recent research by Kline and Odum (1966) has shown that epiphytes on foliage were apparent accumulators of fallout radionuclides in tropical forests; and that when radionuclide tracers (^{134}Cs, ^{85}Sr, and ^{54}Mn) were sprayed on the forest floor at El Verde, there was negligible uptake by roots (Kline 1966). Canopy and understory leaves in lower montane tropical forests might get many of their nutrients by aerial interception, including interception by epiphytic organisms, but chemical budget studies to determine this have been started only recently (Kline 1966, Witkamp, in press). The fraction of the fallout initially washed from the leaves to the forest floor in Fig. 1 might be cycled between the forest floor microorganisms and the litter, because studies in the El Verde forest have shown that radionuclides reaching the forest floor may not be available as income to the underlying soil and ground water (Kline 1966). However, Kline (in press) has also shown that a small fraction of some radionuclides may be taken up by the roots which form a thick mat in the litter layer. Witkamp (in press) has shown that there is a high year-round activity of microorganisms in the forest floor litter at El Verde. This observation may account for the apparent retention of some radionuclides within the forest floor ecosystem of the lower montane forest.

Provisionally, assume that the forest canopy and the forest floor litter function as ecosystem sinks in the highland forest, as shown by the model, and that there is no loss to compartments outside the highland forest ecosystem (e.g., by erosion of litter and soil to the lowlands). Thus, in this hypothetical situation, the fraction of the total fallout that falls in the highland ecosystem would be isolated from any direct inputs to man (unless man foraged there for food). If this ecosystem were to become a sink, it would be of significant importance because it would apply to the area of the continental divide in Panama where the elevation is about 300 m. Whether or not these broad generalizations and extrapolations between El Verde and Panama can be made remains to be determined from field observations. The *first* major question about the pathways diagram is whether there may be significant movement of radioactivity from the upland forests (in organic or inorganic form) to the lowland floodplains where the location of agriculture may be critical. Is it misleading to assume no such transfer, as the diagram implies?

Another feature of the diagram is the apparent potential importance of both ground water and surface water as pathways for radioisotope transfers eventually leading to man. All major ecosystem compartments, except perhaps the highland forest ecosystem, may have incomes from surface water or ground water. The *second* major question, therefore, is whether ground water can be expected to carry significant quantities of radionuclides, in view of the potential sorption of radionuclides in tropical soils.

The diagram has not been expanded at this time, in order to keep the preliminary model as simple as possible for illustrative reasons, and because actual field data were not available to do otherwise. Possibly some compartments should be subdivided into numerous subcompartments to provide a more realistic approximation of the environment. For example, it may be necessary to add one or more soil compartments for the crop, pasture, and forest areas, since minerals in these compartments may provide more-or-less immobilized "sinks" for many nuclides. Alternately, all root compartments may have to be pooled with soil until there is more specific information on roots and soils. In Oak Ridge forest-tagging studies, and in many fallout studies, a major question has been: How soon will critical elements be taken out of the local biogeochemical cycles and transferred into compartments with very slow turnover? This is a *third* major question to be answered for each of the broad ecosystems in Panama or Colombia.

Losses from crop foliage to the soil surface or herbivorous animals are not shown in Fig. 1, but may need to be added to remove the unrealistic simplifying assumption that man consumes all of the crops he grows. The coupling of the crop fruit compartment is an interesting example, because it shows that crop fruit has loss only to man. Thus, the *fourth* major question, or complex of questions, concerns the typical and extreme amounts of production of crops per unit area, activity per unit weight (fresh and dry), and the amounts or fractions of this total which actually enter the digestive tract after wasted and discarded production are taken into account. Some of these numbers are available now (or can be assumed with reasonable accuracy), and others will be provided by field investigations now in progress.

In many cases it may be necessary to deal with a single large compartment measured by an average value rather than to measure the individual small compartments that make up the large compartment. For instance, it would be easier to estimate the average concentration of radioactivity for total plankton than it would be to estimate the concentration of radioactivity for all individual species of plankton.

Uses of Frequency Response Techniques in Exposure Pathway Analysis

Typically a compartment model consists of a set of first-order nonlinear differential equations, each equation accounting for the inventory changes of a radionuclide or biomass. By considering all banana stems, all grass, or all dairy cows in a landscape area as distinct compartments, we imply, for example, that a homogeneous radionuclide distribution can be assumed for an "average" stem, patch of grass, or cow. Thus, the success of the compartment approach depends upon the accuracy with which we can describe and measure transfer rates on a macroecological, rather than microecological, scale. If it were necessary to distinguish between all the different species of grass, whether or not the grass is located under a palm tree or open skies, 10 cm or 100 m from a stream, etc., then compartment model solution would be hopelessly complex.

A major problem in radionuclide transfer is the large uncertainty in the postulated mathematical models and in the parameters used in these models. Hence in this initial phase of the work it is important to gain an understanding of the relative importance of these uncertainty factors in estimating intakes to man. For example, if certain aspects of the dose computation (such as integrated intake and peak intake) are very sensitive to changes in a parameter, this indicates that it would be worthwhile to try to obtain accurate data on this parameter. On the other hand, an insensitive parameter would merit little additional research.

A straightforward method of performing this sensitivity analysis is to compute the time response of a "reference model," then vary in turn each parameter an amount ΔP_i and compute the difference in the response ΔR_i. The quotient $\Delta R_i / \Delta P_i$ is known as the sensitivity, and its limit as ΔP_i approaches zero is $\partial R_i / \partial P_i$. The difficulty with this approach is that it requires considerable computation. If we use linear, constant coefficient models, then methods can be employed to compute sensitivities directly for the model.

A sensitivity analysis using frequency response techniques can be performed for illustrative purposes using a small system such as the one shown in Fig. 2. This hypothetical five-compartment banana plantation is described by the parameters listed in Table 1. We visualize this plantation as a uniform 1 ha plot (no other plants living on area) which has negligible inputs and outputs with environmental systems other than the ones listed in Table 1. Our initial conditions specify a fallout pulse input of 100 μCi/m^2 divided equally between the leaf compartment and soil compartment. The following five system equations are mathematical representations of the incomes and losses shown in Fig. 2.

Compartment 1 — leaves

$$m_1 \frac{dX_1}{dt} = -m_1 \lambda_r X_1 - m_1 \lambda_{1,3} X_1 - m_1 \lambda_{1,5} X_1 + m_3 \lambda_{3,1} X_3$$

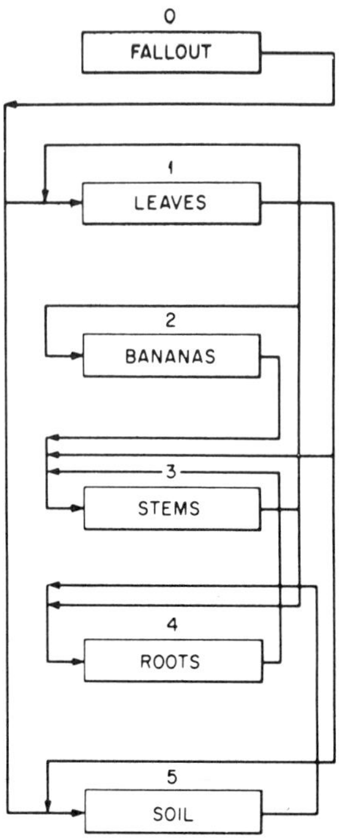

Fig. 2. System diagram of a hypothetical banana plantation.

Compartment 2 – bananas

$$m_2 \frac{dX_2}{dt} = -m_2 \lambda_r X_2 - m_2 \lambda_{2,3} X_2 + m_3 \lambda_{3,2} X_3$$

Compartment 3 – stems

$$m_3 \frac{dX_3}{dt} = -m_3 \lambda_r X_3 - m_3 \lambda_{3,1} X_3 - m_3 \lambda_{3,2} X_3 - m_3 \lambda_{3,4} X_3$$

$$+ m_1 \lambda_{1,3} X_1 + m_2 \lambda_{2,3} X_2 + m_4 \lambda_{4,3} X_4$$

Compartment 4 – roots

$$m_4 \frac{dX_4}{dt} = -m_4 \lambda_r X_4 - m_4 \lambda_{4,3} X_4 + m_3 \lambda_{3,4} X_3 + m_5 \lambda_{5,4} X_5$$

Compartment 5 – soil

$$m_5 \frac{dX_5}{dt} = -m_5 \lambda_r X_5 - m_5 \lambda_{5,4} X_5 + m_1 \frac{\lambda_{1,5}}{2} (X_1 - X_5) + m_1 \frac{\lambda_{1,5}}{2} (X_1)$$

We have specified two pathways which are quantitatively equal for transfer of radioactivity from leaves to soil. The first pathway term, written $m_1 \lambda_{1,5} (X_1 - X_5)/2$, is the case where there is a

36

Table 1. System parameters for a hypothetical banana plantation

Compartment Number	Biomass (g/m^2)	Loss Pathway	Transfer Coefficient ($days^{-1}$)
1	6×10^3	1,3	0.099
		1,5	0.35
2	4.6×10^3	2,3	0.014
		Harvest	0.014
3	9×10^2	3,1	2.8
		3,4	0.099
		3,2	2.8
4	8×10^2	4,3	2.8
5	3×10^{5} [a]	5,4	0.023
		Radioactive decay	0.0128

[a]Only top 15.2 cm (6 in.) of soil is arable.

mass transfer accompanying the radionuclide transfer from compartment 1 to compartment 5, such as in leaf fall. The second pathway term, $m_1 \lambda_{1,5}(X_1)/2$, is used to represent the "wash out" or "fall off" of radioactivity from leaves to soil. Two frequency response techniques developed for analyzing coupled systems such as the banana plantation will now be described.

A computer plot of the radionuclide concentration, X_2, in the bananas vs time for our postulated plantation is shown in Fig. 3. One response parameter of interest is the integrated concentration in the bananas, (ICB), represented by the area under the time-response curve. This can be represented by a new function X_{ICB}, where $dX_{ICB}/dt = X_2$ so that $\int_0^\infty X_2'(t)\, dt = X_{ICB}|_{t=\infty}$, where in this case $X_2'(t)$ is the response to an impulse input from fallout. The impulse response of a system and its frequency response are Laplace transform pairs as shown by

$$\mathcal{L}\,[X_2'(t)] = G_2(S)\,,$$

where

$\mathcal{L}$ is the Laplace transform operator,

S is the Laplace transform variable,

$G_2(S)$ designates the frequency response for concentration in bananas.

Noting this relationship, we can establish the relationship between frequency response and integrated concentration. From the final value theorem,

$$\lim_{t \to \infty} X_{ICB}(t) = \lim_{S \to 0} S \cdot G_{ICB}(S)\,,$$

and it follows that

$$S \cdot G_{ICB}(S) = \mathcal{L}\left[\frac{dX_{ICB}}{dt}\right] = \mathcal{L}\,[X_2'(t)]\,.$$

Thus, the frequency response $G_2(S)|_{S=0}$ is numerically equal to the integrated concentration for infinite time. Both $G_2(S)|_{S=0}$ and the sensitivity of $G_2(S)|_{S=0}$ to parameter changes are calculated directly by the SFR-3 computer code developed at ORNL (Kerlin and Lucius 1966).

37

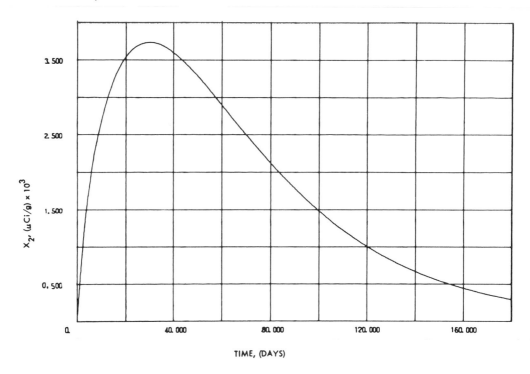

X_2, (μCi/g) $\times 10^3$

TIME, (DAYS)

Fig. 3. Computer plot of radionuclide concentration in bananas as a function of time for an impulse input of 100 μCi/m^2.

A second response parameter of interest is the maximum radionuclide concentration in the banana compartment. An estimate of this peak value, the time at which it occurs, and the sensitivity of its magnitude to parameter changes can all be calculated directly by the SFR-3 code.

A frequency response curve is directly interpreted as the response of a linear system to a sustained sinusoidal input, and many good rule-of-thumb methods are available in standard references for relating frequency response and step response (Grabbe et al. 1958). For example, the magnitude of a peak in the frequency response amplitude is directly related to the peak step response amplitude, and the time-to-peak is inversely related to the peak frequency. In order to interpret frequency response in terms of an impulse, rather than a step input, we can make use of the fact that the step response and $G_2(S)/S$ are Laplace transform pairs. Hence the frequency function $S \cdot G_2(S)$ can be interpreted for an impulse response in the same way that $G_2(S)$ could be interpreted for a step response.

To illustrate some of the results of a typical frequency response analysis, consider the concentration of the radionuclide in the bananas and its sensitivity to changes in the soil-to-root transfer coefficient, $\lambda_{5,4}$. For a 100 μCi/m^2 impulse input, the integrated concentration in bananas (ICB), peak concentration (X_2 max), time-to-peak, and sensitivities of the ICB and X_2 max to changes in $\lambda_{5,4}$ are shown for 180-day time response calculations and frequency response approximations in Table 2. The differences between ICB results are due to the different effective integration times (180 days vs ∞). It should be noted that the frequency response estimate of X_2 max from the maximum amplitude of $S \cdot G_2(S)$ can only be expected to give an approximate estimate. The frequency response estimate of time-to-peak is the reciprocal of the frequency at which the amplitude of $S \cdot G_2(S)$ peaks (radians/day). Sensitivity calculations for all 15 input parameters were made by a single frequency response run, and only the one is shown here, while the time response sensitivities shown were calculated by a separate run with a 10% higher value

Table 2. Comparison of time and frequency response analyses for radionuclide concentration in bananas

	ICB $\left(\dfrac{\mu Ci \cdot days}{g}\right)$	X_2 max $\left(\dfrac{\mu Ci}{g}\right)$	Time to Peak (days)	$\dfrac{\partial ICB}{\partial \lambda_{5,4}}$ $\left(\dfrac{\mu Ci \cdot days}{g}\Big/ days^{-1}\right)$	$\dfrac{\partial X_2 \, max}{\partial \lambda_{5,4}}$ $\left(\dfrac{\mu Ci}{g}\Big/ days^{-1}\right)$
Time response[a]	0.315	0.00351	30	7.22	0.102
Frequency response	0.328	0.00467	33	7.46	0.138

[a]ICB time response calculations for 180 days.

of $\lambda_{5,4}$. As an example of how to interpret the sensitivity results, assume a 10% increase in $\lambda_{5,4}$, i.e., $\lambda_{5,4}$ increases 0.0023 from 0.023 to 0.0253 days^{-1}. The frequency response estimate would give a change in ICB of (0.0023) (7.46) = 0.017, resulting in a new ICB of (0.328 + 0.017) = 0.345 $\mu Ci \cdot days/g$. This shows that a 10% increase in $\lambda_{5,4}$ results in about a 5% increase in ICB, so that the results are quite sensitive to the assumed value of $\lambda_{5,4}$.

It has been shown that there are very powerful mathematical techniques for rapidly analyzing coupled compartments. In order to fully utilize these techniques, ecologists will have to perfect better techniques than are presently available for measuring rate processes in the environment. Furthermore, the environment will have to be reassessed in conjunction with multicompartment rate studies of environmental transfer coefficients to produce flow diagrams that will lead to flexible working models.

Acknowledgments

We wish to thank T. W. Kerlin, Dept. of Nuclear Engineering, University of Tennessee, for his expert advice on the application of techniques for systems analysis of nuclear reactor dynamics to analysis of ecological systems. Many members of the Radiation Ecology Section of the Health Physics Division contributed valuable suggestions for construction of the pathway diagram representing radionuclide movement in the tropical environment.

Literature Cited

Grabbe, E. M., S. Ramo, and D. E. Wooldridge (Eds.). 1958. Handbook of automation, computation and control, Vol. I. Wiley, New York.

Kerlin, T. W., and J. L. Lucius. 1966. The SFR-3 code — a FORTRAN program for calculating the frequency response of a multivariable system and its sensitivity to parameter changes. ORNL-TM-1575.

Kline, J. R. 1966. Cycling of fallout radionuclides in tropical forests. Bull. Ecol. Soc. Amer. 47: 119.

Kline, J. R. Retention of fallout radionuclides of stratospheric origin by tropical forest vegetation. in: A Tropical Rainforest. H. T. Odum (Ed.). In press. USAEC TID-24270.

Kline, J. R., and H. T. Odum. 1966. Comparison of the amounts of fallout radionuclides in tropical forests. Bull. Ecol. Soc. Amer. 47: 121.

Witkamp, M. Aspects of soil microflora in a gamma irradiated rainforest. in: A Tropical Rainforest. H. T. Odum (Ed.). In press. USAEC TID-24270.

Witkamp, M. Mineral retention by epiphyllic organisms. in: A Tropical Rainforest. H. T. Odum (Ed.). In press. USAEC TID-24270.

Editors' Comments
on Papers 4, 5, and 6

EXTENSIONS FROM POPULATION MODELING

To some extent, there has always been a schism between population modelers and systems ecologists. Dichotomies between the two fields include academic affinities, interest in basic versus applied research, and even journals typically used as publication outlets. Recently, there have been welcome signs of synthesis (Levin 1975, 1976).

Because early systems ecologists were interested in material transfer, they were drawn to linear, donor-controlled differential equations since conservation of matter and energy is a necessary consequence of the structure of these equations. There seemed little merit in the autonomous (no inputs from outside the system) and nonconservative (in the sense of Newton's laws) Lotka-Volterra equations.

Although the Lotka-Volterra equations have been roundly criticized over the past several decades (e.g., Smith 1952), they are fundamental to modern theoretical ecology. Population theorists have made tremendous advances using these equations and have little sympathy with mass-balance constraints as important reasons to discard them.

It is unfortunate that the debate between population and ecosystem ecologists has only recently begun to generate light as well as heat. Probably the lack of acceptance of mathematical approaches has, until recently, reduced the number of arenas in which confrontations and the resultant synthesis could take place. We anticipate a great deal of interest and progress over the next decade resulting from cross-fertilization between population and ecosystem ecologists.

In anticipation of these developments, we have included three papers that, relatively early in the development of systems ecology, injected perspectives from population modeling. Paper 4 by Garfinkel is an early attempt at ecosystem simulation by an individual who later was involved in several (Garfinkel 1967a, 1967b; Garfinkel and Sack 1964; Garfinkel et al. 1964) explorations of methods to increase the realism of the Lotka-Volterra equations.

Paper 5 by Holling has been influential because of its lucid explanation of how to develop a model. The concept involves analyzing a complex process into logical components, examining and modeling each component separately, and reassembling a model by the juxtaposition of the component submodels. Paper 6 by Davidson and Clymer is immediately relevant to our topic since it advocates the use of population models (i.e., Lotka-Volterra equations) as components that can be assembled into models of the total ecosystem.

REFERENCES

Garfinkel, D. 1967a. A simulation study of the effect on simple ecological systems of making rate of increase of population density-dependent. *J. Theor. Biol.* **14**:46–58.

Garfinkel, D. 1967b. Effect on stability of Lotka-Volterra ecological systems of imposing strict territorial limits on populations. *J. Theor. Biol.* **14**: 325–27.

Garfinkel, D., R. H. MacArthur, and B. Sack. 1964. Computer simulation and analysis of simple ecological systems. *Ann. N. Y. Acad. Sci.* **115**: 943–51.

Garfinkel, D., and R. Sack. 1964. Digital computer simulation of an ecological system based on a modified mass action law. *Ecology* **45**:502–07.

Levin, S. A., ed. 1975. *Ecosystem Analysis and Prediction.* Society for Industrial and Applied Mathematics, Philadelphia, 337 pp.

Levin, S. A., ed. 1976. *Ecological Theory and Ecosystem Models.* Office of Ecosystem Studies, The Institute of Ecology, Madison, Wisc., 71 pp.

Smith, F. E. 1952. Experimental methods in population dynamics: A critique. *Ecology* **33**:441–50.

4

DIGITAL COMPUTER SIMULATION OF ECOLOGICAL SYSTEMS

By Dr DAVID GARFINKEL

Johnson Research Foundation, University of Pennsylvania, Philadelphia, 4

THE mathematical representation of ecological systems has been investigated for several decades, starting with the pioneer work of Volterra[1] and Lotka[2]. Most of this work has been done with pencil and paper and has necessarily been limited to systems simple enough to permit the analytical solution of the resulting equations. Some work has been done with analogue computers[3], and an application of digital computers to the problem has been made by Watt[4,5]. It is the purpose of this article to describe a general computer method of representing ecological systems of any desired degree of complexity in terms of simultaneous first-order differential equations, with some simple illustrative applications.

The calculations thus far performed are based on the assumption[1,2,6] that the rate at which two species interact is proportional to the product of their numbers. This rate law is identical to the mass action law of chemistry and therefore permits the use of a computer programme[7,8] developed to represent complex chemical systems with the *Univac I* computer. This programme enables the computer, when given as input a set of chemical equations of the form:

$$A + B \rightarrow C + D$$

automatically to write instructions to solve the associated set of differential equations:

$$-\frac{dA}{dt} = -\frac{dB}{dt} = k\,(A)\,(B)$$

and then perform the solution by numerical methods when given a set of initial conditions. It is possible to write and solve sets of equations having no relationship to chemistry, and it would be possible to substitute some other interaction law for the mass action law, or to make allowance for its limitations[5].

The system examined here (represented by the 'chemical' and differential equations shown in Table 1) consists of a prey, grass (GRS), and a predator, rabbits, which may be either adequately fed (FRB) or starved (SRB). The starved rabbits are assumed to die at a moderately rapid rate if unable to feed; the fed ones, to reproduce. In most of the calculations, it has been

Table 1

'Chemical' equations	Rate constants
(1) $GRS \rightarrow GRS$	$k_1 = 24$
(2) $SRB + GRS \rightarrow FRB - SRB - GRS^d$	k_2 varied $\begin{cases} 3 \times 10^{-1} \\ 10^{-2} \end{cases}$
(3) $FRB \rightarrow FRB$	$k_3 = 30$
(4) $FRB \rightarrow SRB - FRB$	$k_4 = 300$
(5) $SRB \rightarrow DUM - SRB^b$	$k_5 = 33$
(6) $FRB \rightarrow DUM - FRB$	$k_6 = 3$
(7) $FRB + GRS \rightarrow DUM - GRS^c$	$k_7 = 0.012$
(8) $RTS \rightarrow GRS^d$	$k_8 = 5 \times 10^{-4}$
(9) $GRS + GRS \rightarrow DUM - GRS^e$	$k_9 = 5 \times 10^{-4}$

Corresponding differential equations

$$\frac{d(GRS)}{dt} = k_8(RTS) + k_1(GRS) - k_9(GRS)^2 - k_7(FRB)(GRS) - k_2(SRB)(GRS)$$

$$\frac{d(SRB)}{dt} = k_4(FRB) - k_5(SRB) - k_2(GRS)(SRB)$$

$$\frac{d(FRB)}{dt} = k_3(FRB) + k_2(GRS)(SRB) - k_6(FRB) - k_4(FRB)$$

a. Substances which disappear in a reaction must be listed again on the right side of the equation.

b. This statement is equivalent to 'SRB disappears'.

c. Fed rabbits continue to eat.

d. RTS = roots. To abolish the roots, k_8 is set equal to zero.

e. This is to limit the amount of grass that may grow in a fixed land area.

assumed that some of the grass is segregated in such a way that the rabbits cannot get at it (that is, roots which they cannot dig up), which is equivalent to a constant influx of grass into the system.

Varying the predative efficiency of the starved rabbits yields results shown in Figs. 1–4. If the rabbits are highly efficient predators with access to all the grass, they will exterminate it and then starve to death (Fig. 1), a situation observed experimentally by Gause[9]. If some of the grass is inaccessible, a steady state results in which the grass is eaten as fast as it appears (Fig. 2). As the predative efficiency of the rabbits is decreased the system goes into an oscillation (Fig. 3) and then a damped oscillation (Fig. 4). An even more inefficient predator effectively starves to death in the midst of plenty.

A given predator may attack more than one prey organism with differing degrees of efficiency[6]. In many cases this may stabilize the system, but it need not do so: when the predators of Fig. 3 were given an alternative prey, which they attacked less efficiently, they first ate approximately all the available preferred prey, and then came to a steady state with the alternative prey.

In an attempt to duplicate a system where a stable predator–prey relationship may be upset with drastic consequences by adding another predator[10], a second predator (foxes) which eats the rabbits was introduced; the foxes are assumed to reproduce (and die of starvation) more slowly than the rabbits. When foxes were introduced into a system (Fig. 4) which would otherwise come to a steady state, it went into wild oscillations (Fig. 5); at the end of the

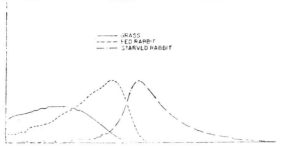

Fig. 1. No. of animals as a function of time with $K_2 = 0.3$ and $K_8 = 0$

Fig. 2. No. of animals as a function of time with $K_2 = 0.3$ and $K_8 = 0.24$

Fig. 3. No. of animals as a function of time with $K_2 = 5 \times 10^{-2}$ and $K_8 = 0.24$

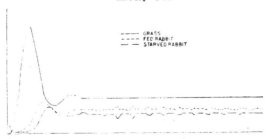

Fig. 4. No. of animals as a function of time with $K_2 = 10^{-2}$ and $K_8 = 0.24$

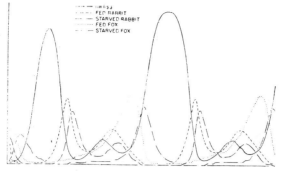

Fig. 5. No. of animals as a function of time with foxes introduced
into the system of Fig. 4

second oscillation the rabbits were approximately
extinct.

Application of this method calculation to ecological
problems offers two distinct advantages: (1) The
computer can be made to deal with nearly all the
mathematics.   (2) The limit of complexity is that
imposed by the speed and memory size of the computer used (the largest now in existence would be
able to handle more than 1,000 species in this manner).
By suitably defining new 'species' and taking advantage of other properties of the programme, it is
possible to represent environmental variations,
disease, different states of nutrition, effect of age,
etc.

This method of calculation can be applied both to
studies in theoretical ecology and to real situations
(such as agricultural pest control). It must, however,
be emphasized that the ability to perform calculations
in this manner is no substitute for detailed knowledge
of the system being studied.

This work was performed during the tenure of a
Research Career Development Fellowship, U.S.
Public Health Service, and supported by the Office
of Naval Research.

[1] Volterra, V., *Leçons sur la theorie mathematique de la lutte pour la vie*
(Gauthier-Villars, Paris. 1931).
[2] Lotka, A. J., *Elements of Mathematical Biology* (Dover, New York,
1956).
[3] Wangersky, P. J., and Cunningham, W. J., *Cold Spring Harbor
Symp. Quant. Biol.*, **22**, 329 (1961).
[4] Watt, K. E. F., *Science*, **133**, 706 (1961).
[5] Watt, K. E. F., *Canad. Entomol.*, **93** (Suppl. 19, 1961).
[6] Slobodkin, L. B., *Growth and Regulation of Animal Populations*
(Holt, Rinehart, Winston, New York, 1961).
[7] Garfinkel, D., Rutledge, J. D., and Higgins, J. J., *Comm. Assoc.
Computing Machinery*, **4**, 559 (1961).
[8] Larson, R., Sellers, P., and Meyer, R., *Comm. Assoc. Computing
Machinery*, **5**, 63 (1962).
[9] Gause, G. G., *Verification Experimentales de la Theorie Mathematique
de la lutte pour la vie* (Herman, Paris, 1935).
[10] Brown, H. S., *The Challenge of Man's Future*, 12 (Viking, New
York, 1954).

5

The Analysis of Complex Population Processes[1]

By C. S. HOLLING

Forest Insect Laboratory, Sault Ste. Marie, Ontario

Introduction

Population ecology promises to advance dramatically in the next few years through the development of a number of new approaches that are beginning to provide fresh insights into the dynamics of animal populations. The need for new approaches and new analytical tools has arisen from the distinctive complexity of biological systems. In my own studies of predation, for example, a number of complex features have emerged which seem to be characteristic not only of predation but of other population processes as well. It is my purpose here to discuss these distinctive features, the analytical problems they pose and some of the possible solutions to these problems.

The Desirable Attributes

The exciting and productive developments in science have been those that have analyzed whole processes realistically and precisely and in such a way that the resulting theories had general application to most examples of the process. The astonishing structure developed by the atomic physicists since Rutherford's time has been so rewarding largely because these qualities have been achieved, and the recent developments in molecular biology promise equal rewards for the same reason. If population ecology is to make the same contribution these qualities should be an integral part of population theory.

Perhaps the most important quality arises from the need to analyze whole processes and not just fragments of them. There are many examples of analyses of discrete fragments of ecological processes. We know, for example, the concentration of bacteria required to kill a fixed proportion of an experimental group of animals, the maximum number of eggs a particular insect can lay, or the caloric requirements of some predators. Each is an isolated fragment of one biological process — of the disease process, of fecundity or of predation — and each — comprised of many such fragments. However intriguing and necessary the analysis of an individual fragment, its value is drastically limited if there is no unifying framework to provide insight into the action and interaction of all the significant fragments of one process or group of processes. This unifying framework is, in fact, an explanatory model that expresses real events in artificial

Contribution No. 990, Forest Entomology and Pathology Branch, Department of Forestry, Ottawa, Canada.

symbols — in words, tables of numbers, graphs or mathematical symbols. The model is an attempt to simulate a whole process or system so that knowledge of isolated fragments can be used in a meaningful way.

Functional wholeness alone, however, is a barren prospect. If the assumptions underlying the unifying structure are unrealistic no amount of wholeness will save it from being a futile exercise. In addition, precision is almost as important as realism. Population theory has been particularly bedevilled by vague, imprecise hypotheses that defy testing. An imprecise, verbally expressed hypothesis is an inevitable stage in the development of a precise and meaningful explanation, but if the analysis stops at this stage the literature becomes smothered by untestable generalities that inhibit rather than enhance understanding.

These three qualities, wholeness, realism, and precision, give depth to an analysis. But breadth is just as important. It would be most disappointing, for example, to develop a precise, realistic model of one example of a process and discover that it was unique and inappropriate for other examples of the same process. There are many examples in the biological literature of precise equations which were thought to describe consistently but subsequently were shown to describe only certain isolated cases. Wigglesworth (1947, pp. 368-372), for example, discusses five different equations that have been proposed to describe the effects of temperature on various physiological processes. None of them describes consistently. Similarly, at least seven different predation models have been proposed since Lotka (1923) and Volterra (1926) tentatively published an extremely simple one in the 1920's. Only two of these models adequately describe a large number of cases (Watt 1959; Holling 1959) and even these do not provide a universal description. This inconsistency of prediction in part reflects the unreality of these models, so that achieving a realistic model will in some cases automatically make it general. It is also possible, however, that specific ecological processes are so unique that only small groups of cases have the same causation even though the end result might be similar. In this event striving for great depth will preclude great breadth.

The qualities promoting depth—wholeness, realism, and precision—similarly might be impossible to obtain simultaneously, and it is for this reason that many biologists criticize mathematical models. Some of these arguments concern semantics more than biology but some are based on an intimate analytic knowledge of biological processes. Labeyrie (1960), for example, after an exhaustive analytical study of parasitism, concluded that the process was so complex and intricate that any attempt to develop a precise mathematical model must inevitably be unrealistic. Important parameters of parasitism would be presumed to be constant when in fact they vary in response to other parameters. These arguments are very compelling, particularly since they arise from a highly detailed insight into the operation of a specific population process. It is still possible, however, that the criticisms apply only to existing population models and the analytical and mathematical tools used to develop these models. Population processes thus might be unique only in that new analytical and conceptual tools are required in order to yield theories with the required wholeness, realism, precision and generality.

The methods used to develop existing mathematical models have been largely borrowed from the physics of 50 years ago. As a result, the models are basically similar to the physicist's model, having the same neat, well-structured form, the same language of expression (largely differential calculus) and the same logical structure. Such similarities arise from the desire to retain simplicity, so that the equations are manageable. Simplicity, however, is clearly a relative concept, for the criterion of simplicity used in describing a simple process must surely be different from that used in describing, with the same realism, a more complex one. Applying a simple concept to a complex process can be done only by sacrificing reality. Hence the logical tools and mathematical language used so effectively in analyzing processes of physics might well be inappropriate for the more complex processes of population biology.

Increasing complexity of systems requires different methods of analysis as well as different techniques of expression. Man's imagination is a severely limited instrument that can proceed only one small step into the unknown. When the unknown is simple then this small step might be sufficient to expand the known to include it. When the unknown is complex, however, the possibilities become so great that imagination requires a crutch to aid it. Since there is little question that population processes are complex, their analysis must be given detailed direction through an intimate wedding of theory and experiment.

If these arguments have any real merit then it would be an entertaining and potentially profitable exercise to develop a realistic and precise model of a population process by largely ignoring the traditional concept of simplicity and emphasizing the need for reality. By conducting an analysis that emphasizes the need for a realistic explanation, it should become apparent if any of the qualities promoting depth and breadth are intrinsically mutually exclusive or if the difficulties arise from the use of inadequate methods of analysis and synthesis.

This has been the aim behind the studies of predation that I have conducted for the past few years. In order to fulfil the aim I used an approach in which the components of a simple example of the process were separately analyzed experimentally. Precise hypotheses were proposed for the action of each component, these were tested experimentally, and when a set of postulates proved adequate they were then expressed in a mathematical form. The resulting series of equations were then synthesized to produce a model of the simple example chosen. This provided a base from which to explore more complex examples where additional components were operating, for these new components could be analyzed in the same way and the model expanded to include them. In this manner a more and more complex structure is built, each progressive step being taken only when an explanation of proven validity has been obtained for the previous step.

In order to demonstrate the problems involved in obtaining depth and breadth I shall discuss the most recent step taken in the analysis of predation. This concerns the effect of prey size on attack, and since the analysis is only partly completed the specific explanations proposed here should be considered as tentative.

The Effect of Prey Size on Attack

Most, if not all, predators are particularly responsive to a specific size of prey and less readily attack prey smaller or larger than this optimum. The ultimate causes underlying this behaviour concern the selective forces that have acted during the predator's evolutionary history. On the one hand, it is advantageous for a predator to develop methods of attacking larger and larger prey, for large prey represent a greater energy source per capture than do small ones. On the other hand, if the prey attacked is too large, an enormous amount of energy must be expended to subdue it and, in fact, the killer faces the possibility of becoming the killed. Similarly, if the dangers associated with the attack of large prey are avoided by selectively attacking small ones, which are likely to be more manageable and numerous, the prey selected could become so small that the energy expended searching for each one would exceed the energy gained by

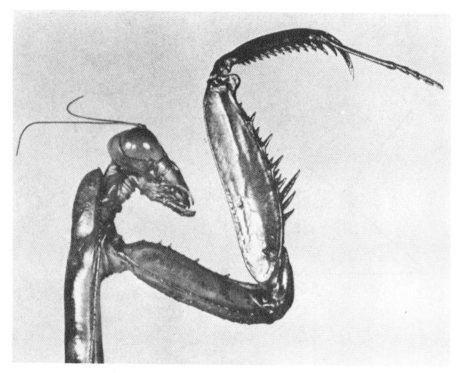

Fig. 1. The grasping foreleg of *H. crassa*.

eating it. Furthermore, interspecific competition, as Lack (1954) suggests, has had a major influence on the evolution of attack behaviours that are specifically directed towards certain types and sizes of prey.

It would be extremely difficult to unravel these ultimate causes in any precise and realistic way. But whatever the ultimate causes that determine the size of prey selected by a particular predator, the proximate causes can be found in physiological and structural features of the predator. Hence Haldane (1954) points out that for many measures of animals the optimum with respect to survival is usually near the mean, thus implying that the population studied is nearly in equilibrium under natural selection. Therefore part of the question of how prey size effects attack can be rephrased in more specific terms. That is, how do the structures that a predator uses to capture prey determine the size of prey captured?

As an example, consider the grasping foreleg of the praying mantid, *Hierodula crassa* Giglio-Tos (Fig. 1). It appears to even the least anthropomorphic of viewers as a most formidable weapon and an insight into its efficiency can be gained by building, in imagination, a grasping mechanism with the same salient feature. In Fig. 2A a very simple grasping mechanism is drawn as two straight arms, each hinged to the other at one end. The forces generated as the arms are brought together, operate at right angles to tangents of the circular object held by the arms, so that a parallelogram of forces can be drawn by considering the forces as acting at the centre of the object. The two forces P_1 and P_2 generate a resultant force, R in Fig. 2A, which tends to push the object out along a line bisecting the angle formed by the two arms. The object will be held in this position so long as the frictional forces are greater than R. If the object held is increased in size, the angle formed between the arms and the resultant force R

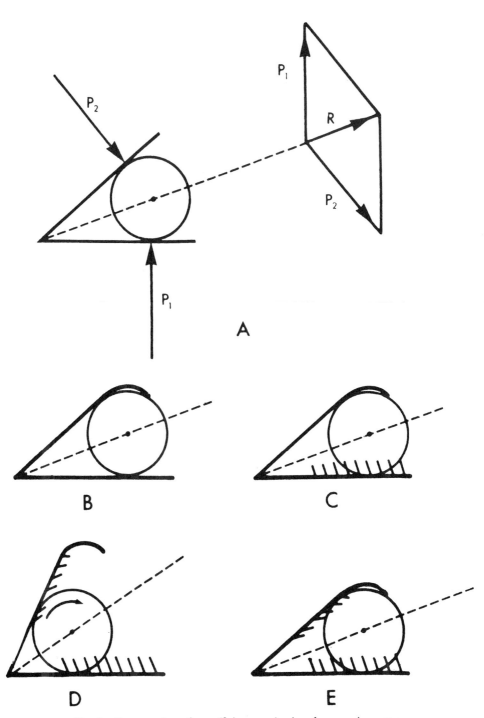

Fig. 2.   Construction of an efficient mechanism for grasping prey.

will also increase. At some point the frictional forces will become too small to balance R and the object will be forced out of the grasp of the arms. It is essentially the same problem that is involved in a person holding a basketball as compared to a baseball. Thus even this simple grasping system is limited to holding only certain-sized objects.

The efficiency of this structure is greatly increased if a hook is formed at the end of one of the arms (Fig. 2B). Now the object is held at three points so that it becomes securely locked in position, the tip of the hook effectively opposing the outward motion of the object. The lock can be made all the more effective by adding spines to the other arm, for the spines will act in the same way as the spikes on a meat carving platter. As shown in the diagram these spines are most efficient if they are angled to oppose the outward movement of the object (Fig. 2C).

The mantid has one more significant refinement lacking in our manufactured grasping structure. This is the line of teeth along the tibia. These teeth are slanted in the same direction as the spines on the femur, i.e. distad, but their function is different. Consider, for example, an object that has been caught well within the two arms, as shown in Fig. 2D. The slanted spines of the femur would so greatly increase the frictional coefficient of the femur relative to the tibia that the resultant force, R, would be translated into an angular moment. Thus the object would tend to roll outward along the femur. This is exactly what should happen in order to take advantage of the locking device offered by the tibial hook. Hence the teeth on the tibia should be slanted so as not to oppose the rolling motion. Moreover, when the object finally touches the hook, the tibial teeth will effectively jam the object between the teeth and the hook (Fig. 2E). They thus

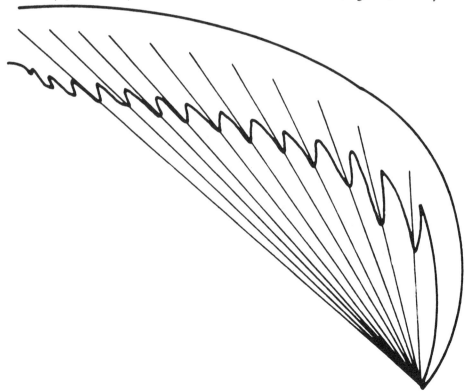

Fig. 3. Tracing of a projected image of the tibia of the foreleg of a mantid to show the angle of the tibial teeth.

act very much like the teeth on a ratchet wheel. In order to function most efficiently the tibial teeth should not only slant towards the opening formed by the two arms but should point towards the tip of the tibial hook, following the line of force generated between the tip of the hook and the part of the tibia touching the object. In this way, different-sized objects which touch the tibia at different points all become readily impaled on the teeth. Fig. 3 was prepared from a tracing of the projected image of a tibia of an adult *H. crassa*, and the drawing shows that the tibial teeth actually do tend to point towards the tip of the tibial hook.

The fundamental feature of this grasping structure is the locking device formed by the femur, the tibia and the tibial hook. It should therefore be possible to predict the optimum size of prey from the geometry of this locking device. If the optimum size is defined as the maximum size that can be securely held without violating the locking principle, the optimum will be that size of prey (1) which touches the tip of the tibial hook, the tibia and the femur and (2) in which a line drawn from the tip of the hook perpendicular to the femur runs through the centre of the prey, as in Fig. 4. If the prey is only a little larger

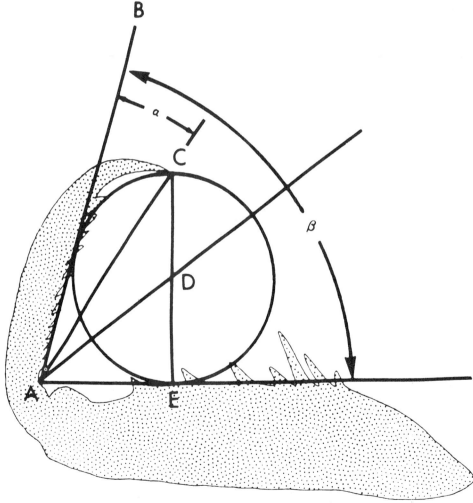

Fig. 4. Geometry of the mantid grasping foreleg.

than the one drawn, its centre (D in Fig. 4) would lie outside the line (i.e. to the right of CE) and the tibial hook would no longer lock the prey in position.

The specific size of prey that is optimal can therefore be calculated knowing the length of the tibia (AC) and the tibial-hook angle (α). From the relationship in the triangle CAE

$$\tan (\beta - \alpha) = \frac{CE}{AE} \tag{1}$$

where β = the angle BAE.

Since the lines AB and AE form tangents to the circle, the line AD through the centre of the circle bisects the angle BAE or β. Hence from the relationships in the triangle DAE

$$\frac{\tan \beta}{2} = \frac{DE}{AE} \tag{2}$$

Since   CE = 2r

where r is the radius of the circle, and

$$DE = r$$

equations (1) and (2) may be written as, respectively,

$$\tan (\beta - \alpha) = \frac{2r}{AE} \tag{3}$$

$$\text{and} \quad \frac{\tan \beta}{2} = \frac{r}{AE} \tag{4}$$

Therefore from (3) and (4)

$$\tan (\beta - \alpha) = 2 \frac{\tan \beta}{2} \tag{5}$$

The optimum angle between tibia and femur can be calculated iteratively from equation (5), if α is known. Once β is calculated, it is then possible, given the length of the tibia to the tip of the tibial hook (AC), to determine the optimum size of prey in terms of its radius, r. Thus from the triangle CAE in Fig. 4,

$$\sin (\beta - \alpha) = \frac{CE}{AC}$$

Since   CE = 2r

$$\text{therefore } r = \frac{T \sin (\beta - \alpha)}{2} \tag{6}$$

where T = AC, the length of the tibia.

Equations (5) and (6) have received support from two sources. First, the angle of the femoral spines seems to be specifically adjusted to hold a prey of the size predicted as optimal. It was mentioned earlier that these spines impede the force pushing out the object and this force, R, acts along a line that bisects the angle formed between the tibia and femur. Hence the femoral spines would most efficiently immobilize the optimum size prey if they were placed at right angles to this line of force. That is, their angle on the femur should equal ($90° - \beta/2$) if equation (6) is correct. Table I summarizes the pertinent data obtained from 29 mantids, and shows that the observed femoral spine angle is remarkably close to the angle predicted as that most efficient for immobilizing an optimum-sized prey.

TABLE I

Observed angle of the femoral spines of 29 mantids compared with the angle predicted

Observed tibial-hook angle (mean ± 1 S.E.) a	β calculated from equation (5)	Angle of femoral spines	
		Predicted $(90° - \beta/2)$	Observed (mean ± 1 S.E.)
17.2° ± .6	73.3°	53.3°	53.7° ± 1.2

As a further test, the optimum size predicted by equation (6) was compared to the actual optimum determined experimentally with five female *H. crassa*. These five mantids were offered model prey of different sizes and their reactions were recorded either as an attack response (striking at the model) or no response. The models were cut from black paper in the shape of a rectangle with rounded ends; the width was ⅓ of the length, and the sizes ranged from 8.5 mm. long (about the length of a housefly) to 69 mm. long (about the length of a mantid). Each model was attached to a stiff wire and in each test was rotated for a maximum of 30 seconds within striking distance of the mantid. The five mantids were fed to satiation and were tested once every two hours thereafter for 12 hours. The next eight hours was a period of darkness so that there were no further tests until the beginning of the next day, when two additional tests were conducted, separated by two hours. Thus each of the five mantids was tested eight times over a 24-hour period to give a total of 40 tests for each prey size.

The proportion of tests resulting in attack are plotted in Fig. 5 against prey size, measured as ½ the width of the model. The mantids were clearly least responsive to very small and very large prey, and were most responsive to a prey with a radius of about 4 mm. (i.e. a model length of 24 mm. and width of 8 mm.). In order to compare this observed optimum with the optimum predicted by equations (5) and (6), the tibial length and the tibial hook angle of the five mantids were measured. They were, respectively, 10.23 mm. (range 9.18-11.22 mm.) and 17.67° (range 15.75-20.40°). Hence, from equation (5), $\beta = 74.1°$ and from equation (6), r = 4.26 mm., very close to the optimum actually observed.

While the explanation proposed above for the size of prey that is optimal seems logical, and has excellent predictive powers, it is only one aspect of the problem of the effect of prey size on predation. The other aspects, particularly the way responsiveness of a predator is affected by prey size, are now being explored using the same type of analysis. There is little doubt that the full explanation will be of the same kind as the partial one outlined here, and will differ only in degree. The present analysis therefore provides a useful basis to discuss the problems involved in achieving depth and breadth in an analysis of a typical population process.

Achieving Depth and Breadth

The above analysis shows only what should have been self-evident — i.e. that a realistic explanation of one fragment of a process can be obtained and expressed in a precise mathematical form. Realism and precision alone are not enough, however, for the analysis of one isolated fragment cannot be much more than an entertaining exercise. It can gain meaning only in combination with other frag-

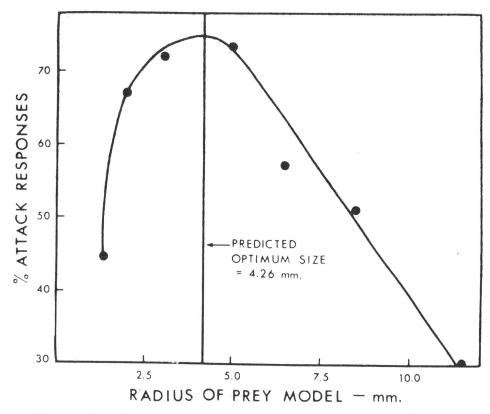

Fig. 5. Effect of size of prey model on the proportion of encounters that initiated attack.

ments so that its significance in the whole process, or a major part of it, can be explored.

The effect of prey size is actually the latest fragment to be studied. A number of other components have already been analyzed, and have been combined into a model of a functional response to prey density. This model simulates attack by a predator exposed to different densities of prey and includes twenty-two different parameters, each associated with some biologically meaningful feature of attack. One, for example, is the rate of digestion, another represents speed of movement of the predator and still another measures the predator's success in capturing prey. The cryptic and defensive qualities of the prey, the size and shape of the predator's reactive field, the maximum food capacity of the gut, and the rate of pursuit and eating are each represented by one or more parameters. Since a general description of the model has already been presented (Holling 1963) and a more detailed version is now being prepared for publication, further description here is unnecessary. It is necessary only to point out that the model can be readily expanded to include the effects of prey size. A number of parameters will be affected. The size of the reactive field of the mantid, for example, is partly determined by the distance it can see an object. This distance, and the associated parameters, is related to a linear measure of prey size. Weight of the prey is another parameter of the model which can be expressed as a linear function of a volumetric measure of prey size. Finally, for some uses of the model, it might be advantageous to express prey density as a function of prey size by exploiting the size — density relations demonstrated by the pyramid of numbers. In this way the isolated effects of prey size can be combined with other components to simulate a complex and major part of the predation process. The resulting model will show not only how prey size affects the number of prey destroyed but will also show how the predator's energy input and output are affected and the conditions under which net energy is optimized.

The analysis of predation has extended far enough now to show that a useful degree of wholeness can be achieved without sacrificing reality or precision. In the process of achieving this depth, it has become clear that a number of significant features of predation are shared by a large number of population processes, and that the methods of analysis and synthesis used here can be profitably applied elsewhere.

First, thresholds and limits are a common feature of the components of predation. In the example already discussed, the predicted optimum prey diameter is itself a threshold, for the locking device formed by the tibia, tibial hook, and femur becomes inoperative for prey larger than the optimum. There are many other examples of thresholds and limits affecting various components of predation. A predator, for example, attacks only when its hunger exceeds a certain value, and when prey exceed a certain size only part of the prey is eaten and the rest is discarded. Thresholds and limits are distinctive features not only of predation but of other biological processes. A nerve impulse, for example, is propagated only when the intensity of a stimulus exceeds a certain value and a population of animals persists in an area only when its density attains a certain level. It is true, of course, that thresholds affect non-biological processes, as well, but the large number of components affecting biological systems and the commonness of thresholds gives a distinctive character to these systems.

Time is also a dominant feature of many of the components of predation, for it takes time for a predator to search for a prey, to pursue, subdue, eat, and digest it. The dominating influence of time did not really become apparent, however, until the various fragments were synthesized into a whole attack model. It then became obvious that the various time-consuming activities produced lag effects so that events occurring at any moment depended not only on conditions at that moment but on conditions in the past as well. This historical character has long been recognized as an integral part of genetics and of evolutionary theory but it is an equally important feature of population processes.

Finally, in order to combine realistically the various components, it seemed essential to retain the basically discontinuous feature of predation. When a predator begins a day of attack, its hunger is usually high because of a preceding non-feeding period. As time passes, the hunger continues to increase until a prey is finally discovered, attacked, and eaten. The hunger is thereby abruptly lowered to a new level, after which the same cycle is repeated. This step-like, discontinuous process is an essential feature of predation that should be preserved. This feature is not restricted to predation, however, since parasitism, reproduction, and competition have important discontinuous elements arising from the relatively few contacts animals have with one another.

These three prime features — the prevalence of thresholds, and the historical and discontinuous features — can be demonstrated experimentally in a precise and realistic way. It proved particularly difficult, however, to retain these features in a mathematical model. The individual fragments themselves could be readily modelled using fairly straightforward mathematical expressions. The equation predicting the optimum prey size is a particularly simple example. Synthesizing all these equations posed a more difficult task. Since differential equations seemed quite inappropriate because of the discontinuous nature of attack, the model was formulated using difference equations (Goldberg 1961), which generated each attack successively. Thus, given a starting level of hunger, one equation predicts the searching time required to contact the first prey, another predicts the pursuit time, another the time spent eating and a final one the time spent in a "digestive-pause", during which the predator is too satiated to search. The sum of all these times represents the time for one step or cycle, and at this point another cycle can be generated. This continues until all the available time for attack elapses.

A digital computer proved to be ideally suited to handle this complex model. Because of its speed, the significant history could be actually duplicated step by step, until a stable condition prevailed. In a typical case, for example, it took 20 minutes to simulate 10 days of attack using the moderately fast I.B.M. 1620 digital computer. Moreover, the language used to program the computer, I.B.M.'s Fortran, was admirably suited to cycle operations economically through one attack after another and to handle the threshold problem. Thus the combination of an experimental component analysis which gives direction to theory and the digital computer which effectively copes with the distinctive complexity of biological processes, produces an analysis of great depth.

There is one potentially crushing criticism of this approach. Because it emphasizes the need to unravel the intricate action and interaction of the components of the process, there is a real danger that the analysis will become so enmeshed in detail that a broad synthesis becomes unlikely. And it is breadth of synthesis, more than any other feature, that gives significance to an analysis.

It seems inevitable that a variety of proximate causes will underlie any given biological phenomenon. Selection acts on end results and if many paths produce the same end these will all be exploited depending on the restrictions imposed by existing organization. In the optimum prey size example discussed earlier, the causes were related to the geometry of the grasping foreleg of the mantids. The optimum size of prey selected by other predators that have exploited the same type of capturing mechanism can presumably be explained in basically the same manner. Hence, with only minor modifications, the equations should apply to bellastomatids, to phymatids, and even to some vertebrates like snakes, for all capture prey with a two-armed structure similar to that of the mantids.

Other predators, however, have exploited quite different methods to capture prey. The ant-lion digs a pit in sand, the baleen whale has developed a sieve, the bee fly immobilizes bees with poison, and the spider traps prey with a web. Each has developed a different method to capture prey so that the detailed explanation proposed for the mantid can hardly be generalized to include these cases. Nevertheless, the ultimate causes arising from natural selection have been the same so that the end result might basically be similar in all cases. Hence, despite the differences in capturing technique, all methods would tend to be most efficient in capturing but one prey size and prey larger or smaller would be less readily captured. It is therefore possible that the specific equation appropriate for the mantid might adequately describe other, apparently different situations. If this proves to be the case then the parameters will have different meanings. The parameter representing tibial length, for example, might in the case of the ant-lion represent the slope of the pit, in the case of the whale the size of the sieve openings, in the case of the bee fly, the LD 50 per gram of organism and in the case of the spider, the tensile strength of the silk in the web. In this way generality of a high order might be possible. It is, however, premature to assume this is true and we will never know for sure until more models of great depth are developed and tested against a variety of examples. The possible rewards justify the great effort involved.

To conclude, I shall summarize the main point developed in this paper. The analysis of predation has shown that a whole process can be realistically and precisely analyzed by ignoring the traditional concept of simplicity, by establishing an intimate feed-back between experiment and theory, and by proceeding in gradual steps with experimental evidence providing direction at each step. Biological theoreticians have long pleaded, and rightly, for experimental evidence upon which to base theories. I wonder, however, if it is possible for them to obtain the necessary data from others. Many biological processes are so complex that an extremely intimate union is required between experiment and theory, with experiments dictating theory and theory suggesting experiments in successive, repetitive steps. Some aspects of physics could develop with considerable separation between experimentalist and mathematical theoretician — between Faraday and Maxwell or Rutherford and Bohr, for example. The same separation, however, might not be possible in many areas of biology.

Acknowledgments

I am particularly indebted to P. J. Pointing, R. F. Morris and K. E. F. Watt, for their many helpful discussions during the development of the predation studies, and to H. Mittelstaedt for providing the mantids.

References

Goldberg, S. 1961. Introduction to difference equations. John Wiley and Sons, Science Editions, Inc., New York.

Haldane, J. B. S. 1954. The statics of evolution. *In* J. Huxley, A. C. Hardy and E. B. Ford (eds.) Evolution as a process, Allen and Unwin, London.

Holling, C. S. 1959. Some characteristics of simple types of predation and parasitism. *Canad. Ent.* 91: 385-398.

Holling, C. S. 1963. An experimental component analysis of population processes. *Mem. ent. Soc. Can.* 32: 22-32.

Labeyrie, V. 1960. Contribution à l'étude de la dynamique des populations d'insectes: I Influence stimulatrice de l'hôte *Acrolepia assectella* Z. sur la multiplication d'un hymenoptère Ichneumonidae (*Diadromus* sp.). *Entomophaga Mem.* 1: 1-193.

Lack, D. 1954. The natural regulation of animal numbers. Clarendon Press, Oxford.

Lotka, A. J. 1923. Contribution to quantitative parasitology. *J. Wash. Acad. Sci.* 13: 152-158.

Volterra, V. 1926. Variazioni e fluttuazioni del numero d'individui in specie animali conviventi. *Mem. Acad. Lincei* (6) 2: 31-113.

Watt, K. E .F. 1959. A mathematical model for the effect of densities of attacked and attacking species on the number attacked. *Canad. Ent.* 91: 129-144.

Wigglesworth, V. B. 1947. The principles of insect physiology. 3rd ed. Methuen and Co. Ltd., London.

6

Copyright © 1966 by New York Academy of Sciences

Reprinted from *N.Y. Acad. Sci. Ann.* **128**:790–794 (1966)

THE DESIRABILITY AND APPLICABILITY
OF SIMULATING ECOSYSTEMS

R. S. Davidson

Battelle Memorial Institute, Columbus, Ohio

and

A. B. Clymer

Consulting Analytical Engineer, Columbus, Ohio

Introduction

Ecosystems are concerned with living species: plants, animals, and humans interacting with each other and with the physical features of environment. This paper briefly explores the desirability and feasibility of applying computer simulation to ecosystems particularly in the area of environmental problems. Emphasis is given also to preliminary work done by the authors, since the monograph itself is devoted to recent or current studies.

Simulation of Plankton Population Dynamics

In 1964, the second named author directed a simulation of the annual cycle of estuarine phytoplankton and zooplankton, using an analog computer and later the digital computer at Battelle Memorial Institute. In formulating the mathematical model the assumptions were:

1. The locale is the Middle Atlantic seaboard of the United States
2. All variables were uniformly distributed through the volume of water of concern
3. All variables are averages over 24 hours, since diurnal frequencies were not of interest
4. All species present can be lumped into two classes: phytoplankton and zooplankton, each having constants averaged over all species, ages, and times of year
5. There is only one critical nutrient
6. Growth and reproduction need not be distinguished
7. Illumination and water temperature have sinusoidal annual cycles.

These assumptions gave rise to the following mathematical model:

$$\frac{1}{P}\dot{P} = K_1 N^{LIM} IT - K_2 Z - K_3 - K_4 T$$

$$\frac{1}{Z}\dot{Z} = K_5 P - K_6$$

$$\dot{N} = K_7 - K_8 P, N \geqslant 0$$

59

$$T = K_9 - K_{10} \cos \frac{\pi t}{6}$$

$$I = K_{11} - K_{12} \cos \frac{\pi t}{6}$$

where:

P = phytoplankton population density, millions of cells per cubic meter

Z = zooplankton population weight density, milligrams per cubic meter

N = nutrient concentration, milligram-atomic-weights per cubic meter

t = time, months, zero at January 1

T = water temperature in °C

I = incident solar and sky radiation, gram-calories per cm^2-min.

K_i = constants (see below)

N^{LIM} = N limited from above (also nonnegative)

Superscript dot = first derivative with respect to time.

The K_1 term denotes phytoplankton growth rate due to photosynthesis, K_2 the death of phytoplankton eaten by zooplankton, K_3 all other contributions to the death or disappearance rate of phytoplankton, K_4 debits the energy necessary for respiration, K_5 represents zooplankton growth resulting from phytoplankton consumption, K_6 is the death rate of zooplankton, K_7 is the rate of nutrient influx (due to tides, bacterial action, and marine animal excreta), and K_8 is the nutrient bound by phytoplankton. K_9 is the average annual temperature, K_{10} is the maximum summer elevation of temperature, K_{11} is the annual average illumination, and K_{12} is the maximum summer elevation of illumination. Most of the following values for the constants were based on values available from pertinent literature; the remaining constants $(K_1, K_2, K_5, K_6,$ and $K_7)$ were obtained by curve fitting on the computer.

$K_1 = 6.4$	$K_5 = 0.021$	$K_8 = 00.08$	$K_{11} = 0.22$
$K_2 = 0.0015$	$K_6 = 4.0$	$K_9 = 15.0$	$K_{12} = 0.10$
$K_3 = 3.0$	$K_7 = 6.0$	$K_{10} = 9.0$	N^{LIM} upper limit $= 0.5$
$K_4 = 0.15$			

The results obtained are shown in FIGURE 1. The "spring algal bloom" obtained had, as expected, a maximum value of the order of 1000 times the winter minimum. The small peaks during the summer and fall growth period could not be expected to be as regular as shown in FIGURE 1. However, this regularity results from the fact that cloudy days were not included in the illumination function used. The simulation indicated that the spring algal bloom peak is reached at exactly the time when the nutrient supply is depleted to zero. That is, phytoplankton growth is limited by the nutrient available.

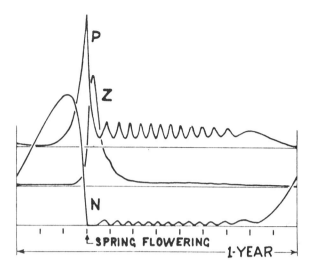

FIGURE 1. Simulated annual cycles.

The zooplankton peak occurs slightly after the phytoplankton peak, as expected.

The analog computer simulation in this preliminary study was not completely satisfactory for representing the reduced winter populations, because the voltages required were down in the noise level. To avoid this difficulty, the problem was also programmed for a digital computer, where sufficient dynamic range is available. More realistic winter population results were obtained by means of this program; the results for the remainder of the year were in agreement with the analog computer.

Applications for Large Ecosimulations

The appended bibliography lists those ecosimulations known to the authors. They are small in number but cover a wide variety of ecosystems: insect pest control, the housefly and a pupal parasite, two competing species of flour beetle, flux of radiocarbon, the vole population cycle, waste treatment, a fox-rabbit-grass system, habitat selection by animals, the effect on lynx population of the snowshoe hare cycle, crop yield and animal price, epidemics, water pollution, fish populations, and microbial growth. We can reasonably expect that many more such simple ecosimulations will be developed in the near future and that they can be put together to form much larger ecosimulations.

The feasibility of larger and more complex ecosimulations has been well demonstrated in simulation applications which are quite similar in their essential considerations to ecosimulation. Among these systems which have been treated by simulation techniques are: physiological processes, chemical

kinetics, industrial chemical processes, operations research systems, and socioeconomic systems.

Desirability of Large Ecosimulations

Large ecosimulations could have many important roles in: Interpreting
ecological systems, teaching ecology, ecological research, extending test data,
model identification, ecosystem disaster diagnosis, prediction, design of controls, and facilitating management decisions.

Consider, for example, the role of prediction. It is quite conceivable that
some man-caused disturbance could destroy an entire ecosystem as a result of
forcing the ecosystem beyond its limited range of stability. It is a certainty
that man has been responsible for the extinction of many individual species.
This automatically results in a loss to the ecosystem of which the species was
a member. Ecosimulation enables us to explore and hopefully prevent or
control such possibilities.

Ecosimulation is of paramount importance to many investigators dealing
with problems of biomedical simulation. The following selected applications of
ecosimulation are to be found in the biomedical area:

1. The spread of epidemics and plagues in a population, including the roles
   of all vectors (e.g., rats and fleas).
2. The dynamics of a disease syndrome in the human body, in which the
   populations of pathogens, antibodies, toxin concentrations, body temperature, etc., must be related.
3. Public health aspects of water pollution, such as toxic wastes, infected
   fish and shellfish, bacteria from sewage, worm eggs from farm
   animals, etc.
4. The growth and reproduction of parasites in an animal.
5. Public health aspects of the dissemination, spread and concentration of
   radionuclides in an ecosystem, including the food web of man.

Of lesser interest in biomedicine, but of greater socioeconomic concern to
man, are numerous ecological problems for the study of which ecosimulation
would be valuable, such as the worldwide food shortage, the population explosion, the conservation of all natural resources (water, air, soil, forests, etc.), life
support in space travel, pest control, and many others.

Opportunities in Ecosimulation

Thus the authors believe that ecosimulation is highly desirable and quite
feasible. It would seem that the conditions are right for a discipline in which
this demonstrated tool may be applied to investigations of ecological problems.

The field of ecosimulation is growing rapidly from a rather recent start
Papers reporting ecosimulations are doubling in number approximately every
two years. It is expected this growth rate will increase in the next few years.

Ecosimulation offers much. It is rich in humanistic values, probably equalling if not surpassing biomedical simulation in this respect. The rapid rate of development of ecosimulation promises opportunity for a variety of professional disciplines to join or form teams for the development of new ecosimulations. Development of proficiencies in ecosimulation can be derived from the fields of biomedical simulation, ecology, and several branches of engineering (chemical, civil, mechanical, sanitary, etc.).

Acknowledgments

The authors wish to express their appreciation of the work of K. N. Braun and M. Edwards in the programming of the plankton simulations.

Bibliography

1. WANGERSKY, P. J. & W. J. CUNNINGHAM. 1957. Time lag in population models. Cold Spring Harbor Symposia on Quantitative Biology 22: 329-338.
2. WANGERSKY, P. J. & W. J. CUNNINGHAM. 1957. Time lag in prey-predator population models. Ecology 38: 136-139.
3. WATT, K. E. F. 1961. "Use of a computer to evaluate alternative insecticidal programs. Science 133: 706-707.
4. WATT, K. E. F. 1961. Canad. Entomol. Suppl. 19. 93: 202-220.
5. WATT, K. E. F. 1964. Computers and the evaluation of resource management strategies. Amer. Scientist 52: 408-418.
6. WATT, K. E. F. 1964. Canad. Entomol. 96:
7. ODUM, H. T. 1960. Ecological potential and analogue circuits for the ecosystem. Amer. Scientist 48(1): 1-9.
8. NEEL, R. B. & J. S. OLSON. 1962. Use of analog computers for simulating the movement of isotopes in ecological systems. Report ORNL-3172. Oak Ridge National Lab.
9. MEIER, R. L., E. H. BLAKELOCK & H. HINOMOTO. 1964. Simulation of ecological relationships. Behavioral Science 9(1): 67-76.
10. GARFINKEL, D. 1962. Digital computer simulation of ecological systems. Nature 194(4381): 856-857.
11. EISGRUBER, L. M. 1965. Farm operation simulator and farm management decision exercise. Purdue Univ., Agr. Exp. Sta., Research Progress Report 162.
12. BLACK, M. L. & I. D. GAY. 1965. Some kinetic properties of a deterministic epidemic confirmed by computer simulation. Science :981-985.
13. THOMANN, R. V., et al., 1965. Delaware estuary comprehensive study. Reports 1-6, Public Health Service.
14. ANONYMOUS. 1965. Desk-Top Computer Keeps Track of our Fish Population. Computers and Automation. Feb. : 29.
15. NEUGROSCHL, E. J. 1963. Computer control of an activated sludge treatment plant. Case Inst. of Tech., Systems Research Center, Report SRC 25-C-63-8.
16. RAMKIRSHNA, D., A. G. FREDERICKSON & H. M. TSUCHIYA. Models for the dynamics of microbial growth. J. Theror. Bio. (in press).
17. GARFINKEL, D., R. H. MacARTHUR & R. SACK. 1964. Computer Simulation and Analysis of Simple Ecological Systems. Annals N. Y. Acad. Sci. 115 Art. 2: 943-951.

Editors' Comments
on Papers 7 and 8

7   **VAN DYNE**
Excerpts from *Ecosystems, Systems Ecology, and Systems Ecologists*

8   **REICHLE and AUERBACH**
Analysis of Ecosystems

SYSTEMS ECOLOGY AS AN APPROACH TO
LARGE-SCALE PROBLEMS

As mentioned in the introduction to Part I, a significant precursor of systems ecology was the "systems approach." This is basically a holistic methodology that solves individual problems in the context of the total system of interest. While the holistic epistemology was certainly nothing new, systems engineering (e.g., Shinners 1967) was a departure from approaches taken earlier in this century. The systems engineering approach and related developments in cybernetics (Weiner 1948) showed spectacular success during World War II in the solution of military problems and promised applicability to a wide range of disciplines.

The postwar period also saw a growing interest in studying the dynamics of ecosystems. Early success in quantifying ecosystem energy flow (Odum 1957; Tilly 1968; Teal 1962; Golley 1960) showed the feasibility of addressing total ecosystem problems. Developments also began in experimental manipulation of ecosystems (Auerbach et al. 1964; Likens et al. 1970) that involved large, complex studies by teams of researchers.

Total ecosystem studies were greatly stimulated by the advent of the International Biological Programme. Following a long series of organizational meetings at the national and international levels (Worthington 1975), the United States initiated a series of ecosystem studies through the Biome Programs. These programs proposed major interdisciplinary studies that would involve hundreds of scientists and dozens of research specialties.

It was natural that the Biome Programs should be strongly attracted to the systems approach and early documentation (Van Dyne 1972; Auerbach 1971) clearly shows a cognizance of the concept. From the beginning, modeling was envisioned as playing a major role in these studies. Indeed, the IBP Biome Programs provided the major financial and intellectual stimulation for the development of systems ecology.

In the context of the systems approach, modeling was conceived as having a complex and multiple role. The initial specification of significant components and interactions in a model was viewed as a systematic and mathematically rigorous approach to program planning. The model provided a framework from which program management could decide which areas needed emphasis. From the "Box and Arrow Diagram," one could determine in a fairly precise manner which processes had to be measured.

The model could also serve as a framework for analysis and synthesis of data. Since specific parameters would be needed to quantify the model, field and laboratory data could be structured toward the derivation of the required parameter values. Sensitivity analysis of preliminary versions of the model could be used to adjust program goals and funding allocation so that critically needed areas could be emphasized.

At the completion of the study, synthesis of the total program could be achieved through the model. With all data synthesized as model parameters, the final model would, in some sense, represent a total picture of the dynamics of the system under study. In this synthesized form, the model would then be applied to resource management or environmental impact assessment problems.

This ambitious and optimistic viewpoint dominated the late 1960s and early 1970s and is well reflected in the first of the papers in this section (Paper 7). However, the scenario presented above was considered overly ambitious by almost all the Biome Programs. The plan underestimated personnel problems (Van Dyne 1972), modeling management problems (O'Neill 1975), and a host of other existential difficulties (Auerbach et al. 1977). As actually implemented, the programs were more conservative and less optimistic about the possibilities for sweeping breakthroughs within the context of any single program. Nevertheless, this vision of the systems approach and the role of mathematical modeling was an important factor when the Biome Programs were first formulated.

Paper 7 by Van Dyne is a comprehensive presentation of the systems approach and reflects the forward-looking approach that characterized early applications. It should be noted that the term "systems ecology" is used to designate the entire interdisciplinary venture. This sense of the term is retained in some later presentations (Van Dyne 1969). In most contexts today, the term system ecology designates the mathematical modeling components of the program, rather than the total approach.

Paper 8 by Reichle and Auerbach appeared six years later and attempted an assessment of the systems approach following initial experiences in the Biome Programs. The viewpoint is still optimistic, if somewhat less ambitious. The broad role envisioned for modeling and the ultimate utility of the models for addressing practical problems is still strong in this article.

REFERENCES

Auerbach, S. I. 1971. The Deciduous Forest Biome Programme in the United States of America, pp. 677–84. In P. Duvigneaud (ed.), *Productivity of Forest Ecosystems.* UNESCO, Paris, 707 pp.

Auerbach, S. I., J. S. Olson, and H. D. Waller. 1964. Landscape investigations using cesium-137. *Nature* **201**:761–64.

Auerbach, S. I., R. L. Burgess, and R. V. O'Neill. 1977. The Biome Programs: Evaluating an experiment. *Science* **195**:902–04.

Golley, F. B. 1960. Energy dynamics of a food chain of an old-field community. *Ecol. Monogr.* **30**:187–206.

Likens, g. F., F. H. Bormann, N. M. Johnson, D. W. Fisher, and R. S. Pierce. 1970. Effects of forest cutting and herbicide treatment on nutrient budgets in the Hubbard Brook Watershed Ecosystem. *Ecol. Monogr.* **40**:23–47.

Odum, H. T. 1957. Trophic structure and productivity of Silver Spring, Florida. *Ecol. Monogr.* 27:55–112.

O'Neill, R. V. 1975. Management of large-scale environmental modeling projects, pp. 251–82. In C. S. Russell (ed.), *Ecological Modeling in a Resource Management Framework.* Resources for the Future, Washington, D.C., 394 pp.

Shinners, S. M. 1967. *Techniques of Systems Engineering.* McGraw-Hill, New York, 498 pp.

Teal, J. M. 1962. Energy flow in the salt marsh ecosystem of Georgis. *Ecology* **43**:614–24.

Tilly, L. J. 1968. The structure and dynamics of Cone Spring. *Ecol. Monogr.* **38**:169–97.

Van Dyne, G. M., ed. 1969. *The Ecosystem Concept in Natural Resource Management.* Academic Press, New York, 383 pp.

Van Dyne, G. M. 1972. Organization and management of an integrated ecological research program, pp. 111–72. In J. M. R. Jeffers (ed.), *Mathematical Models in Ecology.* Blackwell Scientific Publications, Oxford, 398 pp.

Wiener, N. 1948. *Cybernetics.* John Wiley & Sons, New York, 295 pp.

Worthington, E. B. 1975. *The Evolution of IBP.* Cambridge University Press, Cambridge, 257 pp.

7

Reprinted from pages 1–17 and 26–31 of *Ecosystems, Systems Ecology and Systems Ecologists*, ORNL-3957, Oak Ridge, Tenn.: Oak Ridge Natl. Lab., 1966, 40pp.

ECOSYSTEMS, SYSTEMS ECOLOGY, AND SYSTEMS ECOLOGISTS

George M. Van Dyne

ABSTRACT

This paper defines and discusses ecosystems, systems ecology, and systems ecologists, in that order. Some properties of ecosystems and the ecosystem concept are given as a basis for defining the area of study called systems ecology. Problems, methods, tools, and approaches of systems ecology are considered in defining tasks, problems, and training of systems ecologists. The interdisciplinary nature of systems ecology research and the importance of computers in this research are considered. Examples of methods, concepts, and applications are drawn from a diverse body of ecological, natural resource management, and mathematical literature, which further illustrate the interdisciplinary nature of systems ecology. Advantages and limitations, with respect to total-ecosystem problems, of research by ecologists in universities, in state and federal experiment stations, and in national laboratories are compared. An example is given wherein, possibly under International Biological Program support, the skills and resources of these three groups of ecologists might be combined for integrated attack on nationally important ecosystem problems.

ACKNOWLEDGMENTS

In any organization in which there is frequent and free interchange it is difficult to clearly identify the origin of concepts and ideas. Much of the material herein is a product of cross-fertilization of ideas with S. I. Auerbach, J. S. Olson, and B. C. Patten. Generally, we agree in principle, although our approaches differ in practice. Several other colleagues in the Radiation Ecology Section also have provided constructive criticisms of this paper. Acknowledgment also is extended to J. W. Barrett, J. E. Cantlon, F. B. Golley, A. M. Schultz, E. G. Struxness, and K. E. F. Watt, who have criticized the manuscript. G. C. Battle is thanked for his editorial assistance. The author assumes responsibility for any misconceptions or errors in this paper and would appreciate comments and criticisms. Part of this paper is based on a lecture given to the Radioecology Institute, Oak Ridge Institute of Nuclear Studies, June 1965.

INTRODUCTION

The Radiation Ecology Section of the ORNL Health Physics Division has initiated a new program of studies in "Systems Ecology." Neither is this area of work in ecology clearly defined nor do all ecologists view it equally. As with any new field, systems ecology is beset with vociferous skeptics (largely those who have done well under the old conditions) but supported primarily by lukewarm champions (largely those who may do well under the new conditions). It is desirable to examine the subject more closely as a further basis for clarification of objectives of work in this area.

Often we feel that our own work and interests are of extra importance, but I do not propose systems ecology to be a new panacea, nor that we neglect more conventional approaches in ecology. However, I feel that a systems approach has much to offer in various phases of ecology and especially in renewable resource management. This leads to a brief review of some ecological concepts and to suggestions about the tools, training, and work of systems ecologists. It is in this context that this essay is offered.

The purposes of this paper are twofold:

(1) To discuss important properties of ecosystems and to consider application of recent techniques from systems analysis and other fields. This is done not only to help define systems ecology and to help prevent semantic confusion, but also to help define some of the needs, training, and perspective of systems ecologists.

(2) To provide an introduction to a selected part of the large and diverse literature, through 1965, which encompasses the interdisciplinary area of systems ecology.

ECOSYSTEMS

Definitions

In 1935 Tansley (*106*) introduced the term ecosystem, which he defined as the system resulting from the integration of all the living and nonliving factors of the environment. Webster now defines the term as a complex of ecological community and environment forming a functioning whole in nature. An ecosystem is a functional unit consisting of organisms (including man) and environmental variables of a specific area (*3*). Macroclimate has an overriding impact on the other components, each of which is interrelated at least indirectly (Fig. 1). The term "eco" implies environment; the term "system" implies an interacting, interdependent complex.

Russian ecologists use the term ecosystem less frequently than the term biogeocoenosis, which Sukachev (*105*) defines as "any portion of the earth's surface containing a well-defined system of interacting living (vegetation, animals, microorganisms) and dead (lithosphere, atmosphere, hydrosphere) natural components...." Biogeocoenosis is derived from the Greek "bio" or life, "geo" or earth, and "koinos" or common. Sukachev and others feel that biogeocoenosis is the more accurate and descriptive term from an etymological viewpoint, but Sukachev recognizes that ecosystem is the older term and that the two terms are widely used as synonyms.

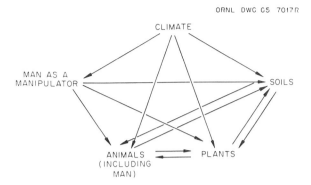

ORNL DWG 65 7017R

ORNL DWC C5 7017R

Fig. 1. An ecosystem is an integrated complex of living and nonliving components. Each component is influenced by the others, with the possible exception of macroclimate. And now man is on the verge of exerting meaningful influence over macroclimate.

For practical purposes and to avoid semantic argument, ecosystem and biogeocoenosis will be considered synonymous in this paper. Further discussion of ecosystem terminology is given by Schultz (*100*) and Maelzer (*73*).

The Ecosystem Concept: Unlimited Size and Complexity

A system is an organization that functions in a particular way. The functions of an ecosystem include transformation, circulation, and accumulation of matter and flow of energy through the medium of living organisms and their activities and through natural physical processes. Some specific functional processes include photosynthesis, decomposition, herbivory, carnivory, parasitism, and symbiosis (*31*). The ecosystem must be studied as a whole in order to understand energy transformations, the hydrologic cycle, or cycles of carbon, nitrogen, phosphorus, or other elements (*66, 68, 100*).

The ecosystem is the fundamental unit of study in "pure" and "applied" ecology (*31, 81, 106*). Directly or indirectly the ecosystem concept is useful in the management of renewable resources such as forests, ranges, watersheds, fisheries, wildlife, and agricultural crops and stock (*16, 26, 64, 72, 89, 91*). Understanding the ecosystem concept is required in the disposal of radioactive wastes and in analyses of environmental pollution (*108*). It has even found a place in medical studies of the digestive tract (*22*).

The ecosystem as a unit is a complex level of organization. It contains both abiotic and biotic components. The order of increasing complexity is: cell < tissue < organ < organism < population < community < ecosystem. Although the ecosystem is the most complex level, the study of a given ecosystem is less complex in many instances than the study of the lower levels (*33*).

We can consider steppe grasslands, deciduous forests, or oceans as examples of macroeco-systems, as well as considering a given small plot or a spaceship and its contents as a micro-ecosystem.

The term ecosystem is also used to describe a concept or approach of studying biotic-abiotic complexes. In applying the ecosystem concept there is no limit to size and complexity (*31, 34*). We delineate boundaries of ecosystems chiefly for convenience of study, although some natural boundaries may occur (e.g., shore lines and air-water or soil-water interfaces for aquatic systems), and man often introduces distinct boundaries, such as fences and field edges. Most ecosystems are bounded in nature by gradual and indistinct boundaries.

In the sense that the term ecosystem implies a concept and not a unit of landscape or sea-scape, the emphasis is that the biologist must look beyond his particular biological entity (e.g., cells, tissues, organs, etc.) and must consider the interrelationships between these components and their environment. For example, in applying the ecosystem concept to the tissue level of organization (Fig. 6) the environment of a specific tissue would include the fluids, such as blood and lymph, that may bathe or pass through the tissue as well as surrounding tissues. Part of the environment, therefore, consists of other parts of the same organism as well as ex-ternal components. Generally, the ecosystem concept is used in situations where at least several organisms are being considered.

Trends in Ecological Research

Although the concept of the ecosystem and many methods for studying ecosystems have been available for some time, only recently have many ecologists given more than lip service to the idea. Recently it has been suggested that the ecosystem is the rallying point for ecologists. There has been a gradual but distinct shift in emphasis in ecological studies and training from the description or inventory of ecosystems, or parts thereof, to the study of energy flow, nutrient cycles, and productivity of ecosystems. More workers are extending knowledge from the "anatomy" to the "physiology" of the environment. This requires different concepts, tools, and methods. The gradual change in emphasis from inventory to experimentation also requires more use of scientific methodology; this will be discussed below in the section, "Systems Ecology."

Ecosystem Components

Jenny (*54–55*) discusses dependent and independent variables in ecosystems and shows their relationships by the following equations:

$$l, \ s, \ v, \ \text{or} \ a = f(L_0, \ P_x, \ t) \ ,$$

where the internal properties are l = ecosystem property, s = soil, v = vegetation, and a = animals. The external properties are L_0 = initial state of the system, P_x = external flux poten-tials, and t = the age of the system. External flux refers to the flux of nutrients, energy, etc. in from and out to adjoining systems and can be defined by

$$P_x = \left(\frac{P_{out} - P_{in}}{dX}\right) m ,$$

where P_{out} and P_{in} are the flux out and in over a boundary thickness dX which has a permeation parameter m.

The controlling factors of the ecosystem are macroclimate, available organisms, and geological materials, where the last term includes parent material, relief, and ground water. Time is considered as a dimension in which the controlling factors operate, rather than as an environmental factor. The controlling factors are partially or entirely independent of each other. Each of the controlling factors is a composite of many separate elements, and each element is variable in time or space. Operationally, we may consider each controlling factor as a multiple-dimensioned matrix. Each change in a controlling agent in the ecosystem produces in time a corresponding change in the dependent elements of the ecosystem. In space and time there is a continuum of ecosystems.

Internal properties of ecosystems, such as rate of energy flow, might be considered as dependent factors which vary through time under the influence of a series of independent controlling factors. The dependent factors of the ecosystem are soil, the primary producers (vegetation), consumer organisms (herbivores and carnivores), decomposer organisms (bacteria, fungi, etc.), and microclimate. Each of these factors is dynamically dependent on the others (Fig. 1), and each is a product of the controlling agents operating through time.

Producers, consumers, and decomposers are not distributed at random in the abiotic part of an ecosystem. To maintain either dynamic equilibrium or ordered change in an ecosystem requires that a tremendous number of ordered interrelations exist among its dependent elements (82). To function properly ecosystems must process and store large amounts of information concerning past events, and they must possess homeostatic controls which enable them to utilize the stored information. This information may be expressed in amino acid and nucleotide sequences in genetic codes which have developed over evolutionary time, or it may be expressed in spatial or temporal patterns (20). For example, the changing patterns of plant populations and communities in secondary succession can be considered as expressions of genetically coded information. One species, population, or community is replaced by another with greater genetic potential for utilizing the resources of the changing environment.

Dynamics

Ecosystem changes may be caused by fluctuations in internal population interactions or by fluctuations of the controlling factors. Such changes may be cyclical or directional (14). Directional change from less complex to more complex communities may be considered as progression or succession; directional change from more complex to less complex communities may be considered as regression or retrogression; both are shown in Fig. 2.

Autogenic succession occurs when the controlling factors are stable and change is due to the effect of the system or some part of it on the microhabitat. Clements (15) formalized this

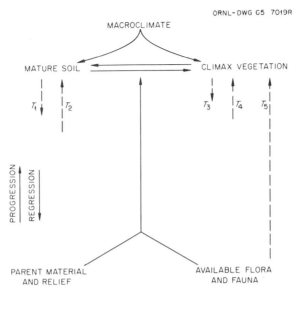

ORNL-DWG 65 7019R

$T_3 < T_4 \simeq T_1 < T_2 \lll T_5$

Fig. 2. Ecosystems develop through time, under climatic control, from the original flora and fauna under a given set of relief and parent material conditions. A final dynamic equilibrium is reached in which there exist a mature soil and climax plant and animal populations.

process as migration, ecesis, competition, reaction, and stabilization. This type of primary succession produces changes which are usually gradual and continuous. Allogenic succession occurs when there is a change in the controlling factors. Most changes in the ecosystem are products of both allogenic and autogenic successions. Most macroecosystems can be said to be polygeneic and are the result of several climatic changes and erosion cycles. Purposeful alterations, such as disruption by man, in the controlling and controlled factors of the ecosystem may induce relatively permanent changes in the ecosystem.

Because ecosystems vary both temporally and spatially, and to prevent ambiguity, it is important to specify at least semiquantitative time and space scales. The importance of specifying a time scale is illustrated in Fig. 2, where the time for primary succession (see T_5 for progression in Fig. 2) is shown as much greater than the time requirement for man to disrupt the system and alter soil or vegetation (T_1 and T_3 in Fig. 2). In the process of retrogression, changes take place in the vegetation more rapidly than they do in the soil. Generally, the ecosystem will recover towards the stable state through a progressive process called secondary succession (T_2 and T_4). Again, the rate of progressive changes of soil properties is usually lower than that for vegetation. Recovery of the vegetation to the climax state may take an amount of time similar to that required for deterioration of the soil. Change in a given ecosystem component or property may be negligible in T_3 but considerable in T_5.

During progressive succession there is usually an increase in productivity, biomass, relative stability and regularity of populations, and diversity of species and life forms within the ecosystem (*74*). Finally, the ecosystem reaches a steady state or equilibrium, which is characterized by dynamic fluctuation rather than by directional change. This steady state of the ecosystem is referred to as climax (*119*). At climax the dependent factors are in balance with the controlling factors; the climax is an open steady state (*101*). A diversity of species and life forms occupies every available ecological niche at climax and, because there is a maximum number of links in the food web, the stability of the system is maximized (*63*). A maximum amount of the entering energy is used in maintenance of life. Fosberg (*34*) considers "that climax communities [are those] in which there is the greatest range and degree of exploitation or utilization of the available resources in the environment." There is no net output from an ecosystem in the climax state (*86*). Three states of ecosystems exist with regard to energy or nutrient balance: steady state or climax, positive balance or succession occurring, and negative balance or decadence and senescence (*99*).

There is continual interchange of matter and energy among contiguous ecosystems. This interchange or flux is an essential property of ecosystems. The fluxes in and out of an ecosystem may be difficult to measure accurately, but there is relatively less error in measuring flux in a macroecosystem than in a microecosystem, because usually the error in measurements is inversely proportional to the magnitude of the object, rate, or processes being measured. Also, the relative amount of relevant surface or area around an ecosystem decreases as its size increases; many of the measurement errors or biases occur at such interfaces because of subjective decisions in defining boundaries. Still, we may find it convenient to study microecosystems such as a sealed bottle containing nutrients, gases, organisms, and water. Essentially, this is the type of system we need to study in preparing for interplanetary travel. But even such discrete microcosms are not adiabatic with their environment, and ultimately they are dependent upon their environment for a continuing energy input.

When flux of some element in and out of a given ecosystem is negligible for a defined period of time we consider that ecosystem to be stable with regard to that element. The equilibrium is referred to as climax only if it is reached naturally. Other equilibria, or disclimaxes, can be maintained by man's intervention. Here is the essence of renewable resource management: maintaining disclimaxes at equilibrium for the benefit of man.

Manipulation of Ecosystems

Man is a vital part of most major ecosystems, and there is an increasing human awareness of man's part in them and his influence on them (*108*) (Fig. 3). Traces of his pesticides probably can be found in living organisms throughout the world. Humans are both parts of and manipulators of ecosystems. Induced instability of ecosystems is an important cause of economic, political, and social disturbances throughout the world. In altering his environment in order to overcome its limitations to him, man learns that he often is faced with undesirable consequences

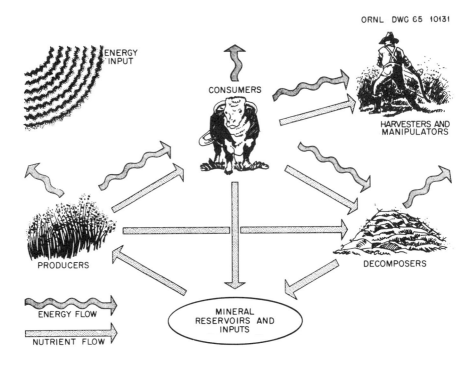

ORNL DWG 65 10131

Fig. 3. Man is both a spectator of and a participant in the functioning of ecosystems. He has manipulated ecosystems to maximize the flow of nutrients and energy to him from the producers and primary consumers. He has attempted to minimize the respiratory losses of energy from producers, consumers, and decomposers.

of the environmental change (*13, 38, 39*). In manipulating his environment (e.g., felling forests, burning grasslands or protecting them from fire, and draining marshes), seldom has he foreseen the full consequences of his action (*104*).

Most ecosystems in our country were in climax states when civilized man began to affect them, but the economy of civilized man demanded that the ecosystems produce a removable product under his domination. In order to reach this goal he disrupted the climax ecosystems, perhaps by shortening food chains or by altering the diversity of life forms of primary producers. He has altered the rate of and amount of nutrients cycling through the system by such means as fertilization, both in aquatic and terrestrial systems (Fig. 3). In some instances the fertilization has been excessive and has led to undesirable side effects, such as algal blooms caused by excesses of organic wastes. In other instances man has altered the structure of ecosystems by simplifying them and diverting the flow of energy into his food products, such as in replacing a grassland and wild animals with a wheatfield. Eventually he has produced changes in some ecosystem properties, which in some instances has led to new quasi-stable levels. In other instances such changes have led to desertification, such as the result of centuries of overgrazing in the Middle East.

Man has also encountered difficulties when he attempts to return ecosystems to their native state or to preserve vegetation by the development of national parks or by control of predators (*104*). In several instances ungulate populations have multiplied rapidly, outstripped the natural control by predators, exceeded the carrying capacity of their ranges, and severely damaged their habitat. Examples include the classical Kaibab mule deer problem (*94*) and the elk problem in Yellowstone National Park (*64*). Man himself has had a direct and profound effect on some ecosystems he has attempted to maintain in a natural state, such as in Yosemite National Park (*39*).

Exploitation of ecosystems is still occurring throughout the world, although the consequences are yet unknown. Systems ecology may contribute to a better understanding of ecosystems and ecosystem processes, which will help civilized man produce new and useful quasi-stable equilibria. We still need to know the long-term effects and profits of ecosystem manipulation, such as even further shortening of food chains, as human populations continue to increase exponentially and impose greater stresses on our world ecosystem. Knowledge about the entire ecosystem has become so important that ecologists can no longer be satisfied to be concerned with specific individuals, species, or populations in the ecosystem. In addition to plant ecologists, animal ecologists, microbial ecologists, etc., we must now train more and more young ecologists to confront the entire complexity of the ecosystem.

We are still in the process of developing technology and scientific knowledge that will enable us to better perceive the influences of our manipulations of ecosystems. Intelligent manipulation of ecosystems is increasing, and there is increased interest in scientifically defining the carrying capacity of ecosystems, as in the proposed International Biological Program, for example. Scientific ecology has clarified many cause-and-effect relationships in environmental change. An example of the value of basic ecological knowledge may be found in studies relating community stability, diversity, and biological control practices (*116*).

SYSTEMS ECOLOGY

Definitions

The July 1964 issue of *Bioscience* contained perspective articles by several noted ecologists, and the term systems ecology was used. E. P. Odum (*83*), then president of the Ecological Society of America, used the term "systems ecology" as follows:

> "... the new ecology is thus a systems ecology — or to put it in other words, the new ecology deals with the structure and function of levels of organization beyond that of the individual and species."

We have been taught that ecology is the study of the relationships between organisms and their environment (*51*) and that ecology may be subdivided into autecology (of individuals or species), population ecology, and community ecology (*80*). Systems ecology in a sense approximates communities ecology. The terms system and ecology both imply a holistic viewpoint. Just as ecology and ecosystems are considered by some people, systems ecology may be not so much an independent branch of study, but a point of view, a way of looking at things and explaining them, a concentration of pertinent concepts, facts, and data from various fields (61).

Some workers consider the realm of systems ecology to be that of using mathematics to study ecological systems. Although application of mathematical techniques to study ecosystems is an important part of systems ecology, it is far from being all of it. Systems ecology can be broadly defined as the study of the development, dynamics, and disruption of ecosystems. I consider systems ecology to have two main phases — a theoretical and analytical phase and an experimental phase.

Earlier I stated that for studying function in ecology we need methods and concepts which are different from those for studying structure. Essentially, study of problems in systems ecology requires three groups of tools and processes: conceptual, mechanical, and mathematical.

Study of Ecosystems

The tools and processes required for systems ecology are different from those for conventional phases of ecology because of the complexity of the total ecosystem as compared with a segment of it. When we consider the totality of interactions of populations with one another and with their physical environs — i.e., ecosystem ecology — we face a new degree of complexity (10). Other than some recent papers (e.g., ref. 41) only a few reasonably adequate functional analyses of natural ecosystems exist (80).

One of the major problems in systems ecology is that of analyzing and understanding interactions. Events in nature are seldom, if ever, caused by a single factor. They are due to multiple factors which are integrated by the organism or the ecosystem to produce an effect which we observe (45). To further complicate the matter, various combinations of factors and their interactions may be interpreted and integrated by the ecosystem to produce the same end result.

Conceptual Requirements

A first conceptual requirement in systems ecology is clearer definition of problems. It is axiomatic that ambiguous use of terminology and an ambiguous statement of the problem lead to ambiguities of thought as well (19). These statements apply to many fields, but, particularly here, clear definitions are required because of the type of people systems ecologists will be and the types of people with whom they will work (discussed further below). Furthermore, in using computers, which are essential tools for systems ecologists, it is necessary to formulate the problems precisely and to clearly delineate the factors involved.

A second conceptual requirement in systems ecology is more and better use of logic and scientific and statistical methods. Essentially, we can define scientific method as the pursuit of truth as determined by logic and experimentation. In scientific method we use the approach of systematic doubt to discover what the facts really are. Experimentation is one of several tools of scientific method used to eliminate untenable theories, that is, to test hypotheses (*32*). Other experiments may be conducted to determine existing conditions, or to suggest hypotheses, etc. The conclusions from experiments may be criticized because the interpretation was faulty, or the original assumptions were faulty, or the experiment was poorly designed or badly executed (*88*). Experimental design and statistical inference are aids in testing hypotheses.

Much past ecological research has not tested a hypothesis. There is a tendency for ecologists to pass over the primary phase of analysis. The lack of understanding of what is known already (inadequate knowledge of the literature, in part) is understandable because of the volume of material to be covered (*58*). Glass (*40*) has clearly stated this dilemma — "the vastness of the scientific literature makes the search for general comprehension and perception of new relationships and possibilities every day more arduous." But inadequate examination of facts and data and inadequate formulation of hypotheses lead to uncritical selection of experiments testing poorly formulated hypotheses, and ecologists are often at fault here (*51*). The experimental design is, essentially, the plan or strategy of the experiment to test clearly certain hypotheses (*32*). Statistical methods are especially important in experimentation with ecosystems, because not all factors influencing the system can be controlled in the experiment without altering the system (*29*). These uncontrollable factors lead to error or "noise" in our measurements, and inferences to be made from the results of experiments should be accompanied by probability statements (*32*).

Eberhardt (*27*) has discussed many of the problems ecologists encounter in sampling, and has stressed the importance of statistical techniques in analysis of such problems. Methods of statistical inference are also useful in suggesting improvements in our mathematical models and in suggesting alterations in the design of future experiments. Some of the work initiated and developed by the late R. A. Fisher on partial correlation and regression is invaluable to us in evaluating independent and interaction effects in complex ecosystems where experimental control is neither possible nor desirable (*45*).

The first two conceptual needs for systems ecology, mentioned above, lead naturally to the third, the approach of modeling (Fig. 4), with models which are mathematical abstractions of real world situations (*17, 107*). In this process some real world situation is abstracted into a mathematical model or a mathematical system. Next, we apply mathematical argument to reach mathematical conclusions. The mathematical conclusions are then interpreted into their physical counterparts. In some instances we are able to proceed from the real world situation via experimentation to reach physical conclusions. In other instances, however, we cannot experiment with a situation that does not exist but may become real; examples are such situations as thermonuclear war and wide-scale environmental pollution (*50*). In many cases we find it too costly to experiment; therefore mathematical modeling or mathematical experimentation may be especially useful.

ORNL-DWG 65-7016

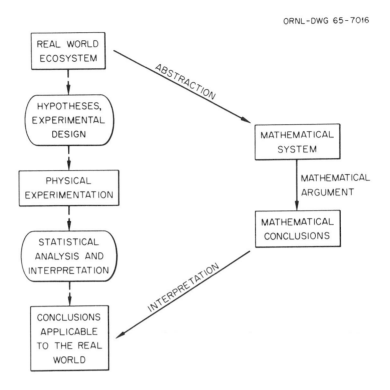

Fig. 4. Two ways of experimenting with ecosystems. One involves the conventional process of formulating hypotheses, designing and conducting experiments, and analysis and interpretation of results. The second involves the abstraction of the system into a model, application of mathematical argument, and interpretation of mathematical conclusions.

Mathematical modeling is somewhat new to many conventional ecologists and, in part, is just as much an art as a science. To ensure that the model will be valid, the mathematical axioms must be translations of valid properties of the real world system. The application of mathematical argument gives rise to theorems which we hope can be interpreted to give new insight into our real world system. However, the value of these conclusions should, where possible, be verified by experimentation. We must then accept the conclusions or reject them and start over again. This procedure of modeling, interpretation, and verification is used in many engineering and scientific disciplines. The success of the procedure, however, depends on the existence of an adequate fund of basic knowledge about the system. This knowledge permits predictive calculations. Hollister (50) outlines some of the problems to be encountered in modeling ecological phenomena.

Tools for Study of Ecosystems

The above conceptual tools should provide a framework in which to attack the complex problems of systems ecology. To implement these methods in studying ecosystems we will

need both physical and mathematical tools, including digital and analog computers and electrical, mechanical, and hydraulic simulation devices, and artificial populations (*44, 75, 85*). The act of expressing and testing biological problems with numerical, electrical, or hydraulic analogs often reveals some unsuspected relationships and leads to new approaches in investigation.
In conducting experiments in systems ecology, more refined chemical analytical equipment will be needed, such as gas chromatographs, infrared gas analyzers, and recording spectrophotometers. Physical analytical equipment required includes micro-bomb calorimeters, biotelemetric equipment, and other electronic equipment useful for rapid, nondestructive sampling and measuring of plant and animal populations and parameters under field conditions.

The importance of these chemical and physical tools is apparent when one considers the amount and variety of apparatus required to construct and maintain even the simplest aquatic ecosystems or to transplant and manipulate naturally occurring ecosystems for detailed measurements (e.g., ref. *2*). A major reason for the scarcity of detailed studies of entire terrestrial ecosystems is that many ecologists are not trained to use many of the required, diversified tools. In other instances these tools may not be available to the ecologist. The systems ecologist cannot be an expert with each of these tools, but he must be aware of their applications and limitations in the study of components and processes in ecosystems. He will need to be conversant with the specialists in other disciplines who make increasing use of these modern and complex tools. For example, in the last 12 years there has been a sixfold increase in the use of large, expensive, and complex instruments in chemistry (*118*).

One of the important tools is tracing with radioisotopes. This is valuable for identifying food chains, for determining the mass of nutrients in various compartments of an ecosystem, and for determining the time and extent of transfer of matter and energy among compartments within the ecosystem. Possibly we can use two, three, or more tags simultaneously in many ecological experiments if we select tracers which have appropriate physiological properties and types of radiation. Ionizing irradiation and selective poisons are other tools or treatments that can enable us to learn more about the function of ecosystem components without physically dismantling the entire ecosystem.

One interesting feature is that many of the new methods and instruments are simpler to use, although more expensive, because much laboratory skill in manipulation has been eliminated by instrumentation (*40*). The systems ecologist will still require the tools of conventional ecology, but he cannot rely on them alone. An important caution for the systems ecologist is that the problem should dictate the tools to be used; the opportunity to use a complex tool should not dictate the problem to be studied.

Operations Research and Systems Analysis Applications

Mathematical analysis will become increasingly important in providing advances in systems ecology. Large, fast digital computers have become available in the last 15 years and have allowed the development of special methods of analyzing and studying complex systems in industry and government. Most of these newer mathematical tools were developed in and are used primarily in two loosely defined and somewhat overlapping fields, operations research and systems analysis.

Operations research may be defined as the application of modern scientific techniques to problems involving the operation of a system looked upon as a whole (77). Included therein are any systematic, quantitative analyses aimed at improving efficiency in a situation where "efficient" is well understood (103).

Systems analysis is more difficult to define. Perhaps it can best be defined by opposites. The opposite of a systems approach is unsystematic or piecemeal consideration of problems; intuition may be taken as the opposite of analysis (46). Essentially systems analysis is any analysis to suggest a course of action arrived at by systematically examining the objectives, costs, effectiveness, and risks of alternative policies – and designing additional ones if those examined are found to be insufficient (93).

It is easily seen that operations research and systems analysis are both alike and different. They both contain elements from mathematical, statistical, and logical disciplines. In operations research, however, there usually is an unambiguous goal to be achieved, and the operations researcher is interested in optimization. The systems analyst faces a multiplicity of goals, a highly uncertain future, a frequent predominance of qualitative elements, and an exceedingly low probability of building an accurate and satisfactory model for his total problem (103). Because of the methods and techniques he can use effectively, the systems engineer has much to offer in study of ecosystems but he will need considerable guidance. In systems ecology he will be facing a collection and coupling of "green, pink, and brown" boxes (plants, animals, and physical environment) rather than the black boxes with which he is familiar (56). The interconnections between these boxes may be known only imperfectly, and the functional significance of the boxes will need to be established.

Some of the mathematical tools to be employed and examined in systems ecology include scientific decision-making procedures, theory of games, mathematical programming, theory of random processes, and methods of handling problems of inventory, allocation, and transportation (77).

Linear, nonlinear, and dynamic programming, which are especially important to the operations researcher, already show promise in ecology (4, 112, 115) and in management of renewable resources (12, 60, 69). Mathematical programming has already been used widely in agro-ecological problems, such as crop or yield prediction (97), in formulation of least-cost rations for livestock (110), and in farm management decisions (5). Game theory has been applied to decisions in cultivated-crop agriculture (113) and appears to have potential in dealing with wildland resources. Queuing theory and network flow appear to offer much in looking at problems of flow rates in ecosystems (90). Margalef (74) has discussed and indicated some important applications of information theory in ecology. Cybernetics principles and techniques are also useful in studying biological systems (37). Simulation is another important tool in operations research, although not limited to it. Mathematical simulation models have been used to study important resource problems, such as salmon population biology (59, 95), and abstract systems (36).

Importance of Digital Computers

Probably most systems ecology problems will be attacked first with deterministic models as first approximations (70, 71). However, to increase their usefulness and their realism, stochastic elements will be involved in most models or an indeterministic point of view will be taken; for example, see Leslie (65), Neyman and Scott (79), and Jenkins and Halter (53). This will require extensive use of digital computers, not only in simulation but also in analysis. Most stochastic models in ecology to date have been concerned with only one or two species rather than populations or ecosystems (6). Stochastic simulation of biological models or processes has been a useful process in some problems (109, 111). Many problems of modeling and analysis will require study and examination of the underlying statistical distributions (28). In addition to the normal distribution, other distributions which will need examination and use in systems ecology problems include the Poisson, the exponential, and the log-normal. Monte Carlo methods will be especially valuable in developing, testing, and using stochastic or probabilistic models (30, 67). Computers are essential in studying and using these statistical techniques in systems ecology. Other statistical aspects are discussed by Eberhardt (29).

Compartment model methodology, implemented with both analog and digital computers, has proven its value in theoretical studies and is beginning to be put to use in analysis and extension of real data in medical (9) and ecological fields (87, 90). Thus far, however, compartmental simultation models have been restricted to relatively simple ecological and agricultural situations, because most investigators have worked with analog systems of limited capacity (1, 35), although simulation systems have been developed for and used with digital computers in the study of renewable resources (e.g., ref. 42). Most systems ecologists will find it surprisingly easy to express many problems in the pseudoalgebraic languages, such as the many dialects of FORTRAN, used to communicate with digital computers.

An ecosystem might be depicted, as in Fig. 5, as composed of trophic levels represented as three-dimensional matrices. PRODUCER (I,J,K), CONSUMER (I,J,K), and DECOMPOSER (I,J,K) are matrices of species, individuals, and parts. The ranges of I, J, and K in each matrix are variable and depend upon the study. Matrices of transfer functions, depicted and simplified by the arrows in Fig. 5, are concerned with movement of matter or energy within individuals, between individuals, or between species. The latter two types of transfers may be between or within trophic levels. Also included in the figure is the fact that flux among contiguous ecosystems may be considered in matrix representation. Some of the transfer functions themselves may contain random noise and may be functions of a driving variable, such as macroclimate, acting on the system over time. Models or functions for macroclimatic influences may be constructed from actual data or may follow some prescribed hypothetical statistical distribution.

ORNL-DWG 65-7018

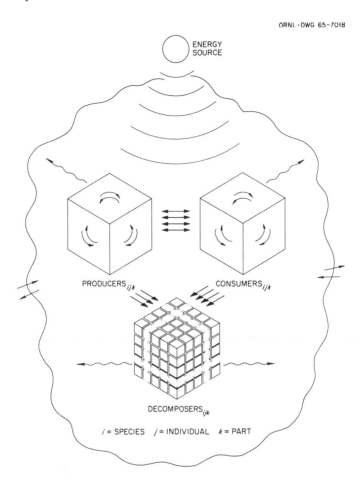

ENERGY SOURCE

PRODUCERS$_{ijk}$ CONSUMERS$_{ijk}$

DECOMPOSERS$_{ijk}$

i = SPECIES j = INDIVIDUAL k = PART

Fig. 5. A matrix representation of ecosystems adapted to pseudoalgebraic computer languages. Each trophic level is represented as a three-dimensional matrix. The arrows, wavy for energy and straight for matter, represent matrices of transfer functions interconnecting parts within individuals, individuals of a species to each other and to other species, etc., on up to connections between contiguous ecosystems. The transfer functions may contain probabilistic components and may be probabilistic functions of external variables such as macroclimate.

Consider the simplified case (Fig. 5) with only three parts per individual, three individuals per species, three species per trophic level, three trophic levels per ecosystem, and three eco-systems per problem. This leads to 3^5 microcompartments to be accounted for in addition to the many transfer functions interrelating the compartments. Many of the transfers will be zero, but this simplified model exceeds the capacity of most analog computers even if the problems of using various random function generators with an analog computer are bypassed. This example does not indicate that analog computers will not play an important role in systems ecology, but only that they may be of limited value in many realistically complex situations. Their major role may be as teaching (and learning) tools and as components of hybrid (digital-analog) systems. The capabilities and versatility of digital computers in general are far greater than those of analog computers (62).

Maximum use of most of the above mathematical tools and others by systems ecologists depends upon access to fast digital computers with large memory capacities (115). Such access will be especially important in working with large complicated models where remote-console access to large central computers will be essential for efficient and rapid progress. Computer technology is approaching the point where the rate of debugging of programs is the limiting factor.

The role of computers in the future of systems ecology is too readily underestimated. Computers in tomorrow's technology will have larger and faster memories, remote consoles, and time-sharing systems. Some may accept hand-written notes and drawings, respond to human voices, and translate written words from one language into spoken words in another (96). There will be vast networks of data stations and information banks, with information transmitted by laser channels over a global network. This network will be used not only by researchers but also by engineers, lawyers, medical men, and sociologists as well as government, industry, and the military. Computers could become tomorrow's reference library used by students in the university; they are already starting to revolutionize our present approaches to certain kinds of teaching. To utilize computers effectively in ecology we will have to state precisely what we know, what we do not know, and what we wish to know. Also, it will be necessary to assemble, analyze, identify, reduce, and store our ecological data and knowledge in a form retrievable by machine.

A Systems Approach

Systems ecology will call for an interdisciplinary team of systems ecologists, systems analysts and operations researchers (if they can be separated), conventional ecologists, mathematicians, computer technologists, and applied ecologists, including agriculturalists and natural resource managers of various disciplines. Systems ecologists studying ecosystems will devote at least as much time to delineating the problem as they will to solving it. This gives a hint as to the nature of the work of the systems ecologist.

The physical and mathematical tools to be used by this team are impressive, even though the list in the preceding sections is only partial. It serves to show that the systems ecologist will have to have more types of specialized training than did his predecessors. That different tools and methods may be needed to solve some of today's complex ecological problems is emphasized by the fact that many important contributors to advances in ecology in recent years may not be identified as ecologists (92). This will be especially true of systems ecology in the future, even though ecologists must be generalists and systems ecologists also will, in part, have to be generalists. Still, there are probably few if any authentic generalists or truly great minds who are not firmly grounded in a specialty (18). Most systems ecologists will serve their apprenticeship in basic fields. The conventional plant, animal, and aquatic ecologists will not be acceptable as systems ecologists, because they will lack the depth required in many specialties (78).

[*Editors' Note:* Material has been omitted at this point.]

LITERATURE CITED

(*1*) Arcus, P. L. 1963. An introduction to the use of simulation in the study of grazing management problems. Proc. New Zealand Soc. Animal Prod. 23: 159–168.

(*2*) Armstrong, N. E. and H. T. Odum. 1964. Photoelectric ecosystem. Science 143: 256–258.

(*3*) Bakuzis, E. V. 1959. Structural organization of forest ecosystems. Proc. Minn. Acad. Sci. 27: 97–103.

(*4*) Barea, D. J. 1963. Analisis de ecosistemas en biologia, mediante programacion lineal. Archivos de Zootecnia 12: 252–263.

(*5*) Barker, R. 1964. Use of linear programming in making farm management decisions. New York Agr. Exp. Sta. Bull. 993. 42 pp.

(*6*) Bartlett, M. S. 1960. Stochastic population models in ecology and epidemiology. Methuen and Co., Ltd. London. 90 pp.

(*7*) Bellman, R. 1961. Mathematical experimentation and biological research. Rand Corp. P-2300. 12 pp.

(*8*) Berman, M. 1963. A postulate to aid in model building. J. Theoret. Biol. 4: 229–236.

(*9*) Berman, M. 1963. The formulation and testing of models. Ann. New York Acad. Sci. 108: 182–194.

(*10*) Blair, W. F. 1964. The case for ecology. BioScience 14: 17–19.

(*11*) Briegleb, P. A. 1965. The forester in a science-oriented society. J. Forestry 63: 421–423.

(*12*) Broido, A., R. J. McConnen, and W. G. O'Regan. 1965. Some operations research applications in the conservation of wildland resources. Manage. Sci. 11: 802–814.

(*13*) Caldwell, L. K. 1963. Environment: a new focus for public policy? Public Administration Review 23: 132–139.

(*14*) Churchill, E. D. and H. C. Hanson. 1958. The concept of climax in arctic and alpine vegetation. Bot. Rev. 24: 127–191.

(*15*) Clements, F. E. 1916. Plant succession: an analysis of the development of vegetation. Carnegie Inst. Wash. Publ. 242. 512 pp.

(*16*) Cole, L. C. 1958. The ecosphere. American Scientist 198: 83–92.

(*17*) Coombs, C. H., H. Raiffa, and R. M. Thrall. 1954. Some views on mathematical models and measurement theory. *In*: Thrall, R. M., C. H. Coombs, and R. L. Davis (eds.). Decision Processes. John Wiley & Sons, Inc. New York. 332 pp.

(*18*) Dansereau, P. 1964. The future of ecology. BioScience 14: 20–23.

(*19*) Davis, C. C. 1963. On questions of production and productivity in ecology. Arch. Hydrobiol. 59: 145–161.

(*20*) Deevey, E. S. 1964. General and historical ecology. BioScience 14: 33–35.

(*21*) Dubos, R. 1964. Environmental biology. BioScience 14: 11–14.

(*22*) Dubos, R. and R. W. Schaedler. 1964. The digestive tract as an ecosystem. Amer. J. Med. Sci. 248: 267–271.

(*23*) Duckworth, W. E. 1962. A guide to operational research. Metheun and Co., Ltd. London, England. 145 pp.

(*24*) Duren, W. L., Jr. (Chr.). 1964. Tentative recommendations for the undergraduate mathematics program of students in the biological, management and social sciences. Mathematical Association of America, Committee on the Undergraduate Program in Mathematics. 32 pp.

(*25*) Duren, W. L. (Chr.). 1964. Recommendations on the undergraduate mathematics program for work in computing. Mathematical Association of America, Committee on the Undergraduate Program in Mathematics. 29 pp.

(*26*) Dyksterhuis, E. J. 1958. Ecological principles in range evaluation. Bot. Rev. 24: 253–272.

(*27*) Eberhardt, L. L. 1963. Problems in ecological sampling. Northwest Sci. 37: 144–154.

(*28*) Eberhardt, L. L. 1965. Notes on ecological aspects of the aftermath of nuclear attack. pp. 13–25 *in*: Hollister, H. and L. L. Eberhardt. Problems in estimating the biological consequences of nuclear war. U. S. Atomic Energy Commission TAB-R-5.

(*29*) Eberhardt, L. L. 1965. Notes on the analysis of natural systems. pp. 27–40 *in*: Hollister, H. and L. L. Eberhardt. Problems in estimating the biological consequences of nuclear war. U. S. Atomic Energy Commission TAB-R-5.

(*30*) Elveback, L., J. P. Fox, and A. Varma. 1964. An extension of the Reed-Frost epidemic model for the study of competition between viral agents in the presence of interference. Amer. J. Hygiene 80: 356–364.

(*31*) Evans, F. C. 1956. Ecosystem as the basic unit in ecology. Science 123: 1127–1128.

(32) Feibleman, J. K. 1960. Testing hypotheses by experiment. Persp. Biol. and Med. 4: 91–122.

(33) Feibleman, J. K. 1965. The integrative levels in nature. pp. 27–41 *in*: Kyle, B. (ed.). Focus on information and communication. ASLIB. London.

(34) Fosberg, F. R. 1965. The entropy concept in ecology. pp. 157–163 *in*: Symposium on Ecological Research in Humid Tropics Vegetation, Kuching, Sarawak, July 1963.

(35) Garfinkel, D., R. H. MacArthur, and R. Sack. 1964. Computer simulation and analysis of simple ecological systems. Ann. New York Acad. Sci. 115: 943–951.

(36) Garfinkel, D. and R. Sack. 1964. Digital computer simulation of an ecological system, based on a modified mass action law. Ecology 45: 502–507.

(37) George, F. H. 1965. Cybernetics and biology. Freeman and Co., San Francisco, California. 138 pp.

(38) George, J. L. 1964. Ecological considerations in chemical control: Implications to vertebrate wildlife. Bull. Entomol. Soc. Amer. 10: 78–83.

(39) Gibbens, R. P. and H. F. Heady. 1964. The influence of modern man on the vegetation of Yosemite Valley. Calif. Agr. Exp. Sta. Manual 36. 44 pp.

(40) Glass, B. 1964. The critical state of the critical review article. Quart. Rev. Biol. 39: 182–185.

(41) Golley, F. B. 1965. Structure and function of an old-field broomsedge community. Ecol. Monogr. 35: 113–137.

(42) Gould, E. M. and W. G. O'Regan. 1965. Simulation, a step toward better forest planning. Harvard Forest Paper 13. 86 pp.

(43) Gross, P. M. (Chr.). 1962. Report of the committee on environmental health problems. Public Health Service Publ. 908. 288 pp.

(44) Harris, J. E. 1960. A review of the symposium: models and analogues in biology. Symp. Soc. Exp. Biol. 14: 250–255.

(45) Hasler, A. D. 1964. Experimental limnology. BioScience 14: 36–38.

(46) Hitch, C. 1955. An appreciation of systems analysis. Rand Corp. P-699. (Symposium on Problems and Methods in Military Operations Research, pp. 466–481.)

(47) Hoag, M. W. 1956. An introduction to systems analysis. Rand Corp. RM-1678. 21 pp.

(48) Holling, C. S. 1963. An experimental component analysis of population processes. Mem. Entomol. Soc. Canada. 32: 22–32.

(49) Holling, C. S. 1965. The functional response of predators to prey density and its role in mimicry and population regulation. Mem. Entomol. Soc. Canada. 45: 3–60.

(50) Hollister, H. 1965. Problems in estimating the biological consequences of nuclear war. pp. 1–11 *in*: Hollister, H. and L. L. Eberhardt. (same title). U. S. Atomic Energy Commission TAB-R-5.

(51) Hughes, R. D. and D. Walker. 1965. Education and training in ecology. Vestes 8: 173–178.

(52) Jehn, K. H. 1950. The plant and animal environment: a frontier. Ecology 31: 657–658.

(53) Jenkins, K. B. and A. N. Halter. 1963. A multi-stage stochastic replacement decision model. Ore. Agr. Exp. Sta. Tech. Bull. 67. 31 pp.

(54) Jenny, H. 1941. Factors of soil formation. McGraw-Hill Book Co., New York. 281 pp.

(55) Jenny, H. 1961. Derivation of the state factor equation. Soil Sci. Soc. Amer. Proc. 25: 385–388.

(56) Jones, R. W. 1963. System theory and physiological processes: An engineer looks at physiology. Science 140: 461–464.

(57) Kennedy, J. L. 1956. A display technique for planning. Rand Corp. P-965.

(58) Kramer, P. J. 1964. Strengthening the biological foundations of resource management. Trans. N. Amer. Wildlife and Natural Resources Conf. 29: 58–68.

(59) Larkin, P. A. and A. S. Hourston. 1964. A model for simulation of the population biology of Pacific salmon. J. Fish Res. Bd. Canada 21: 1245–1265.

(60) Leak, W. B. 1964. Estimating maximum allowable timber yields by linear programming. U. S. For. Serv. Res. Paper NE-17. 9 pp.

(61) Lebrun, J. 1964. Natural balances and scientific research. Impact of Sci. on Society 14: 19–37.

(62) Ledley, R. S. 1965. Use of computers in biology and medicine. McGraw-Hill Book Co., Inc. New York. 965 pp.

(63) Leigh, E. G. 1965. On the relation between the productivity, biomass, diversity, and stability of a community. Proc. Nat. Acad. Sci. 53: 777–783.

(64) Leopold, A. S., S. A. Cain, C. H. Cottam, I. N. Gabrielson, and T. L. Kimball. 1963. Wildlife management in the national parks. Amer. Forests 69: 32–35, 61–63.

(65) Leslie, P. H. 1958. A stochastic model for studying the properties of certain biological systems by numerical methods. Biometrika 45: 16–31.

(66) Lewis, J. K. 1959. The ecosystem in range management. Proc. Ann. Meet. Amer. Soc. of Range Manage. 12: 23–26.

(67) Lloyd, M. 1962. Probability and stochastic processes in ecology. *In*: H. L. Lucas (ed.) The Cullowhee Conf. on Training in Biomath. Institute of Statistics, N. Car. St. U., Raleigh, N. C.

(68) Lotka, A. J. 1925. Elements of physical biology. Williams and Wilkins Co. Baltimore. 460 pp. (Also published by Dover Publications, Inc. New York. 1956.)

(69) Loucks, D. P. 1964. The development of an optimal program for sustained-yield management. J. Forestry 62: 485–490.

(70) Lucas, H. L. 1960. Theory and mathematics in grassland problems. Proc. Intern. Grassland Cong. 8: 732–736.

(71) Lucas, H. L. 1964. Stochastic elements in biological models; their sources and significance. pp. 355–383 *in*: Gurland, J. (ed.). Stochastic models in medicine and biology. U. Wisc. Press, Madison, Wisc. 393 pp.

(72) Lutz, H. J. 1963. Forest ecosystems: their maintenance, amelioration, and deterioration. J. Forestry 61: 563–569.

(73) Maelzer, D. A. 1965. Environment, semantics, and system theory in ecology. J. Theoret. Biol. 8: 395–402.

(74) Margalef, R. 1957. Information theory in ecology. Mem. Real Acad. Ciencias y Artes de Barcelona 23: 373–449.

(75) Margalef, R. 1962. Modelos fiscos simplificados de poblaciones de organismos. Mem. Real Acad. Ciencias y Artes de Barcelona 24: 83–146.

(76) Margalef, R. 1963. On certain unifying principles in ecology. The Amer. Nat. 97: 357–374.

(77) Miller, I. and J. E. Freund. 1965. Probability and statistics for engineers. Prentice-Hall Inc., Englewood Cliffs, N. J. 432 pp.

(78) Miller, R. S. 1965. Summary report of the ecology study committee with recommendations for the future of ecology and the Ecological Society of America. Bull. Ecol. Soc. Amer. 46: 61–82.

(79) Neyman, J. and E. L. Scott. 1959. Stochastic models of population dynamics. Science 130: 303–308.

(80) O'Connor, F. B. 1964. Energy flow and population metabolism. Science Prog. 52: 406–414.

(81) Odum, E. P. 1959. Fundamentals of ecology. W. B. Saunders Co. Philadelphia, Pa. 2nd Ed. 546 pp.

(82) Odum, E. P. 1963. Ecology. Holt, Rinehart, and Winston. New York. 152 pp.

(83) Odum, E. P. 1964. The new ecology. BioScience 14: 14–16.

(84) Odum, H. T. 1963. Limits of remote ecosystems containing man. The Amer. Biol. Teacher 25: 429–443.

(85) Odum, H. T. 1965. An electrical network model of the rain forest ecological system. U. S. Atomic Energy Commission PRNC 67.

(86) Odum, H. T. and R. C. Pinkerton. 1955. Time's speed regulator: the optimum efficiency for maximum output in physical and biological systems. Amer. Sci. 43: 331–343.

(87) Olson, J. S. 1965. Equations for cesium transfer in a *Liriodendron* forest. Health Physics 11: 1385–1392.

(88) Ostle, B. 1963. Statistics in research. Iowa St. Univ. Press. 2nd Ed. 585 pp.

(89) Ovington, J. D. 1960. The ecosystem concept as an aid to forest classification. Silva Fennica 105: 73–76.

(90) Patten, B. C. 1964. The systems approach in radiation ecology. Oak Ridge National Laboratory Technical Memorandum 1008. 19 pp.

(91) Pechanec, J. F. 1964. Progress in research on native vegetation for resource management. Trans. N. Amer. Wildlife and Natural Resources Conf. 29: 80–89.

(92) Platt, R. B., W. D. Billings, D. M. Gates, C. E. Olmsted, R. E. Shanks, and J. R. Tester. 1964. The importance of environment to life. BioScience 14: 25–29.

(93) Quade, E. S. (ed.). 1964. Analysis for military decisions. Rand Corp. R-387. Rand McNally & Co., Chicago.

(94) Rasmussen, D. I. 1941. Biotic communities of Kaibab Plateau, Arizona. Ecol. Monogr. 11: 229–275.

(95) Royce, W. F., D. E. Bevan, J. A. Crutchfield, G. J. Paulik, and R. L. Fletcher. 1963. Salmon gear limitation in northern Washington waters. U. Wash. Publ. in Fisheries (N. S.) 2: 1–123.

(96) Sarnoff, D. 1964. The promise and challenge of the computer. Amer. Fed. Infor. Process. Soc. Conf. Proc. 26: 3–10.

(97) Schaller, W. N. and G. W. Dean. 1965. Predicting regional crop production: an application of recursive programing. USDA Tech. Bull. 1329. 95 pp.

(98) Schmitt, O. H. and C. A. Caceres (eds.). 1964. Electronic and computer-assisted studies of bio-medical problems. C. C. Thomas. Springfield, Illinois. 314 pp.

(99) Schultz, A. M. 1961. Introduction to range management. U. California. Ditto notes. 116 pp.

(100) Schultz, A. M. 1965. The ecosystem as a conceptual tool in the management of natural resources. *In*: Parsons, J. J. (ed.). Symposium on quality and quantity in natural resource management (in press), manuscript 33 pp.

(*101*) Sears, P. B. 1963. The validity of ecological models. pp. 35–42. *In*: XVI International Congress of Zoology. Vol. 7. Science and Man Symposium – Nature, Man and Pesticides.

(*102*) Slobodkin, L. B. 1965. On the present incompleteness of mathematical ecology. Amer. Scientist 53: 347–357.

(*103*) Specht, R. D. 1964. Systems analysis for the postattack environment: some reflections and suggestions. Rand Corp. RM-4030. 34 pp.

(*104*) Stone, E. C. 1965. Preserving vegetation in parks and wilderness. Science 150: 1261–1267.

(*105*) Sukachev, V. N. 1960. Relationship of biogeocoenosis, ecosystem, and facies. Soviet Soil Science 1960: 579–584.

(*106*) Tansley, A. G. 1935. The use and abuse of vegetational concepts and terms. Ecology 16: 284–307.

(*107*) Thrall, R. M. 1964. Notes on mathematical models. U. Mich. Engin. Summer Conf. – Foundations and Tools for Operations Research and Management Sciences. Multilith 97 pp.

(*108*) Tukey, J. W. *et al.* (Environmental Pollution Panel, President's Science Advisory Committee). 1965. Restoring the quality of our environment. Superintendent of Documents, U. S. Govt. Printing Office, Washington, D. C. 317 pp.

(*109*) Turner, F. B. 1965. Uptake of fallout radionuclides by mammals and a stochastic simulation of the process. pp. 800–820 *in*: Klement, A. W. (ed.). Radioactive fallout from nuclear weapons tests. U. S. Atomic Energy Commission Symp. Ser. 5. 953 pp.

(*110*) van de Panne, C. and W. Popp. 1963. Minimum-cost cattle feed under probabilistic protein constraints. Manage. Sci. 9: 405–430.

(*111*) Van Dyne, G. M. 1965. Probabilistic estimates of range forage intake. Proc. West. Sect. Amer. Soc. Animal Sci. 16 (LXXVII): 1–6.

(*112*) Van Dyne, G. M. 1965. Application of some operations research techniques to food chain analysis problems. Health Physics 11: 1511–1519.

(*113*) Walker, O. L., E. O. Heady, and J. T. Pesek. 1964. Application of game theoretic models to agricultural decision making. Agronomy J. 56: 170–173.

(*114*) Watt, K. E. F. 1962. Use of mathematics in population ecology. Ann. Rev. Entom. 7: 243–260.

(*115*) Watt, K. E. F. 1964. The use of mathematics and computers to determine optimal strategy and tactics for a given insect pest control problem. Canad. Entomol. 96: 202–220.

(*116*) Watt, K. E. F. 1965. Community stability and the strategy of biological control. Canad. Entomol. 97: 887–895.

(*117*) Watt, K. E. F. 1965. An experimental graduate training program in biomathematics. BioScience 15: 777–780.

(*118*) Westheimer, F. H. (Chr.). 1965. Chemistry: opportunities and needs. Nat. Acad. Sci.– Nat. Res. Counc. Publ. 1292. Washington, D. C. 222 pp.

(*119*) Whittaker, R. H. 1953. A consideration of climax theory: the climax as a population and pattern. Ecol. Monogr. 23: 41–78.

(*120*) Yambert, P. A. 1964. Is there a niche for the generalist? Trans. N. Amer. Wildlife and Natural Resources Conf. 29: 352–372.

8

Reprinted from pages 260–280 of *Challenging Biological Problems: Directions Towards Their Solutions,* J. A. Behnke, ed., Arlington, Va.: Am. Inst. Biol. Sci., 1972, 502pp., with the permission of the authors and the American Institute of Biological Sciences

Analysis of Ecosystems

DAVID E. REICHLE
STANLEY I. AUERBACH
ENVIRONMENTAL SCIENCES DIVISION
Oak Ridge National Laboratory[1]

> Actually, as should be apparent, the general problem of ecosystem analyses is, with the exception of sociological problems that are also ecological, the most difficult problem ever posed by man.
> (P. Handler, 1970)

Introduction

The landscape, when viewed as an abstraction, represents different things to different people: source of food to the agriculturalist; land for the developer; places for cities, towns, factories, and parks to the general population. To the general biologist, the landscape may appear as a complex of organisms influenced by a complex of climatic, edaphic, genetic, and other interactive factors. To the ecologist, it also appears as a pattern of interdependent units, or ecosystems, all derived through evolutionary processes and mediated by climate, substrate, and geography. Examples of ecosystems are the various types of forests, grasslands, streams, lakes, deserts, tundras, and oceans which comprise the biosphere. These ecosystems, although part of a larger continuum, possess specific characteristics; some may be unique, others differing between systems only in degree and not in kind.

What is the goal or utility of "understanding the structure and func-

Research supported in part by the U.S. Atomic Energy Commission under contract with Union Carbide Corporation and in part by the Eastern Deciduous Forest Biome, US-IBP, funded by the National Science Foundation under Interagency Agreement AG-199, 40-193-69 with the Atomic Energy Commission—Oak Ridge National Laboratory.

Contribution No. 34 from the Eastern Deciduous Forest Biome, US-IBP.

1. Operated by the Union Carbide Corporation for the U.S. Atomic Energy Commission.

tion" of ecosystems? Similar to other organisms, man interacts with the other components of ecosystems, affects them, and is affected by them. Attempts to interpret the ecological effects of man's population growth and technological development are showing that simplistic, narrow approaches to assessing environmental impacts are, at best, inadequate. For man to learn to live in and to maintain the quality of his environment, it will be essential to understand ecosystems, their driving forces, their dynamics, and the mechanisms by which they are regulated.

Based upon an historical review of the scientific development of ecosystem analysis and the basic principles of the ecosystem concept, the following sections will illustrate, by means of examples, the potential of the systems approach when applied to the analysis of ecosystems.

Historical Perspective

Recognition of the functional ecosystem approach to environmental problems is commonly credited to Lindeman (1942) and the trophic-dynamic school of ecology (Allee, et al., 1949; Odum, 1957). Perhaps the greatest impetus of this concept as a way of addressing and understanding ecological processes at the landscape level came from the writings of E. P. Odum (e.g., Odum, 1971). Odum and colleagues have demonstrated that ecosystem ecology has an intrinsic intellectual tie to our understanding of nature and the environment. The trophic approach to ecosystem analysis soon associated the flow of materials between system components with the overall metabolism of the ecosystem. It soon became apparent that system metabolism was a measure of the collective processes which served to maintain the integrity and stability of the system.

This analogy to cellular and organism metabolism provided a basis for the application of similar methodologies to ecosystem analysis. Of these, the most powerful were the radiotracers. The theoretical and operational principles had been previously established in physiology (Sheppard, 1962), and the advent of radioecology, with its applied mission for determining the fate of radioactive elements released to the environment, provided both the impetus and the tools for testing these concepts. The principles, techniques, and analytical power of radiotracer methods to quantify trophic dynamics were demonstrated by Davis and Foster (1958), Auerbach (1958), Crossley and Howden (1961), and other early radioecologists. Application of transfer coefficient matrices in ecosystem modeling by Olson (1965) illustrated the potential for applying advancements in systems analysis, using the electronic computer to provide solutions to problems of complex ecosystem interactions.

Computer development in the late 1940s was followed by a growing recognition among biologists of the power of the computer for manipulating large data sets which are characteristic of ecosystem research. By the early 1960s, a small number of ecologists had begun to apply this data-handling capability of the computer to the solution of long-standing (but unresolved) problems, such as predator-prey interactions (Holling, 1966), populations and their exploitation (Watt, 1963), and ecosystem dynamics (Olson, 1965).

The potential of the computer in large-scale ecosystem studies awaited recognition of the need for a new level of formal theory. Based on an ecosystem level of analysis, an understanding of mathematical and numerical analysis, and an appreciation of the magnitude of the problem, ecologists (Olson, 1963; Van Dyne, 1966; Watt, 1966; Patten, 1971) began to write convincingly of the need for application of systems analysis procedures to ecology and, in fact, developed the field of Systems Ecology. Currently, the Biome programs of the International Biological Program (IBP) have further stimulated the application of systems analysis to the understanding of complex ecosystems (Reichle, 1970).

Ecosystem Ecology

The goal of ecosystem analysis is to develop a quantitative ecosystem science which may provide new theoretical insights into the organization *and* function of natural systems at their most complex level. One of the basic hypotheses being tested is whether the landscape is organized in logical patterns of self-sustaining, dynamic, and internally regulated components. Because of the complexity of interactions and variables involved, the question is: "How can the ecologist approach the quantitative analysis and interpretation of ecosystems?" From development of conceptual, and then mathematical, models and their subsequent validation, the ecologist strives to acquire a formalistic basis for dealing with complex sets of interacting variables. Subsequent simulations which involve manipulation of model parameters may be used to test hypotheses of system behavior and regulation—and to derive inferences at different levels of scale.

Holistic models are often required to deal with a scale of the landscape

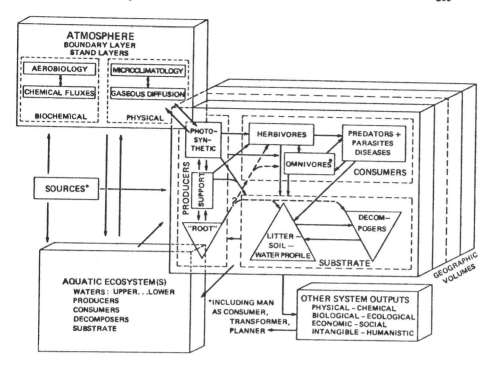

LAND ECOSYSTEM MODULE.

FIGURE 1. A diagramatic representation of a land ecosystem module, interfaced with atmospheric and aquatic systems to represent an integral segment of the landscape. This concept of the environment recognizes the coupling between ecosystems in time and space. The hierarchical organization of ecosystem attributes reflects the interaction of system components at several scales by resolution. (modified after Olson, 1970).

considerably more broad and generalized than the precise levels of resolution necessary to understand the underlying ecological mechanisms. The biosphere may be examined at various levels of resolution, depending upon the complexity of the problem and the kinds of couplings between ecosystems (Fig. 1). Specific ecosystems also may be subdivided into a hierarchy of subsystems based on internal mechanisms (or processes). These may be basic processes of energy transformation, such as production, mortality, and decomposition of green plants, or related functional processes, such as nutrient cycling (Likens et al., 1970).

The ecosystem is conceptually defined (Figs. 1 and 2) as a functional entity with internal homeostasis, identifiable boundaries, and recognizable

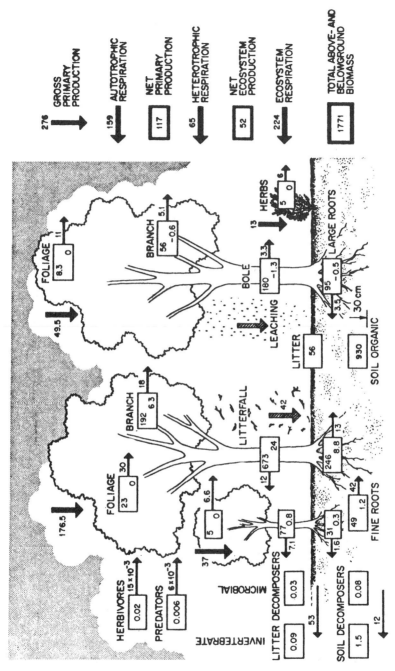

FIGURE 2. Pictorial presentation of the carbon cycle in a mesic hardwood forest ecosystem at Oak Ridge National Laboratory. Structural components of the ecosystem have been abstracted as compartments, with the major fluxes through the system illustrated by arrows. Values in compartments are annual standing crops and those in parentheses are annual increments. Units of measure are 10^5 grams per hectare and 10^5 grams per hectare per year for fluxes on arrows; e.g., value for litter is equivalent to 5600 kg ha^{-1} or 560 g m^{-2}. Summaries of ecosystem metabolism are shown to the right of the figure. (constructed from Sollins, 1972).

94

relationships between subcomponents. System boundaries are established to minimize or maximize transfers across them (to optimize analytical evaluation) and must contain a complete set of ecosystem processes and their interactions (Smith, 1970). Let us consider a specific example by examining the biological cycle of carbon in the biosphere, proceeding from global budgets to ecosystem cycles and related ecological processes.

Among the ways of functionally linking the world's regional and local systems to one another has been the unifying consideration of the circulation of carbon through a common atmospheric pool. Rather simple mechanistic models (e.g., Olson, 1970) have been sufficient to assess recent trends of carbon exchange and to extrapolate a few years into the future. Even now, we are not certain how important terrestrial ecosystems, especially forests, are in modifying or stabilizing these cycles, but

TABLE I. *Net Annual Carbon Fixation in Primary Production and Standing Crop of Carbon in Biomass for Terrestrial Ecosystems of the World.* *

	Area 10^4 km^2	Net Primary Production of Carbon NPP $ton/km^2/yr$	10^9 tons	Live Carbon Pool		
				Tons km^2	10^9 tons	NPP pool
Woodland or forest						
temperate "cold-deciduous"	8	1,000	8	10,000	80	0.10
conifer: boreal and mixed	15	600	9	8,000	120	0.075
rainforest: temperate	1	1,200	1.2	12,000	12	0.10
rainforest: tropical, subtropical	10	1,500	15	20,000	200	0.075
dry woodlands (various)	14	200	2.8	5,000	70	0.04
Subtotal	48		36		482	
Wetlands, thickets	1	2,000	2	5,000	5	0.40
Nonforest						
agricultural	15	400	6	1,000	15	0.4
grassland	26	300	7.8	700	18.2	0.43
tundra-like	12	100	1.2	600	7.2	0.17
other "desert"	32	100	3.2	287	9.2	0.29
glaciers	15		0			
Subtotal	100		18.2		49.6	
Average/continents	149	380	56.2	3,700	544.6	
Average/ocean	362	61	22	8	3	
Total Biosphere	511		78		548	

* Carbon data and geographic estimates of ecosystem area updated by Olson from Table 3 (Olson, 1970). (Courtesy Springer-Verlag)

data suggest that their importance may have been underestimated in the past (Table I). Current information is insufficient to assess the atmospheric impact (and indirect effects on local ecosystems as well as the biosphere) of excess carbon dioxide released from the burning of fossil fuels and land clearing which releases much carbon from humus and dead trees. Better understanding of the internal processes of aquatic and terrestrial ecosystems is essential before we can extend predictions to the decades ahead.

Systems Approach to Studying Ecosystems

Systems ecologists' heritage in trophic dynamics and the flow of energy through ecosystems has resulted in a concern primarily for ecosystem processes rather than for species components. Research on transport mechanisms, such as food chains, relates not only to system components (organisms) and their interactions but also to those physical and biological factors which control the rates at which ecosystem functions occur. This approach has resulted in an interpretation of ecosystem dynamics in terms of the structural and compositional attributes affecting internal transport of materials (carbon, mineral elements, etc.). Ultimately, the ecosystem ecologist is concerned with both system dynamics and system strategies which provide a basis for synthesis of the contributions from both functional and structural fields of ecology.

Since environmental factors themselves are part of ecosystems, the distinction between dependent and independent variables becomes obscure. This feature necessitates varied and innovative approaches to the analysis of ecosystems (Goodall, 1970). *Experimentation*, to cover a broad range of relevant component combinations, may be equally as valuable as the usual effort devoted to replicate sampling for statistical precision. *Observation* of the behavior of components of ecosystems under different conditions gives breadth to the standard experimental approach. In addition, historical information may show that key variables and infrequent events have greater control over the long-term behavior of the system than the normal ranges of obvious variables.

Morton (1964) has suggested that systems analysis is, in effect, the scientific method itself which ensures an organized approach to examination of complex systems. In the study of ecosystems, the modeling activity is of great importance since a model which accurately represents the real system can itself be utilized for experimentation as well as to provide early identification of key variables, parameters, and linkages between subsystems. In the analyses of ecosystems, five sequential operations (objective, conceptual, modeling, empirical, and analytical) are fundamental to both the research and modeling efforts (Dale, 1970; Reichle et al., 1971).

The carbon cycle in deciduous forests can be used to illustrate these facets of environmental systems analyses. From examination of the carbon pools and turnover rates in deciduous forest ecosystems (Fig. 2), the relative contribution of various system components to carbon dioxide exchange with the atmosphere is apparent. The greater percentage of aboveground respiration resulting from autotrophic (green) plants comes from metabolically active tissues, such as leaves and bark of branches and twigs rather than the bulk of the biomass represented in wood. Well over one-third of the total ecosystem respiration of 224×10^5 grams per hectare results from the decomposition of detritus in the soil carbon pool. The subsequent model of decomposition will serve to illustrate one aspect of ecosystem dynamics and continue our analysis of the carbon cycle.

To examine the dynamics of the soil and litter carbon pool, Sollins (1972) developed five linear differential equations describing the organic detritus component of the forest carbon model (Fig. 2). Conceptually, the forest floor was partitioned into two litter layers (0_1 and 0_2) and two mineral soil layers (0-10 cm and 10-60 cm depth). Modeling the system consisted of specifying the mechanisms by which the decomposition processes occurred and writing the system equations. Concise and realistic definitions had to be made for each compartment and for all transfer pathways in the model. As an instructive example, let us examine the equation describing changes in compartment size of the fast decomposing component of 0_1 litter:

$$O_1 = (L_o + L_u + L_g) + C(1\text{-}a) - [O_1(R_{o_1} + \beta)K] + \pi$$

where $L_o = \lambda_1 Q_1 =$ (overstory leaf fall) (kg m^{-2} yr^{-1})

$L_u = \lambda_2 Q_2 =$ (understory leaf fall) (kg m^{-2} yr^{-1})

$L_g = \lambda_3 Q_3 =$ (ground flora leaf fall) (kg m^{-2} yr^{-1})

and λ_i's are the respective litterfall rates (yr^{-1}) which operate only during the dormant season and Q_i's are the respective foliage biomasses (kg m^{-2})

where $C(1 - a) =$ frass influx (kg m^{-2} yr^{-1}) to forest floor and

$C =$ insect feeding flux (kg m^{-2} yr^{-1}) and

$a =$ proportionality constant for digestive assimilation

where $O_1(R_{o_1} + \beta)K$ = decomposition losses of litter and

O_1 = weight of O_1 litter (kg m^{-2}) and

R_{o_1} = respiratory carbon losses (yr^{-1}) and

β = transfer rate of O_1 litter to O_2 horizon and

K = a dimensionless temperature-moisture coefficient

= 0.2 (° C × % moisture in litter)

where π = the input of the rapidly decomposing component of litter other than leaves (e.g., reproductive parts) (kg m^{-2} yr^{-1}).

After such mathematical representations of system processes have been developed, the important analytical aspects of quantifying coefficients and investigating model responses remains. Mean standing crop of litter in the forest was 540 g m^{-2} (Fig. 2). Total annual litterfall averaged 420 g m^{-2} yr^{-1} plus 100 g m^{-2} yr^{-1} branch and bole wood. Insect frass (and aphid honeydew) amounted to 10 g m^{-2} yr^{-1}. Respiration accounted for a total litter (O_1 and O_2) weight loss of 530 g m^{-2} yr^{-1}. Soil organic matter has been estimated to accrue at 100 g m^{-2} yr^{-1} (Sollins, 1972). Consequently, estimated root dieback of ~350 g m^{-2} yr^{-1} was found to be a significant input to soil and litter detritus, heretofore not fully appreciated.

Since models are often constructed from existing data bases, their realism needs to be tested by comparison with other independent data bases to avoid circular reasoning. A high degree of realism implies that the model resembles the real system for the range of system variables relevant to the problem at hand (Walters, 1971). Accuracy of model simulations is not necessarily synonymous with model complexity (O'Neill, 1972). Many mathematical and systems analysis techniques are available for evaluating model performance, but the ecologist must supply the objective criteria to be tested. The model remains not a product in itself but an experimental tool for data synthesis and interpretation. In this final stage, the results of model analyses are applied as the best assessment of the performance of the system. (The entire sequence is iterative.)

Analysis of an Ecosystem

Just as the organism is more than its component tissues and cells, so the whole ecosystem is more complex than the sum of its constituent parts. In ecosystem analysis, however, the challenges lie in interfacing processes to construct total systems. New system properties arise from interactions, feedbacks, and synergisms between individual components.

Total system attributes and perspectives, forthcoming from analysis of ecosystem studies, can be exemplified by again considering the example of energy flow and productivity in a forest ecosystem (Fig. 2).

In this analysis, net primary production (NPP) is obtained by direct measurement of ecosystem components, but evaluation of autotrophic respiration (R_{sA}) is necessary to calculate gross primary production (GPP) and assess the metabolic "costs" of radiant energy conversion via the photosynthetic process. Net ecosystem production (NEP) or the accumulation of organic biomass by the system, however, requires that total ecosystem respiratory losses (including heterotrophic respiration, R_{sH}) be known. New properties for the ecosystem begin to emerge which are aggregates (but not necessarily simple algebraic sums) of its components and presumably relate to the strategy by which the *system* has adapted to its variable environment.

Consider the relationship among parameters of the whole ecosystem (Fig. 2). Can these attributes contribute toward a general set of ecosystem principles? Do the ratios of R_{sA}/R_{sE} of 0.71 and R_{sH}/R_{sE} of 0.29, which compare the autotrophic respiratory energy losses, provide a clue to the efficiencies of anabolism and catabolism in the ecosystem (Reichle et al., 1972)? How do the metabolic costs of ecosystem production ($NEP/R_{sE} = 0.23$) and the size of ecosystem increments relative to standing crop ($NEP/NPP = 0.44$) relate to patterns of structure, composition, and age of different ecosystems (Jordan, 1971; Sollins, 1972)? Are these attributes always proportioned in a fixed 1.0 to 0.35 ratio between above- and below-ground components as in the forest ecosystem example (Fig. 2), or do they vary in accordance with unappreciated patterns of ecosystem dynamics? While pattern is beginning to unfold for components, little is known of the quantitative similarities and differences in function between ecosystems. There is a need to explore what these analytical measures contribute toward our understanding of the living environment. They reflect a deeper and more complex nature of the ecosystem, and pragmatically, they lead to examination of an entirely new series of questions.

The methods of systems analysis have been developed to apply to man-designed systems; e.g., transportation networks, economic systems, and industrial production. Ecosystems are the product of biological evolution and display inherent strategies and mechanisms for self-perpetuation and reproducibility. Ecological concepts derived from the analysis of ecosystems may offer significant new contributions to our knowledge about systems theory.

SYSTEM DYNAMICS

There are many alternative routes for the cycling of materials via the biogeochemical cycle. The importance of any particular pathway in the ecosystem at any single point in time (e.g., transport at what rates and along which one of the myriad of food chains) is an intricate and dynamic property of the system. Hence, the seemingly unpredictable dispersion of many toxic substances in the environment. This is a problem area particularly suitable for examination by critical pathway analysis (Booth et al., 1972).

REGULATING MECHANISMS

Some species in ecosystems that are either rare or contribute insignificantly to cycling of elements or flows of energy may have considerable import in the regulation of system processes. What kinds of negative feedback loops, such as seed consumption (Janzen, 1971) or limiting nutrients (Whittaker, 1970), exist as regulatory mechanisms? How many unanticipated system attributes might be exposed through use of systems analysis techniques, such as control theory and optimization?

STABILITY OF SYSTEMS

Analysis of ecosystems can contribute knowledge of the homeostatic mechanisms which ensure self-perpetuation. What are the structural and functional characteristics of ecosystems which contribute to stability (Woodwell and Smith, 1969)? Here the methodology of sensitivity analysis (Smith, 1970) becomes a powerful analytical tool. The means by which ecosystems persist may differ from those of species components and may even occur at the expense of certain species.

ADAPTIVENESS OF SYSTEMS

If ecosystem evolution, as well as species evolution, has been shaped by selection pressures for existence, then the system strategies which have occurred to effect these objectives may provide significant insights for ecosystem analyses. Is the diversity in ecosystem structure unexplainable or simply an array of alternatives for exploiting dissimilar environments with similar strategy? What is the ecological significance of evergreenness (Monk, 1966)? Are many constituents of ecosystems relatively insignificant in the steady-state dynamics of the system? Does this diversity represent the "buffer" mechanism which permits ecosystems to respond to change? Perhaps ecosystems have evolved structures with concomitant internal dynamics which are geared to the most efficient transfer of energy rather than to maximum energy transfer (Odum and Pinkerton, 1955). The strategies of ecosystems may not always be inferred from the strategies of species; optimization of system properties does not necessarily follow from maximization of internal processes. Is man consistent with these principles in his concepts of management of the environment (Van Dyne, 1969)?

Future Role of Ecosystem Analysis

The world-wide recognition that environmental problems transcend, not only political, but also intellectual boundaries which have segregated many scientific disciplines, is resulting in the emergence of the environmental sciences as a broad, interdisciplinary field of endeavor. Within this context, ecosystem analysis may provide both a conceptual and an operational framework for linking the research of various environmental disciplines, such as meteorology, geophysics, soil chemistry, and hydrology. Each of these disciplines has much to contribute toward an unraveling of the complex environmental processes that must be understood if we are to solve such pressing problems as the maintenance and conservation of renewable resources and pollution, as well as to provide the predictive capability for assessing the consequences of manipulating these systems.

ENVIRONMENTAL IMPACT OF TECHNOLOGY

The National Environmental Policy Act (NEPA) of 1969 requires assessment of both the short- and long-term environmental impacts of technological developments. This Act requires consideration of alternatives in design, engineering, construction, operation, application, or even

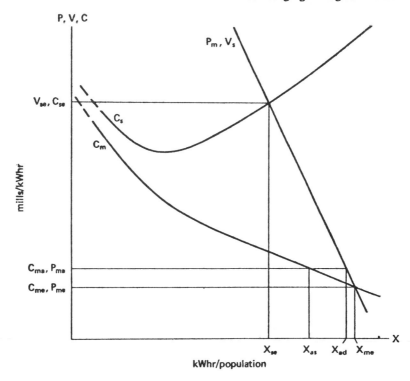

FIGURE 3. Market cost and social equilibria associated with energy production by stationary power plants. Vertical axis represents power costs, and horizontal axis represents per capita power production. (after Chapman and O'Neill, 1970).

C_m = Market cost function
P_m = Market demand function
C_{me} = Market equilibrium cost
P_{me} = Market equilibrium price
C_{ma} = Cost of actual supply
X_{me} = Market equilibrium output
X_{ad} = Output demand at price P_{ma}
C_s = Marginal social cost function
V_s = Marginal social demand function
C_{se} = Social equilibrium cost
V_{se} = Social equilibrium marginal value
P_{ma} = Price of actual supply
X_{se} = Social equilibrium output
X_{as} = Output actually supplied

new location. Preparation of these statements has required unprecedented effort by ecologists in analysis and synthesis of environmental data. Weaknesses which have become evident are a lack of relevant data and adequate capability to analyze and synthesize for assessment of potential risks. Ecosystem analysis, with its built-in requirement for the structured synthesis of multiple parameters associated with large bodies of quantitative environmental data, will lend itself to effective interpretation of research on ecological processes. Information derived from these efforts is needed to develop ecological criteria as a basis for impact evaluations.

The conflict between demands for increased power production and improved environmental quality can be approached by integrating economic and ecological theory in a functional systems analysis. Consider the current problem of "blackouts" or "brownouts" in major cities (Fig. 3), as evaluated through a socioeconomic model by Chapman and O'Neill (1970). Intersection of the market supply and demand functions defines the market equilibrium, X_{me}, with price P_{me}. Environmental impact of power generation, as perceived by society, causes a new (social) equilibrium, X_{se}, defined by the intersection of the marginal social cost and marginal social value functions. Differences between equilibria create conflict between construction plans of power companies and conservation groups. At the total power production from plants actually·built, X_{as}, the utility incurs market costs, C_{ma}. The regulatory agency requires the utility to set its market price, P_{ma}, at or near C_{ma}. With price, P_{ma}, public consumption demands quantity, X_{ad}, which is substantially in excess of the amount, X_{as}, that can be supplied. This study illustrates how the energy production/utilization system is coupled in such a way that demand can exceed supply and cause brownouts.

Ecosystem analysis, dealing as it does with the structure and function of landscape components, will find valuable application to other aspects of technological development. The orderly arrangement of hierarchical data sets describing the landscape (Fig. 1) must be coupled with human ecology (socioeconomic, demographic, and political aspects) to form arsenals of models which can be used to analyze the consequences of new technologies on large landscape regions. Ecosystem analysis will be a powerful tool for evaluating and selecting sites for stationary power plants, industrial facilities, and new cities. The need for this capability is already rising from the nation's developing energy crisis. The location of the large number of electric generating stations that will be required in the next few decades is already posing serious questions of land use allocation and environmental quality.

ENVIRONMENTAL QUALITY

As the earth becomes more populated and as societies move toward global energy-dependent economies, it will be necessary to establish internationally accepted criteria for dangerous pollutants. A concomitant need will be prediction of the behavior and long-term fates of toxic materials in the environment. There is already an urgent need to identify sensitive ecological parameters which can give early warning of ecosystem degradation.

An example is the simulation model of DDT and DDE movement in human food chains developed (O'Neill and Burke, 1971) for the Environmental Protection Agency, Advisory Committee on DDT, to forecast pesticide concentrations in man under alternative assumptions for application levels in the United States. The model consisted of a set of ordinary differential equations to express pesticide concentration changes through time.

$$F(t) = 54 - 7t$$

$$\frac{dx_1}{dt} = a_1 F(t) - b_1 x_1 \qquad b_1 = 1.014$$

$$\frac{dx_2}{dt} = a_2 F(t) - b_2 x_2 \qquad \begin{aligned} a_2 &= 0.0247 \\ b_2 &= 0.0507 \end{aligned}$$

$$\frac{dx_3}{dt} = a_3(x_1 + x_2) - b_3 x_3 \qquad \begin{aligned} a_3 &= 0.1430 \\ b_3 &= 0.4740 \end{aligned}$$

where x_1 and x_2 are food items contaminated directly by spraying and indirectly from the environment, respectively, and x_3 is the concentration in human adipose tissue. The constants a_1 relate rate of input to magnitude of the source and the b_1's are biological elimination coefficients for DDT/DDE. After 1966, no direct spraying of the food supply was permitted, and no value for a_1 could be estimated.

Source terms used in the model and comparison of model simulations and actual pesticide concentrations in man are given in Table II. If DDT usage had continued at the 1966 level, the model extrapolates that concentrations in man approach an asymptote at some value above 6.7 ppm around 2022. Simulation with DDT usage, decreasing at the present rate of about 7 million pounds per year, shows decreasing levels (1 ppm by

TABLE II. *Source Terms Utilized for Modeling DDT and DDE Movement Through the Human Food Chain and Comparison of Model Output with Residue Levels Monitored in Human Adipose Tissue.*

Year	DDT Usage (10^6 lb)[a] $F(t)$	DDT + DDE in Diet (mg/kg)[b] $(x_1 + x_2)$	DDT + DDE in Human Adipose Tissue (ppm)[c]	
			x_3 observed	x_3 model predicted
1965	53	0.031		
1966	46	0.040		
1967	40	0.026	4.65	5.02
1968	33	0.019	5.61	5.57
1969		0.016	5.22	5.42
1970		0.015	5.27	5.14

Data sources: a) USDA Pesticide Review, 1969; b) FDA analyses of food at the market place; c) Human monitoring survey.

2006) but with measurable concentrations still remaining after 50 years.

Traditional concepts of simple biological indicators are not sufficient to detect effects on ecological systems or on their internal processes until we know how species are coupled within ecosystems and how changes wrought in the species composition or population levels reflect damage to the total system. Additional information will be required about the fundamental processes (food chain mechanisms and related biogeochemical cycles) which transform and disperse these materials in the biosphere. Assessment of potential long-term environmental degradation from pollutants requires an understanding of ecologic processes at the ecosystem level. Our stewardship of the biosphere may depend on sophisticated application of the principles of ecosystem analysis.

LAND USE

A national land use policy will require a profound understanding of the landscape and its component processes. The capacity of natural and managed ecosystems to produce food and fiber, with optimized inputs of energy and material resources, needs to be seriously examined. Similarly, ecosystems need to be studied at a number of levels of resolution to derive a more fundamental understanding of how much, if any, of man's indus-

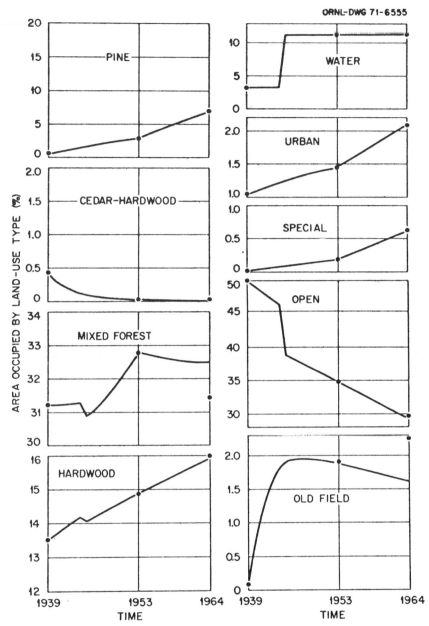

ORNL-DWG 71-6555

FIGURE 4. Results of a land-use simulation model predicting the time response of land use categories (curves) compared to observed land use patterns (solid points). The nine categories are: open (primarily agricultural); old field (abandoned farm land); pine (plantation); mixed forest (conifer and hardwood); cedar-hardwood forest; hardwood forest; water (impoundments and rivers, but excluding small streams); urban; and special (e.g., quarries, power line right-of-ways, etc.). (after Hett, 1971).

106

trial, domestic, and agricultural by-products can be accommodated, re-cycled, or otherwise disposed of without detriment to the environment. There is growing recognition that current land use patterns throughout the country do not always represent the best mix of urban, rural, and recreational allocations and that these uses often undergo rapid and dramatic changes. This capability to forecast land-use changes will in-volve a substantial input from ecologists since the landscape is a mosaic of natural and managed ecosystems, each influenced by ecological events as well as human activities.

Recently, collaborating scientists from the International Biological Pro-gram and the Tennessee Valley Authority (TVA) examined the chang-ing land-use patterns in eastern Tennessee (Hett, 1971). A simulation model was developed, based upon known plant succession for the region and land use trends derived from aerial photographs spanning 25 years. Predictions were made of the rates of change of various landscape cate-gories for a five-county region (Fig. 4). The rates of change for different vegetation types are related to the rates of plant succession and can be evaluated using empirical data from the field. The annual loss of non-forested land for the five-county area was 5.28 km^2 per year, of which 2.39 km^2 per year represented an increase in forested land (primarily due to the influence of a paper mill on the establishment of pine planta-tions) and 2.85 km^2 per year was irreversibly lost to reservoirs (TVA dams in the Valley) and urban development.

Expansion of simplified man-made agricultural ecosystems has renewed the question of whether or not these systems are increasingly vulnerable to biological insults and increasingly costly to maintain in terms of direct energy expenditures or requirements for energy-dependent products, such as fertilizers and insecticides. The future will see more research on land-scape dynamics and interactions to determine how best to organize nat-ural and managed ecosystems. Agricultural systems of the future might be mixes of large and small monocultures intermixed with heterogeneous crop and forest ecosystems to maintain the gene pools necessary to ensure genetic variability and landscape viability. The data and understanding required to lead logically to this development will draw heavily on ad-vancing knowledge of ecological processes operative within these systems. Ecosystem analysis can provide the conceptual and mechanistic frame-work to integrate these findings into meaningful syntheses and predictions of environmental phenomena at landscape levels of resolution.

References

Allee, W. C., A. E. Emerson, O. Park, T. Park, and K. P. Schmidt. 1949. *Principles of Animal Ecology.* W. B. Saunders Co., Philadelphia. 837 p.

Auerbach, S. I. 1958. The soil ecosystem and radioactive waste disposal to the ground. *Ecology,* **39**: 522-529.

Booth, R. S., S. V. Kaye, and P. S. Rohwer. 1972. A systems analysis methodology for predicting dose to man from a radioactivity contaminated terrestrial environment. *In*: D. J. Nelson (ed.). *Symposium on Radioecology.* USAEC CONF-710501. (In press.)

Chapman, D. and R. V. O'Neill. 1970. Ecology and resource economics: An integration and application of theory to environmental dilemmas. ORNL-4641, Oak Ridge National Laboratory.

Crossley, D. A. and H. F. Howden. 1961. Insect vegetation relationships in an area contaminated by radioactive wastes. *Ecology,* **42**: 302-317.

Dale, M. B. 1970. Systems analysis and ecology. *Ecology,* **51**: 2-16.

Davis, J. J. and R. F. Foster. 1958. Bioaccumulation of radioisotopes through aquatic food chains. *Ecology,* **39**: 530-535.

Goodall, D. W. 1970. Studying the effects of environmental factors on ecosystems. p. 19-28. *In*: D. E. Reichle (ed.). *Analysis of Temperate Forest Ecosystems.* Springer-Verlag, Berlin-Heidelberg-New York. 304 p.

Handler, P. 1970. *Biology and the Future of Man.* Oxford University Press, Oxford. 471 p.

Hett, J. M. 1971. Land-use changes in East Tennessee and a simulation model which describes these changes for three counties. International Biological Program, Eastern Deciduous Forest Biome. ORNL-IBP-71-8, Oak Ridge National Laboratory, 56 p.

Holling, C. S. 1966. The functional response of invertebrate predators to prey density. *Mem. Entomol. Soc. Can.,* **48**: 1-85.

Janzen, D. H. 1971. Escape of *Cassia grandis* L. beans from predators in time and space. *Ecology,* **52**: 964-979.

Jordan, C. F. 1971. A world pattern in plant energetics. *Amer. Sci.,* **59**: 425-433.

Likens, G. F., F. H. Bormann, N. M. Johnson, D. W. Fisher, and R. S. Pierce. 1970. Effects of forest cutting and herbicide treatment on nutrient budgets in the Hubbard Brook Watershed Ecosystem. *Ecol. Monogr.,* **40**: 23-47.

Lindeman, R. L. 1942. The trophic-dynamic aspect of ecology. *Ecology,* **23**: 399-418.

Monk, C. D. 1966. An ecological significance of evergreeness. *Ecology,* **47**: 504-505.

Morton, J. A. 1964. From research to industry. Int. Sci. Technol. May 1964. p. 82-92, 105.

Odum, E. P. 1971. *Fundamentals of Ecology*. 3rd ed. W. B. Saunders Co., Philadelphia. 574 p.

Odum, H. T. 1957. Trophic structure and productivity of Silver Springs, Florida. *Ecol. Monogr.*, **27**: 55-112.

———— and R. C. Pinkerton. 1955. Times speed regulator, the optimum efficiency for maximum output in physical and biological systems. *Amer. Sci.*, **43**: 331-343.

Olson, J. S. 1961. Analog computer models for movement of isotopes through ecosystems. p. 121-125. *In*: V. Schultz and A. K. Klement, Jr. (eds.). *Radioecology*. Reinhold Publ. Corp., New York. 746 p.

————. 1965. Equations for cesium transfer in a *Liriodendron* forest. *Health Physics*, **11**: 1385-1392.

————. 1970. Carbon cycle and temperate woodlands. p. 226-241. *In*: D. E. Reichle (ed.) *Analysis of Temperate Forest Ecosystems*. Springer-Verlag, Berlin-Heidelberg-New York. 304 p.

O'Neill, R. V. 1972. Error analysis of ecological models. *In*: D. J. Nelson (ed.). *Symposium on Radioecology*. USAEC CONF-710501. (In press.)

———— and O. W. Burke. 1971. A simple systems model for DDT and DDE movement in the human food chain. International Biological Program, Eastern Deciduous Forest Biome, ORNL-IBP-71-9, Oak Ridge National Laboratory. 18 p.

Patten, B. C. (ed.). 1971. *Systems Analysis and Simulation Ecology*. Academic Press, New York. 607 p.

Reichle, D. E. (ed.). 1970. *Analysis of Temperate Forest Ecosystems*. Springer-Verlag, Berlin-Heidelberg-New York. 304 p.

————, R. S. Booth, R. V. O'Neill, P. Sollins, and S. V. Kaye. 1971. Systems analysis as applied to ecological processes. p. 12-28. *In*: T. Rosswall (ed.). Systems Analysis in Northern Coniferous Forests—IBP Workshop. Bull. No. 14, Ecological Research Committee, Swedish Natural Science Research Council, Stockholm.

————, B. E. Dinger, N. T. Edwards, F. W. Harris, and P. Sollins. 1972. Carbon flow and storage in a woodland ecosystem. In: *Carbon and the Biosphere*. Brookhaven Biology Symposium. No. 24. (In press.)

Sheppard, C. W. 1962. *Basic Principles of the Tracer Method*. John Wiley and Sons, New York. 282 p.

Smith, F. E. 1970. Analysis of ecosystems. p. 7-18. *In*: D. E. Reichle (ed.). *Analysis of Temperate Forest Ecosystems*. Springer-Verlag, Berlin-Heidelberg-New York. 304 p.

Sollins, P. 1972. Organic matter budget and model for a Southern Appalachian *Liriodendron* forest. Doctoral Dissertation, University of Tennessee, Knoxville.

Van Dyne, G. M. 1966. Ecosystems, systems ecology and systems ecologists. ORNL-3957, Oak Ridge National Laboratory.

————. (ed.). 1969. *The Ecosystem Concept in Natural Resource Management.* Academic Press, New York. 383 p.

Walters, C. J. 1971. Systems ecology: The systems approach and mathematical models in ecology. p. 276-292. Chapt. 10. *In*: E. P. Odum. *Fundamentals of Ecology.* W. B. Saunders Co., Philadelphia. 574 p.

Watt, K. E. F. 1963. Mathematical models for five agricultural crop pests. *Mem. Entomol. Soc. Can.*, **32**:83.

————. 1966. *Systems Analysis in Ecology.* Academic Press, New York. 276 p.

Whittaker, R. H. 1970. The biochemical ecology of higher plants, p. 43-70. *In*: E. Sondheimer and J. S. Simeone (eds.). *Chemical Ecology.* Academic Press, New York. 336 p.

Woodwell, G. M. and H. H. Smith (eds.). 1969. Diversity and Stability in Ecological Systems. Brookhaven Nat'l Lab. Publ. No. 22, Upton, New York. 264 p.

Editors' Comments
on Papers 9, 10, and 11

TECHNIQUES TRANSFERRED FROM ENGINEERING SCIENCES

Throughout its development, systems ecology has seen new quantitative methods introduced from engineering. Engineers made solid contributions to ecology as early as the 1950s (Kerner 1957; Kerner 1959; Bellman and Kabala 1960), and there have been continuing contributions from several different engineering disciplines:

1. *Nuclear Engineering*—Nuclear engineers have produced papers on elemental transfer in ecosystems (Booth et al. 1971; Booth and Kaye 1971) and applications of simulation techniques to a variety of ecological modeling problems (Goldstein and Elwood 1971; Goldstein and Harris 1973; Goldstein and Mankin 1972). Techniques developed to study reactor control problems (Kerlin and Lucius 1966) have been used in one of the papers included below (Paper 10).

2. *Civil Engineering*—Civil and environmental engineers were involved in early studies on microcosms (Bungay 1968). More recently they have produced works on the simulation of open-water aquatic systems (Chen and Asce 1970; Chen and Orlob 1968; Chen and Selleck 1970; DiToro et al. 1970) and watershed transport processes (Huff 1967; Huff et al. 1973).

3. *Mechanical and Industrial Engineering*—There is a range

of inputs into systems ecology from mechanical and industrial engineers. At one end of this spectrum, we have extremely detailed ecological process simulators (DiMichele and Sharp 1971) and at the other, the hotly debated (Boyd 1972) large-scale global models of environmental problems (Forrester 1971).

4. *Electrical Engineering*—Probably the greatest number of contributions have come from electrical engineering, which is due, in part, to a shared interest in general systems theory and applied mathematics, particularly differential equations. R. H. Boling has worked on the simulation and analysis of stream ecosystems (Boling 1973; Boling et al. 1972, 1975a, 1975b). Electrical engineers have been involved in methodological developments, including programs for the solution of differential equation models (Mankin and Brooks 1971).

Paper 9 by Martin, Mulholland, and Thornton and Paper 10 by Shugart et al. emphasize control engineering approaches. The reader should note that the engineering background carries with it an interest in applied problems reflected in both of these papers. Paper 11 by Webster, Waide, and Patten is an excellent example of the application of engineering mathematics to ecosystem theory.

REFERENCES

Bellman, R., and R. Kalaba. 1960. Some mathematical aspects of optimal predation in ecology and boviculture. *Proc. Nat. Acad. Sci. (USA)* **46**: 718–20.

Boling, R. H. 1973. Toward state-space models for biological populations. *J. Theor. Biol.* **40**:485–506,

Boling, R. H., P. Cota, D. Curtice, P. Fraleigh, E. D. Goodman, G. Kroh, S. R. Reice, J. A. Van Sickle, S. Weiss, and J. Zimmer. 1972. Modeling developments in aquatic ecosystems, pp. 1–23. In Koenig, H. E., W. E. Cooper, and R. Rosen (eds.), *Design and Management of Environmental Systems*, vol. 2, NSF GI-20, Second Ann. Rep. Nat. Sci. Found., Washington, D.C. Sect. 4.

Boling, R. H., E. D. Goodman, J. A. Van Sickle, J. O. Zemmer, K. W. Cummins, R. C. Petersen, and S. R. Reice. 1975. Toward a model of detritus processing in a woodland stream. *Ecology* **56**:141–51.

Boling, R. H., R. C. Petersen, and K. W. Cummins. 1975. Ecosystem model for small woodland streams, pp. 183–204. In Patten, B. C. (ed.), *Systems Analysis and Simulation in Ecology*, vol. 3. Academic Press, New York, 601 pp.

Booth, R. S., O. W. Burke, and S. V. Kaye. 1971. Dynamics of the forage-cow-milk pathway for transfer of radioactive iodine, strontium and cesium to man. *Proc. Nuclear Methods Environ. Res.* Univ. of Missouri, Columbia.

Booth, R. S., and S. V. Kaye. 1971. *A Preliminary Systems Analysis Model of Radioactivity Transfer to Man from Deposition in a Terrestrial Environment*, ORNL/TM-3135. Oak Ridge National Laboratory, Oak Ridge, Tenn.

Boyd, R. 1972. World dynamics: A note. *Science* **177**:516–19.

Bungay, H. R. 1968. Analogue simulation of interacting microbial culture. *Al. Chem. Eng. Symp. Ser.* **64**:19.

Chen, C. W., and A. M. Asce. 1970. Concepts and utilities of ecologic model. *ASCE J. Sanit. Eng. Div.* **96**(SAS):1085–99.

Chen, C. W., and G. T. Orlob. 1968. *A Proposed Ecological Model for an Eutrophying Environment*, Report to the Federal Water Pollution Control Agency from Water Resources Engineers. Walnut Creek, Calif.

Chen, C. W., and G. T. Orlob. 1972. *Ecologic Simulation for Aquatic Environments*, Final Report for the Office of Water Resources Research. Washington, D.C., 156 pp.

Chen, C. W., and R. E. Selleck. 1969. A kinetic model of fish toxicity threshold. *J. Water Pollut. Control Fed.* **41**(8):R294–R308.

DeMichele, D. W., and P. Sharpe. 1971. *A Morphological and Physiological Model of the Stomata*, EDFB Memo. Rep. 71-27. Oak Ridge National Laboratory, Oak Ridge. Tenn., 15 pp.

DiToro, D. M., D. J. O'Connor, and R. V. Thomann. 1970. *A Dynamic Model of Phytoplankton Populations in Natural Waters*, Internal Report. Environmental Engineering and Science Program, Manhattan College, Bronx, New York.

Forrester, J. W. 1971. *World Dynamics*. Wright-Allen Press, Cambridge, Mass., 142 pp.

Goldstein, R. A., and J. W. Elwood. 1971. A two-compartment, three-parameter model for the absorption and retention of ingested elements by animals. *Ecology* **52**:935–39.

Goldstein, R. A., and W. F. Harris. 1973. SERENDIPITY—A watershed level simulation model of tree biomass dynamics. *Proc. 1973 Summer Comput. Simul. Conf., AICHE, ISA, SHARE, SCI, AMS*. Montreal, Can.

Goldstein, R. A., and J. B. Mankin. 1972. PROSPER: A model of atmosphere-soil-plant water flow, pp. 1176–81. *Proc. 1972 Summer Comput. Simul. Conf., AICHE, ISA, SCI, AMS*. San Diego, Calif.

Huff, D. D. 1967. The chemical and physical parameters in the hydrologic transport model for radioactive aerosols. In *Proc. Int. Hydrol. Symp.* Fort Collins, Colo.

Huff, D. D., J. F. Koonce, W. R. Ivarson, P. R. Weiler, E. H. Dettman, and R. F. Harris. 1973. Simulation of urban runoff nutrient loading and biotic response of a shallow eutrophic lake, pp. 33–55. In Middlebrooks, E. J., D. H. Falkenborg, and T. E. Maloney (eds.), *Modeling the Eutrophication Process*. Utah Water Res. Lab., College of Engr., Utah State Univ., Logan, 228 pp.

Kerlin, T. W., and J. L. Lucius. 1966. *The SFR-3 Code: A FORTRAN Program for Calculating the Frequency Response of a Multivariate System and its Sensitivity to Parameter Changes*, ORNL/TM-1575. Oak Ridge National Laboratory, Oak Ridge, Tenn.

Kerner, E. H. 1957. A stastistical mechanics of interacting biological species. *Bull. Math. Biophys.* **19**:121–46.

113

Kerner, E. H. 1959. Further considerations on the statistical mechanics of biological associations. *Bull. Math. Biophys.* **21**:217–55.

Mankin, J. B., and A. A. Brooks. 1971. *Numerical Methods for Ecosystem Analysis*, ORNL/IBP-71-1. Oak Ridge National Laboratory, Oak Ridge, Tenn., 99 pp.

9

G. D. MARTIN

School of Mechanical Engineering
and Center for Systems Science.
Presently,
Union Carbide,
Charleston, West Va.

R. J. MULHOLLAND

School of Electrical Engineering
and Center for Systems Science.

K. W. THORNTON

Department of Zoology and Center
for Systems Science. Presently,
Waterways Experiment Station,
Army Corps of Engineers,
Vicksburg, Miss.

Oklahoma State University,
Stillwater, Okla.

Ecosystem Approach to the Simulation and Control of an Oil Refinery Waste Treatment Facility[1]

Ecosystems of a series of linked treatment ponds for oil refinery effluent are modeled by state variable techniques and simulated using a digital computer. The model response verifies the observed behavior of the pond series. A control philosophy, using pond series flow rate and feedback aeration, and a measure of system performance, using biological indicators, are both developed. A computer optimization search is conducted in order to maximize system performance.

Introduction

In recent years systems modeling and digital simulation have become important tools in the study of ecology. The modeling of biological populations began with the work of Lotka [1][2] and Volterra [2], and led to relatively recent simulations of ecosystems utilizing the digital computer [3]–[11]. Referring to the analysis of whole ecosystems, Smith [12] argues that "systems oriented techniques with high speed computers offer the only means by which this can be accomplished". This is certainly not, however, the limit of digital simulation capabilities. Many authors have suggested using computer ecosystem models in studying proposed management or control policies [3], [8], [12]. Although the efforts to date have been limited, the ultimate goal of this activity is the optimal design and control of ecological systems [9]. Watt [11] treats resource management as an optimization problem: how to maximize a rate of harvest, or how to minimize the density of a pest. He used a mathematical model of the system to predict the effects of different management policies on yield [13]. Davidson and Clymer [3] express the desirability of applying computer simulation to ecosystems, particularly in the area of environmental problems. An interesting application of this type would

be the optimal control of a pollution abatement ecosystem. An example of this approach is the subject of this work.

The particular system of interest is a series of effluent holding ponds (Fig. 1) located at an oil refinery in north central Oklahoma. Effluent from the refinery flows through the series of ten ponds in approximately eight days, during which time biological and chemical processes tend to improve the effluent water quality. A detailed description of the refinery operations and treatment of effluent waters can be found in Minter [14] and Tubb [15].

The ponds are biologically active, each one causing a degradation of the wastes by natural means. The refinery effluent passes through the pond series until in the final pond the water quality is such that discharging it into a nearby river is no longer a serious pollution problem. As water quality increases the ponds progress through a series of successional stages from pond one to ten, going from a highly polluted ecosystem to one which approximates a natural system. In the first two ponds, anaerobic bacteria metabolize complex organic compounds, and release essential nutrients for algal growth. The next three ponds are dominated by algae, which produce oxygen to maintain an aerobic medium [14], with pond five containing zooplankton. Only algae and zooplankton appear in ponds six and seven in a balanced community with respect to the production and consumption of oxygen. In the final three ponds secondary consumers are present for at least part of the year. The last pond is the most diverse with many different species and at least three trophic levels. It is this biological succession which improves the final effluent water quality.

The holding pond system was installed in 1960 and provided effective effluent treatment for approximately four years. Since then the water quality of the final effluent has diminished. This is probably due to increased flow through the system, which was

[1]This work was supported in part by the National Science Foundation through grants GK-34084 and ENG 75-05341.
[2]Numbers in brackets designate References at end of paper.

Contributed by the Automatic Control Division for publication in the JOURNAL OF DYNAMIC SYSTEMS, MEASUREMENT, AND CONTROL. Manuscript received at ASME Headquarters, November 8, 1975. Paper No. 76-Aut-P.

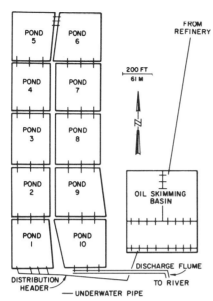

Fig. 1 Holding pond system

designed to accomodate an average flow of 546 m³/hr [16]. Since 1964 the flow has been increased by more than 40 percent.

There has been no attempt to continuously monitor or regulate biological processes in the system. In this study a differential equation model is developed for this ecosystem and a control philosophy adopted. Control possibilities and indicators of effluent quality are reviewed, evolving into a control scheme and performance index. Digital simulation provides verification of the model, and prediction of system performance is carried out by applying a search algorithm to find the minimal cost control policy.

The modeling of the system was carried out in three steps. First, a model for pond ten was formulated as a joint effort between the Center for Systems Science and Department of Zoology of Oklahoma State University. Using the model formulated by this project, models were then developed for two upstream ponds in the system, ponds four and seven. The three models were then linked to form a series similar in function to the actual system, thus simulating the important treatment effect of biological succession. A detailed description of the role of biological succession in the waste treatment system is given by Minter [14].

The Pond Ten Submodel Development

Listed below are the modeling assumptions which were made:

1 All variables were uniformly distributed.
2 Diurnal (daily) frequencies were not of interest.
3 Growth and reproduction need not be distinguished.
4 The species were divided into functional groups.

These assumptions are similar to those of Davidson and Clymer [3].

A *functional group* defines a set of species which have the same or similar function within the ecosystem. This concept is extended through the application of the compartment approach [9], wherein a compartment including several lumped species is represented by a single standing crop, expressed in units of biomass. In this way, the state variables of the ecosystem can be systematically identified in the modeling process.

Interactions between the state variables are indicated as fl connecting together the compartments of the model. These fl arise as a result of trophic considerations, predator-prey fee rates, mortality rates, etc. Depending on the type of model flows can represent biomass, energy, uptake of nutrients, n bers, or any one of a number of flux types [17]. In this work flow values represent biomass flux based on a monthly time sc This monthly temporal reference allows the populations to have as continuous rather than discrete variables, an appro used by Ulanowicz [10].

By considering these flows and average standing crop infor tion, a differential state equation model can be developed for ecosystem. This can follow a variety of established method good reference for these being Patten [9]. The equations wl result take the general form of

$$\dot{y}_i = \sum_{j=1}^{n} a_{ji}y_j - \sum_{j=0}^{n} a_{ij}y_i + b_{0i},$$

for $i = 1, 2, ..., n$, and where the rate coefficient a_{ij} represen flow from compartment i to compartment $j \neq i$. The inputs given by b_{0i}, for $i = 1, 2, ..., n$, and the flows to the environm are indicated by a_{i0}, for $i = 1, 2, ..., n$. Equation (1) represe for each compartment, a mass-balance expression, in which difference between inputs and outputs gives the rate of chang compartmental standing crop. Equation (1) usually implies flows between compartments are linear donor controlled, but notation is also employed when the flows between compartm are nonlinear, and dependent upon the standing crops of t donor and recipient compartments, so that, in general:

$$a_{ji} = a_{ji}(y_i, t),$$

$$a_{ij} = a_{ij}(y_j, t),$$

are obtained.

Assuming that the water's edge would be the ecosystem bou ary for pond ten, application of the functional group and c partment concepts led to compartments representing ab components and trophic levels: producers, herbivores, carniv and decomposers, contained within the pond. After identif the system state, attention now turned to quantification of corresponding average standing crops, flows between comp ments, and flows which cross the ecosystem boundary.

Since light, the primary energy input for the ecosystem, en as a flux through the surface of the pond, average standing c were measured per unit area. The units of measurement ch for the average standing crops were gm/m² (dry weight). Ass ing the pond depth to be an average of two meters, the ave standing crop values were actually double the biomass dens in gm/m³ or mg/l. A similar assumption holds for the ab concentrations.

The next step was to appraise the ecosystem interaction flows between the model compartments. Under the assump that the pond system was in steadystate, the sum of flows a compartment minus the sum out was equal to zero. The p ten submodel flow diagram is presented in Fig. 2, with magnitudes, trophic levels, and average standing crops inclu While it is true bluegill do not naturally appear in the p system, it was determined that sufficient food existed for a s population of these fish. Their introduction into the ecosy model was proposed as an indicator of effluent quality.

Temperature influence was modeled through the Q_{10} facto

$$Q_{10} = \rho^{(T - T_a)/10},$$

in which T designates the pond water temperature in deg Celsius and T_a denotes the time integrated yearly average The biotic state equations were multiplied by this factor w prescribes the nonlinear temperature influence on the gr

116

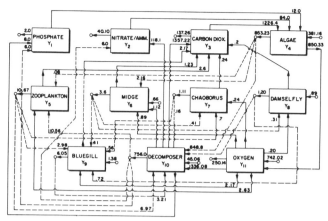

Fig. 2   Pond ten flow diagram with solid flow lines representing donor control and dashed lines recipient control or both

ate of each compartment. It is clear from (3) that, for a ten egree rise in temperature, the Q_{10} factor equals ρ, resulting in a -fold increase in the biotic growth rate. Also, when $T = T_a$, he result is $Q_{10} = 1$. The ρ factor is known to vary with respect o different biological processes. From respiration data for the opulations in the pond communities, values ranging from 1.85 o 2.5 were obtained [18]. A nominal value of $\rho = 2$ was decided pon in order to simplify the analysis of the ecosystem model.

The light influence was introduced by regulating the algal xation of carbon from CO_2 through the photosynthetic process. normalized light parameter (light/avg. light), denoted by L d multiplying the flow of CO_2 to algae, modulated the CO_2 up-ke by algae in the model. As with the previous average value, e light average refers to the time integrated yearly average.

The flows indicated in Fig. 2 between biotic compartments re assumed to be functions only of the donor compartment ntent, while flows between abiotic and biotic compartments re taken to be proportional to both donor and recipient com-rtment levels. The former assumption is standard in ecological deling [9], while the latter allows for limiting the uptake of trients by the producers. It should also be noted that for eral compartments in Fig. 2 the flows do not sum to the con-uent flows to other compartments. This apparent inconsis-cy merely represents losses from the system to its environ-nt. For example, in compartment y_6 the midge larvae emerge the form of flying insects, thus representing a net loss to the system.

The final form of equation (1) for the pond ten submodel is en in the Appendix.

imulations were now performed using the Dynamic Simulation Program (DYSIMP) [19] available at the Oklahoma State University computer facility. The integration routine used by this package is fixed stepsize Runge-Kutta (fourth-order). Due to this fixed stepsize, periodic checks were made throughout this work to insure numerical accuracy. The checks were made by varying the stepsize and noting the corresponding changes in simulation output.

Initial conditions were chosen at the steady-state values, and the simulation was allowed to continue for two years model time. This model replicated the observed pond dynamics, the interesting outcome being the accuracy of the phase relationships displayed among compartments [20].

Formulation of the Upstream Submodels

Consistent submodels for ponds four and seven were obtained using the modeling steps established by the group effort in developing the pond ten submodel. The assumption made here was that within trophic levels all three ponds were similar in structure and function. Under this assumption, the fact that upstream ponds contained progressively fewer trophic levels and were less diverse did not lead to flow determination difficulties. In both instances, the eliminated compartments and respective flows simply discarded while maintaining all other flow interactions recognized in the pond ten submodel.

Measurements were made only for ponds one, four, seven, and ten [18]. Thus, submodels for only these ponds in the series can be constructed based upon collected data. Since the only life forms in pond one are bacteria and algae (only at certain times of the year), measurements made at this pond are viewed as

Table 1 Pond average standing crops[1]						
Compartment	Pond 4	State	Pond 7	State	Pond 10	State	Reference
Phosphate	24.92	21	16.25	12	13.0	1	[18]
Nitrate/Ammonia	32.0	22	27.0	13	22.97	2	[18]
Carbon Dioxide	4.3	23	2.1	14	1.3	3	[30]
Algae	35.21	24	47.43	15	40.0	4	[18], [14]
Zooplankton	0.99	25	3.35	16	3.87	5	[14]
Midges	0.7	26	1.02	17	1.63	6	[15]
Chaoborus					1.4	7	
Damselfly			0.41	18	1.6	8	[16]
Bluegill Fish					9.45	9	[31]
Detritus/Bacteria	1722.0	27	1219.0	19	884.0	10	[32], [9]
Oxygen	4.1	28	11.0	20	10.55	11	[18]

[1](gm/m²).

117

inputs to the pond series. The modeling steps followed for each pond are outlined below:

1. Determine system states
2. Find average standing crop values.
3. Establish balanced flow interactions.
4. Formulate the state equations.
5. Test the homogeneous model (without light and temperature inputs).
6. Include the effects of light and temperature.
7. Verify the final pond submodel.

The submodels constructed were based upon time series of standing crops and estimation of the rate coefficients order to balance the flows. After including the temperature and light data, comparisons of the model outputs with measured time series constituted the validation procedure.

Identification of the significant state variables in the upstream pond submodels demonstrated that three compartments included in the pond ten submodel should be eliminated in one or both of the upstream submodels. Bluegill fish and *Chaoborus* were excluded in both submodels, and the damselfly compartment was eliminated from the pond four submodel. The average standing crops for all three ponds are listed in Table 1, with the corresponding references noted.

The next step in the modeling process was to determine the flows connecting the compartments in each of the submodels. The assumption of similar trophic structure fixed the flow interaction web, but to quantify the flows those in ponds four and seven were scaled proportionately to the ones in ten by a ratio of the respective average standing crops. And by extending the concept of similar trophic structure, the assumption was made that the flow dependence, linear or nonlinear, donor or recipient, were also similar.

Light data for all three ponds were considered to be equivalent, since it was certainly reasonable to assume that light influx per unit area would be identical in each pond. However, this uniformity did not hold true for water temperature, since the flow was characteristically warmer upstream, reaching ambient as it flowed through the series [15]. Thus, independent pond water temperature data was provided for each pond.

Linear interpolation through the data supplied the simulation with causal functions of time for light influx and the pond temperatures. Thus, the light regulator and Q_{10} factors could all be calculated at each integration step. The functional representations for both light and temperature inputs were identical to those used for the formulation of the pond ten submodel.

The final form of the state equations for ponds four and seven appear in the Appendix along with the model for pond ten. Verification of the individual upstream submodels followed [20]. Simulated responses matched the data and expected dynamics. The trajectories oscillated yearly around the average standing crops while the magnitudes of fluctuation seemed to be reasonable. More importantly, phase relationships in the food chain were plausible, with algal blooms occurring when expected.

With the three submodels now complete, linking to form an overall ecosystem model for the holding pond series proceeded.

Linking of the Three Submodels

In choosing the linking method a compromise had to be made between depicting as nearly as possible the pond series ecosystem function and satisfying the limitations imposed by the algorithms used.

Because of the construction of the pond system, with submerged interconnecting pipes generally lying below the euphotic zone, very little coupling exists between the biotic states in the different ponds. To a first approximation, the coupling between ponds depends only upon the abiotic states. Thus, the method chosen was to link the ponds by abiotic flows, neglecting the im-

port or export of biotic states. These flows were delayed in time to produce the correct temporal relationship between the submodels, since it takes approximately eight days for the effluent to flow through the pond series.

At each abiotic compartment of the upstream submodels, there is an exiting flow proportional to the respective standing crop magnitude which is assumed to flow out of the subsystem. Certainly a portion of this flow will enter into the adjacent downstream pond. Thus, a percentage of each of these flows proportional to the upstream standing crops, enters into the downstream pond subsystem, with an associated transport delay.

To illustrate how this action was achieved in the model, consider one of the abiotic flows from one pond to another. The upstream influence on a corresponding downstream compartment is effected by adding a flow term to that compartment which is delayed by an amount consistent with the effluent flow rate through the system. This addition, which is proportional to the upstream standing crop, must be normalized by subtracting a nominal flow from the downstream compartment, this nominal flow being proportional to the upstream average standing crop. This is necessary because merely adding a flow into the downstream compartment would unbalance the total flux at system equilibrium, which must equal zero. Thus, the pond submodel state equations are revised as shown by the following example:

Pond four: $\dot{y}_k = \ldots\ldots - a_{k0}\, y_k$

Pond seven: $\dot{y}_k = \ldots\ldots + K_k\, a_{k0}\, y_k - K_k\, a_{k0}\, y_k{}^*$

Here the kth state is linked between the two ponds with y_k designating the average standing crop of y_k in pond four. Note that the linking will essentially have no effect on the ponds downstream unless the abiotic state magnitudes upstream are reasonably far from their respective average standing crops. This scheme was used to link the abiotic compartments of all three pond submodels. The overall linking mechanism is diagramed in Fig. 3.

Before the completed model could be simulated, the linking

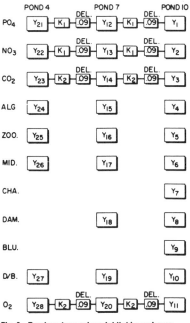

Fig. 3 Pond system submodel linking scheme

percentages and delay duration had to be evaluated. The linking percentages indicate the fraction of abiotic flows reaching the downstream ponds. Remembering that there are actually two ponds between each adjacent pair included in the model, the linking influence would be diminished cubically. In addition to this fact, the possibility of different percentages for different abiotic flows was considered.

Here it was assumed that the two dissolved gases, CO_2 and O_2, could be assigned equivalent percentage rates, since both diffuse out of the pond system. Alternatively, NO_3 and PO_4 were also assigned equivalent but somewhat higher percentage rates. Based upon measured data supplied by the refinery, the parameter values chosen were 0.05 for CO_2 and O_2 and 0.30 for NO_3 and PO_4.

The delay time was assumed to be 0.09 month (approximately 3 days), since the subsystem divided the pond series into three segments, and flow through the entire system took approximately eight days.

Simulation results corresponding to one year of model time revealed that the model behavior closely resembled the actual pond ecosystem dynamics [20]. Linking did have effects on the response of the downstream submodels. The completed model in this sense replicated the pond series behavior. In particular, it has been noted that the dissolved oxygen in pond four is zero during midwinter [21], [18], [22]. This occurrence was predicted by digital computer simulations of the linked pond ecosystem model.

The Control Structure

There are many proven methods of management of waste treatment facilities. These include aeration, activated carbon treatment, microstraining, effluent distribution and coagulation by polymers [23]. Also, the seminatural, engineered pond ecosystem appears to be one of the most inexpensive modes of waste treatment. Although these methods have been in use for several years, testing of control schemes by ecosystem model simulations relatively unexplored. Watt [11] has considered ecosystem management (harvest rate) through simulation while several authors [3], [8], [12] have suggested the need for this type of effort studying many of our environmental problems. Before going to the particular control philosophy adopted in this work, some general remarks will be presented on the requirements the resulting control scheme must meet.

The control scheme must include states or inputs which can be easily measured and economically implemented in the real system, while being directly represented in the model. Controls are desired which will enhance biological action, thus improving the quality of the final effluent. The control scheme was structured with automatic control in mind, limiting direct influence abiotic states with no biotic manipulation (i.e., dredging of detritus or harvesting standing crops).

There were five basic control possibilities which could influence the biological action of the pond series and be directly represented the model. A list of the candidate policies were:

1   Lighting

2   Heating
3   Nutrient (NO_3, PO_4) addition
4   Aeration
5   Flow regulation.

Three of these candidates were removed from consideration as possible controls due to their lack of feasibility. Lighting was ruled out because to effectively influence the photosynthetic process artificially would require that the pond system be virtually covered with high intensity lamps. The cost incurred here would certainly be prohibitive. Both heating and nutrient addition were eliminated due to the nature of the effluent. The flow was both naturally warmer than required for biological influence and nutrient rich as a result of the refinery processes. Thus, an addition of either heat or nutrients as controls to this primarily warm, eutrophic environment would be ineffective. It should be mentioned here that although heat would improve the pond operation during a short period in the winter, at this time additional heated water from the refinery is not available.

Aeration and flow delay were included as controls because both could economically be accomplished and had direct influences on the biological action [21], [18], [14]. Aeration was controlled continuously by feedback monitoring of the oxygen concentration in pond ten. This can be accomplished with existing oxygen level detectors [24]. A feedback structure was constructed by introducing a gain which multiplied the difference between the oxygen level in pond ten and a set point which was arbitrarily chosen. This product represented the feedback aeration rate applied to all three ponds.

Flow delay was accomplished by allowing a variable delay of abiotic flows which linked the submodel series. This control method was also readily attainable through flow regulation in the actual system.

A diagram of the control scheme is shown in Fig. 4. Now that this control scheme foundation has been defined, the performance index development will be presented.

Development of the Performance Index

The optimization of the ecosystem performance with respect to the control scheme of Fig. 4 is approached here via a computer search utilizing the pond performance index to be developed. Applications of this nature have been suggested by Watt [11]. The minimization of the index corresponds to the lowest point on the merit surface (assuming unimodality) [25] generated by the function $PI(K, \tau)$, where K is the oxygen feedback gain and τ the flow delay time. This minimum represents the best combination of the control parameters for the particular problem posed, and the lowest cost attained in the search.

The structure of the performance index included effluent water quality indicators and control cost rates (and penalties). These contributions were summed by the integral form presented below:

$$PI = \int_{t_0}^{t_f} \left[\sum_{i=1}^{n} (w_i u_i)^2 + \sum_{j=1}^{m} (w_j v_j)^2 \right] dt , \qquad (4)$$

where

t_0 = initial time
t_f = final time
u_i = ith control cost (normalized)
w_i = related weight
v_j = the jth system indicator (normalized − reciprocal)
w_j = related weight
n = the number of cost contributions
m = the number of indicator contributions

Since it is desired to minimize the PI of (4), the system indicators are taken as reciprocal in some sense as demonstrated later. The integral is used to provide a measure of the cumulative effects of

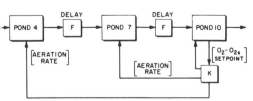

Fig. 4   Control structure for the pond system

cost, etc., while the independent summing emphasizes the individual influence of each constituent (u, v) [26]. The tradeoff involved here is that effluent water quality increase as a result of control effort which is balanced against the costs of that effort (and penalties). The integral was included in the ecosystem model as another state equation, and integrated in time along with the ecosystem simulated response.

The performance index constituents are basically of two types, effluent quality indicators and costs of control effort (or penalties). The notion of indicators of ecosystem quality has been presented in previous studies focused on this effluent holding pond series [14]–[16], [18], [21], [22], [26], [27]. As with the admissible control policies, the indicators were required to be easily measurable in the real system, and applicable in the model. The indicators used refer to pond ten only, since interest was directed to the exiting effluent water quality. Indicators considered for this pond system are listed in Table 2 with respective references.

Many of the proposed indicators of effluent quality were disregarded for the simple reason that there was no accomodation included in the model for their computation. Sulfates, chlorides, dissolved solids, euphotic zone, phenol, alkalinity, chlorophyll/ash free dry weight, boron, oil, pH, and SAR fell into this category. Many of these seem to be quite important in judging the quality of the effluent water, and the lack of their presence in the pond system model might seem unfounded. However, their effects are included implicitly in the biological actions of the model through the flow structure and data incorporated. So their exclusion does not negate the validity of the model or its ability to resolve an optimal control strategy by predicting their effect.

The oxygen demands (BOD, COD, IOD) were excluded from consideration as effluent quality indicators because of their interpretations in the real ecosystem. Because insufficient information exists regarding the chemical reactions which take place within the pond series, attempts to arrive at estimates for these proposed indicators from data provided by the model would produce misleading values. Thus, the oxygen concentration was not used directly as an indicator, but was indirectly included in the P/R measure. Also, oxygen is included as part of the control scheme as shown in Fig. 4.

Bluegill fish, heuristically included in the model for the purpose of indicating effluent water quality, depicted the long term effects of pollution. This longer sensitivity was complemented by the inclusion of zooplankton as an indicator, which possesses a relatively fast turnover rate, resulting in a quick response to pulses of toxic effluent [14]. Both bluegill and zooplankton could easily be sampled in the real system and monitored continuously or sampled in the simulation model.

The P/R ratio was used as a measure of ecosystem stability. This quantity is defined as the ratio of production of oxygen to community respiration including the oxygen demand of inorganic and organic compounds. The P/R ratio approaches unity in a balanced steady state system with no import or export [28]. However, at ecosystem equilibrium, export equals import, and their effects cancel. During periods when sufficient time was allowed for the community to stabilize the pond system was effective in treating the effluent water and the P/R ratio approached unity at the end of holding time in pond ten.

The P/R ratio can be easily measured in the actual ecosystem and is obtainable from the information provided by the model. In the real system P/R could be evaluated by regular sampling or continuous monitoring of oxygen level with application of the diurnal curve method. Eley [29] provides a computer algorithm for this method which, given daily O_2 data, will evaluate the P/R ratio. Note that the control structure already provides the monitoring of oxygen level required here. In the simulation P/R can be calculated by evaluating continuously the ratio of oxygen flow from algae to oxygen (productivity) over the sum of flows of oxygen to the various compartments (respiration).

Details of the procedure are presented below for the pond ten submodel:

$$P/R = 2.015\, y_4/(0.06442\, y_5 + 0.05175\, y_6 + 0.01625\, y_7 + 0.01185\, y_8 + 0.02187\, y_9 + 0.14326\, y_{10}) \quad (5)$$

where the numerator represents the flow from the algae compartment y_4 to the oxygen compartment y_{11} and the denominator is the sum of the flows from oxygen to zooplankton y_5, midge y_6 *Chaoborus* y_7, damselfly y_8 and bluegill y_9. Note that y_{11} has been cancelled from the denominator and numerator terms. These flows which appear in (5) are all included in the equations of the pond ten submodel.

Equation (5) was computed and averaged over one year of model time producing a P/R ratio of slightly less than one. This differs from measured data obtained by Copeland and Dorris [22], in which the average P/R of the final pond in the series was found to be unity. Thus, a small constant (small with respect to the flows involved) was added to (5) in order to correct for this model deficiency. The details of this correction to the P/R ratio are presented in reference [20].

Nitrate (including nitrogen contained in ammonia) was used as an indicator, being easily sampled in both pond ten and its submodel in the simulation. A maximum allowable level for this indicator has been prescribed by the State of Oklahoma [27] see Table 2. It is certainly possible that in the near future, fines will be levied for violation of guidelines such as this. As an outcome of this, the influence of nitrate (or ammonia) on the performance index consisted of introducing a set cost, representing a monthly penalty rate, when the nitrate standing crop magnitude rose above the set standard. In this respect, the nitrate (or ammonia) contribution to the performance index represents penalty, or cost which can indirectly be offset by control effort.

The controls which have been selected are feedback aeration and flow delay adjustment. The associated costs may readily be deduced in the ecosystem application and continuously calculated in the simulation. Aeration cost rates considers the option of mechanical or diffusion aerators and the flow delay cost allows for the burden of a low flow rate on the refinery operations.

Performance Contribution Weighting

With the constituents of the performance index defined, each contributor was normalized so that at ecosystem equilibrium (the average standing crops) its value would equal one or zero. This was necessary so that weighting of the individual performance index contributions could be accomplished [25]. The biotic indicators, fish and zooplankton, were normalized using the following procedure:

$$v_i = y_i^*/y_i \quad (?)$$

where:

y_i^* = ith average standing crop value
y_i = ith standing crop value
v_i = as used in equation (4).

In order to maximize y_i one would minimize the reciprocal, and at equilibrium the indicator value would equal unity.

The ecosystem stability indicator, P/R ratio, was subtracted from one to normalize its value since its equilibrium magnitude was prescribed to be unity, and deviations far from one were undesirable. The structure of the nitrate penalty guaranteed it to equal one or zero in normalized form, depending on whether the limit of allowable nitrate had been exceeded or not. At ecosystem equilibrium, both of these normalized forms equaled zero.

The costs of control, as with the nitrate penalty, were already in a normalized form due to their direct relatibility to expensed dollars.

Table 2   Indicators of effluent water quality

Indicator	Desired Level	Range	Reference(s)
Species diversity	High	5.0–7.0 Species/Cycle	[16], [14], [26]
P/R ratio	~1.0	0.0–4.5	[18], [20], [29], [14], [26]
Bluegill fish	High	0.0–10.0 gm/m³	
Zooplankton	High	0.0–4.0 gm/m³	[14], [26]
Bio. Ox. Demand (BOD)	Low	8.0–28.0 ppm	[18], [32], [26]
Ph	~7.0	6.9–8.6	[32], [14], [26], [27]
Sulfates	Low	6.2–344.0 mg/l	[32], [27]
Chloride	Low	20.0–1700.0 mg/l	[32], [27]
Dissolved solids	Low	157.0–3390.0 mg/l	[27]
Euphotic zone	High	0.5–1.88 m	[14]
Phenol	Low	0.0–1.0 ppm	[18], [14], [26], [15]
Nitrate (available nitrogen, ammonia)	Low	0.0–21.0 mg	[18], [32], [26], [27]
Chem. Ox. Demand (COD)	Low	120.0–130.0 ppm	[18], [32], [26]
Alkalinity	Low	0.0–160.0 ppm	[32], [26]
Chlorophyll/ash free dry weight		0.0–0.015	[18], [14]
Boron	Low	0.0–0.06 mg/l	[27]
Oil	Low	1.0–6.0 ppm	[32]
Inorg. Ox. Demand (IOD)	Low	0.0–3.0 ppm	[32]
Dissolved oxygen	High	2.5–6.5 ppm	[32]
Sodium absorption ratio	Low	5.4–25.0	[27]

With the normalization constructed, weights were determined or the various u's and v's of (4). Weighting of control costs and penalty was done by setting the corresponding weight values qual to the expected costs per month. Aeration costs (mechanical, diffusion) were calculated from data provided by Davis [23]. The cost of flow control and the nitrate penalty were estimated. The biotic weights (P/R, fish, zooplankton) were estimated and re subject to conjecture. The relative value of these ecosystem uality indicators has not been considered previously, but determinations of this kind are required in this type of performance dex application. If the weights arrived at for these indicators em high, one might consider the costs of advertising which uld be offset by the publicity benefits of biotic quality signs .g., fish in the last pond). In this era of ecological emphasis, ch an indication of an unpolluted effluent release can easily be ed to rationalize these weights. The weight values imposed e given in Table 3. The square roots were used for the performance index weighting so that the evaluated integral would rectly represent cumulative cost trade offs.

The form and all of the components of the performance index ve been defined and evaluated. Now, with all of the contribuns and weights included, the index described by equation (4) is

$$= \int_{t_0}^{t_f} \left\{ [w_1/v_1]^2 + [w_2(1.0 - P/R)]^2 \right.$$

$$+ \left[w_3 \left(\frac{9.45}{y_9} \right) \right]^2 + \left[w_4 \left(\frac{3.87}{y_5} \right) \right]^2$$

$$\left. + [w_5 u_1]^2 + [w_6 u_2]^2 \right\} dt , \quad (7)$$

where the weighting factors from Table 3 are

$$w_1 = 170$$
$$w_2 = 10$$
$$w_3 = w_4 = 6$$
$$w_5 = 6$$
$$w_6 = 7,$$

and

$$v_1 = \begin{cases} 1 & \text{if NO}_3 \text{ concentration in pond ten } (y_2) \\ & \text{is above the upper limit of 24.0} \\ 0 & \text{otherwise.} \end{cases}$$

It should be noted that Fig. 4 prescribes the feedback control structure for the pond system optimization, hence the aeration

Table 3   Performance index contributors and weights

Contributor	Normalized form	Notation (Eq. (4))	Weight ($/mo)	Approx. $\sqrt{\text{weight}}$
Aeration cost				
diffusion		u_1	44.7	7.0
mechanical			34.4	6.0
Flow delay cost	Delay/Nominal Delay	u_2	50.0	7.0
Nitrate (ammonia) penalty		v_1	30000.0	170.0
P/R	$(1.0 - P/R)$	v_2	100.0	10.0
Bluegill fish	y_9/y_9^*	v_3	40.0	6.0
Zooplankton	y_5/y_5^*	v_4	40.0	6.0

Table 4   Performance Index contribution sensitivities[1,2]

Contribution	Weight	Range
(a) Original		
Aeration cost	18.0	~700.0
Flow delay cost	7.0	~100.0
Nitrate penalty	170.0	
P/R	10.0	~0.30
Bluegill	6.0	~0.30
Zooplankton	6.0	~0.45

Contribution	Weight	Range	Scale factor
(b) Modified			
Aeration cost	0.90	~35.0	0.05
Flow delay cost	5.0	~72.0	0.715
Nitrate penalty	30.0		0.176
P/R	1000.0	~30.0	100.0
Bluegill	400.0	~20.0	66.7
Zooplankton	400.0	~30.0	66.7

[1]To control.
[2]Oxygen feedback range = 0.0 − 600.0; delay range = 0.090 − 1.0.

was chosen in the following form:

$$u_2 = K\,(y_{11} - y_{11}{}^*)$$

where $y_{11}{}^* = 10.55$, the average oxygen concentration in pond 10. Also, the control $u_1 = \tau$, the pond system delay time.

Preparation of the Model for Optimization

The search procedure selected for the optimization was a grid method [26]. This method seeks an extremum by conducting a series of alternating grid and star patterned performance index evaluations of progressively decreasing size over the control parameter plane (in this two-dimensional case). There were three reasons for selecting this method. First, the nature of the merit surface was totally unpredictable, perhaps the only thing known about it was that it would be continuous. This prevented the application of many of the searching methods with faster convergence requiring knowledge or determination of derivatives. The second reason was that this search will seldom encounter numerical difficulty or stall out. The third reason was that a grid search by its design gives one an overall picture of the merit surface, since individual trial points are spaced relatively far apart.

Before further optimization considerations, the ecosystem model was checked for sensitivity to control effort. These results led to a modification of the linking percentages in which oxygen

and CO_2 were given independent linking parameters. The linking parameter for oxygen was increased to 1.0, while the value of the CO_2 parameter was left at 0.05, and the $NO_3 - PO_4$ parameter was increased to 0.5. Although the controls now had desirable effects on the ecosystem model response (i.e., increasing the standing crop of bluegill in pond ten), the performance index weighting was imbalanced, providing too much influence to the costs and penalty. Thus, the next action taken was to investigate the relative ranges of fluctuation for the performance index constituents.

The weights were modified (or scaled) so that the sensitivities listed in Table 3 would transform into approximately equivalent contribution changes. The modified weights with resulting sensitivity values are given in Table 4.

Simulations were now conducted over a model time period of six months. This was done because for approximately the first six months the oxygen level in the pond ten submodel would not fall below the set point, regardless of delay parameter value. Thus, lack of oxygen feedback led to near duplication of simulated responses for the early months in all parameter variable runs. Therefore, the last six months were used only to save time and money.

Optimization Results

A large two parameter optimization search was now conducted. It indicated conclusively that an optimum does exist in the overall region specified. An illustration of the merit surface evaluations and their locations is provided in Fig. 5.

The minimum performance index value found was 1.9817 × 10⁶ at (66.0, 0.393). Note that the minimum is surrounded by evaluations of higher magnitude. Converging to the optimum by specifying smaller fractional reductions is now possible. It is clear that the location of the minimum is subject to influences by the various weighting factors. Further investigations are not merited because of questions regarding the significance of the numbers used and their accuracy.

These results do show, however, that the minimization technique was successful in finding the optimal control strategy fo the environmental problem. Not only is the minimum performance value bracketed by the fourth evaluation group, but b; considering Fig. 5 there is no implication that the merit surface is other than unimodal. It should be noted that the resultan minimum delay time and oxygen feedback values are not un reasonable for the actual pond system operation. This holds fo biological, physical and economic interpretations of the wast control problem.

Summary and Conclusions

A model has been developed for a particular pollution abat ment ecosystem consisting of a series of oil refinery effluent hol ing ponds. Simulations were performed using a digital compute and the model response was verified with the observed behavi of the pond series.

A control philosophy, using the pond series flow rate and fee back aeration, and a measure of system performance, usi biological indicators, were adopted for an optimization stud The control variables of aeration and flow delay were chosen b cause both could be economically implemented, and they ha a direct influence on the biological action present within t ponds. Aeration was continuously controlled by feedback mor toring of the oxygen level in pond ten. This is easily acco plished using existing oxygen detectors. A feedback structu was assumed by introducing a gain which multiplied the diff ence between the oxygen level in pond ten and a set point chos to reflect system performance. This product represented t aeration rate applied to the ponds. Flow delay was accomplish by allowing a variable delay of the abiotic flows which linked submodels in the series. Assuming the existence of tempor

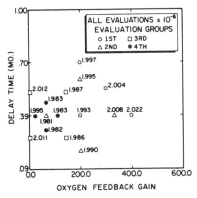

Fig. 5   Optimization results

storage upstream from the pond series, this control method of flow regulation is also readily attainable in the actual system.

Bluegill fish, heuristically included in the model for the purpose of indicating effluent water quality, were chosen to depict the long term effects of pollution as part of the performance measure. This longer sensitivity was complemented by the inclusion of zooplankton as an indicator of system performance. Zooplankton possess a relatively fast turnover rate, resulting in a quick response to slugs of toxic material. The P/R ratio was used as a measure of total biological community performance. Finally, total nitrogen was penalized when the effluent level exceed a maximum as prescribed by the State of Oklahoma.

Several common indicators of effluent quality were disregarded in the construction of the performance index. In particular, the oxygen demands (BOD, COD, and IOD) were excluded from consideration. The claim is that an indirect relationship exists between the measure presented and the more common indicators. This remains an open question.

An optimization search using the computer was programmed in order to maximize the ecosystem performance. The search was conducted over the control parameter plane, defined by the values of the aeration feedback gain and flow delay time, to seek the optimal control policy. Each evaluation of the index determined one point on the merit surface, corresponding to one pair of control variables and a simulation of one half year of model time. The minimization of the performance index indicated the best control policy and most suitable ecosystem behavior.

The application of an ecological performance index is necessary in the evaluation of possible control schemes, and control in the environmental sense is certainly of the utmost value. There are many proven methods of management of waste treatment facilities. Because these schemes are based upon intuitive concepts of system operation, the optimization of performance indicies and testing of control schemes by ecosystem model simulations is a relatively unexplored field. Utilizing models, computer simulations, and optimization to predict ecosystem control policy effects hold promise for future studies concerned with alleviating environmental problems brought on by waste pollution.

APPENDIX

Based upon the data of Fig. 2 and Table 1, the final form of equation (1) for the pond ten submodel is given as

$$\dot{y}_1 = 0.15385\, y_1 - 0.02308\, y_1\, y_4 + 0.01584\, y_{10}$$

$$\dot{y}_2 = -1.74576\, y_2 - 0.9142\, y_2\, y_4 + 0.14039\, y_{10}$$

$$\dot{y}_3 = -105.58462\, y_3 - 23.58462\, y_3\, y_4\, L + 0.17664\, y_3\, y_9$$
$$+ 0.09615\, y_3\, y_8 + 1.18101\, y_3\, y_{10} + 0.13187\, y_3\, y_7$$
$$+ 0.58046\, y_3\, y_6 + 0.51680\, y_3\, y_5$$

$$\dot{y}_4 = (23.58462\, y_3\, y_4\, L + 0.09412\, y_2\, y_4 + 0.02308\, y_1\, y_4$$
$$= -21.33075\, y_4 - 2.01500\, y_4\, y_{11} + 381.16)\, Q_{10}$$

$$\dot{y}_5 = (-2.75711\, y_5 - 0.51680\, y_3\, y_5 + 0.00200\, y_4 + 0.01195\, y_{10}$$
$$+ 0.06442\, y_5\, y_{11})\, Q_{10}$$

$$\dot{y}_6 = (-2.20859\, y_6 - 0.58046\, y_3\, y_6 + 0.05400\, y_4$$
$$+ 0.05175\, y_6\, y_{11} + 0.00127\, y_{10} + 0.66)\, Q_{10}$$

$$\dot{y}_7 = (-0.79286\, y_7 - 0.13187\, y_3\, y_7 + 0.18088\, y_5$$
$$+ 0.25153\, y_6 + 0.01625\, y_7\, y_{11})\, Q_{10}$$

$$\dot{y}_8 = (-0.75000\, y_8 + 0.01185\, y_8\, y_{11} + 0.19018\, y_6$$
$$- 0.09615\, y_3\, y_8 + 0.89)\, Q_{10}$$

$$\dot{y}_9 = (-0.64021\, y_9 + 0.02178\, y_9\, y_{11} + 0.45000\, y_8$$

$$+ 0.40000\, y_7 - 0.17664\, y_3\, y_9 + 0.25153\, y_6$$
$$+ 0.77003\, y_5 + 1.38)\, Q_{10}$$

$$\dot{y}_{10} = (-0.85520\, y_{10} - 0.00127\, y_{10} - 1.18101\, y_3\, y_{10}$$
$$- 0.00582\, y_{10}\, y_2 + 0.11429\, y_7 + 0.14326\, y_{10}\, y_{11}$$
$$+ 1.80103\, y_5 - 0.00070\, y_{10}\, y_1 + 21.22000\, y_4$$
$$+ 0.10000\, y_8 + 45.06)\, Q_{10}$$

$$\dot{y}_{11} = -23.70995\, y_{11} - 0.02177\, y_{11}\, y_9 - 0.06442\, y_5\, y_{11}$$
$$- 0.14326\, y_{10}\, y_{11} - 0.05175\, y_6\, y_{11} - 0.01625\, y_7\, y_{11}$$
$$- 0.01185\, y_8\, y_{11} + 2.01500\, y_4\, y_{11} + 742.02,$$

where the variables $y_1 - y_{11}$, representing standing crops, are defined in Fig. 2.

The submodels for the upstream ponds, obtained using the modeling steps established in developing the pond ten submodel, are presented below. The pond seven submodel is given by

$$\dot{y}_{12} = -0.15385\, y_{12} + 0.0158\, y_{19} - 0.01842\, y_{12}\, y_{15} - 2.57$$

$$\dot{y}_{13} = -1.75185\, y_{13} + 0.14066\, y_{19} - 0.07762\, y_{13}\, y_{15} - 24.77$$

$$\dot{y}_{14} = -201.46714\, y_{14} + 0.05923\, y_{14}\, y_{18} - 14.55779\, y_{14}\, y_{15}\, L$$
$$+ 0.73050\, y_{14}\, y_{19} + 0.35948\, y_{14}\, y_{17} + 0.32125\, y_{14}\, y_{16}$$

$$\dot{y}_{15} = (0.01842\, y_{12}\, y_{15} + 14.55779\, y_{14}\, y_{15}\, L + 0.07762\, y_{13}\, y_{15}$$
$$- 21.50538\, y_{15} - 1.93587\, y_{15}\, y_{20} + 466.40)\, Q_{10}$$

$$\dot{y}_{16} = (0.00200\, y_{15} - 0.32125\, y_{14}\, y_{16} - 2.73731\, y_{16}$$
$$+ 0.06187\, y_{16}\, y_{20} + 0.00820\, y_{19} - 0.946)\, Q_{10}$$

$$\dot{y}_{17} = (-2.21569\, y_{17} + 0.05397\, y_{15} - 0.35948\, y_{14}\, y_{17}$$
$$+ 0.00057\, y_{19} + 0.04991\, y_{17}\, y_{20} - 0.79)\, Q_{10}$$

$$\dot{y}_{18} = (-0.75122\, y_{18} - 0.05923\, y_{14}\, y_{18} + 0.01131\, y_{18}\, y_{20}$$
$$+ 0.19020\, y_{17} + 0.114)\, Q_{10}$$

$$\dot{y}_{19} = (0.85316\, y_{19} + 21.18912\, y_{15} + 1.80597\, y_{16}$$
$$- 0.00496\, y_{13}\, y_{19} - 0.73050\, y_{14}\, y_{19} - 0.00057\, y_{19}$$
$$+ 0.13737\, y_{19}\, y_{20} - 0.00056\, y_{19}\, y_{12} + 0.10000\, y_{18}$$
$$+ 231.009)\, Q_{10}$$

$$\dot{y}_{20} = -23.72727\, y_{20} - 0.01131\, y_{18}\, y_{20} - 0.13737\, y_{19}\, y_{20}$$
$$+ 1.93587\, y_{15}\, y_{20} - 0.06187\, y_{16}\, y_{20} - 0.04991\, y_{17}\, y_{20}$$
$$+ 1095.891,$$

where the variables in these equations are defined in Fig. 3. Note that compartments for *Chaoborus* and bluegill are missing in this upstream pond. The pond four submodel is given by

$$\dot{y}_{21} = -0.15369\, y_{21} + 0.00985\, y_{27} - 0.01208\, y_{21}\, y_{24} - 12.87$$

$$\dot{y}_{22} = -1.75000\, y_{22} + 0.14094\, y_{27} - 0.06568\, y_{22}\, y_{24} - 112.7$$

$$\dot{y}_{23} = -363.76581\, y_{23} - 7.12668\, y_{23}\, y_{24}\, L + 0.35681\, y_{23}\, y_{27}$$
$$+ 0.15621\, y_{23}\, y_{25} + 0.17542\, y_{23}\, y_{26}$$

$$\dot{y}_{24} = (-21.41437\, y_{24} + 0.06568\, y_{22}\, y_{24} + 7.12668\, y_{22}\, y_{24}\, L$$
$$+ 0.01208\, y_{21}\, y_{24} - 5.21609\, y_{24}\, y_{28} - 343.40)\, Q_{10}$$

$$\dot{y}_{25} = (-2.73737\, y_{25} - 0.15621\, y_{23}\, y_{25} + 0.00199\, y_{24}$$
$$+ 0.00172\, y_{27} + 0.16605\, y_{25}\, y_{28} - 0.329)\, Q_{10}$$

$$\dot{y}_{26} = (-2.21429\, y_{26} + 0.05396\, y_{24} - 0.17542\, y_{23}\, y_{26}$$
$$+ 0.13240\, y_{26}\, y_{28} + 0.00029\, y_{27} - 0.682)\, Q_{10}$$

$$\dot{y}_{27} = (-0.85308 \; y_{27} + 21.24396 \; y_{24} - 0.00419 \; y_{27} \, y_{22}$$
$$- 0.35681 \; y_{23} \, y_{27} + 1.79798 \; y_{25} + 0.36713 \; y_{27} \, y_{28}$$
$$- 0.00029 \; y_{27} - 0.00036 \; y_{27} \, y_{21} + 1016.3) \, Q_{10}$$
$$\dot{y}_{28} = -23.78049 \; y_{28} - 0.13240 \; y_{26} \, y_{28} - 0.16605 \; y_{25} \, y_{28}$$
$$+ 5.21609 \; y_{24} \, y_{28} - 0.36713 \; y_{27} \, y_{28} + 1937.554,$$

where the variables in these equations are defined in Fig. 3. It should be noted that compartments for *Chaoborus*, bluegill and damselfly are missing in this upstream pond.

References

1 Lotka, A. J., *Elements of Mathematical Biology*, Dover, New York, 1956.

2 Volterra, V., *Lecons sur la Theorie Mathematique de la Lutte pour la Vie*, Gauthier-Villars, Paris, 1931.

3 Davidson, R. S., and Clymer, A. B., "The Desirability and Applicability of Simulation Ecosystems," *Ann. N. Y. Acad. Sci.*, Vol. 128, 1966, pp. 790–794.

4 Garfinkel, D., "Digital Computer Simulation of Ecological Systems," *Nature*, Vol. 194, 1962, pp. 856–857.

5 Garfinkel, D., "A Simulation Study of the Effect on Simple Ecological Systems of Making Rate of Increase of Population Density-Dependent," *Jour. Theor. Biol.* Vol. 14, 1967, pp. 46–58.

6 Garfinkel, D., MacArthur, R. H., and Sack, R., "Computer Simulation and Analysis of Simple Ecological Systems," *Ann. N. Y. Acad. Sci.*, Vol. 115, 1964, pp. 943–951.

7 Garfinkel, D., and Sack, R., "Digital Computer Simulation of an Ecological System Based on a Modified Mass Action Law," *Ecology*, Vol. 45, 1964, pp. 502–507.

8 Parker, R. A., "Simulation of an Aquatic Ecosystem," *Biometrics*, Vol. 24, 1968, pp. 803–821.

9 Patten, B. C., ed., *Systems Analysis and Simulation in Ecology*, Vol. 1, Academic Press, New York, 1971.

10 Ulanowicz, R. E., "Mass and Energy in Closed Ecosystem," *Jour. Theor. Biol.*, Vol. 34, 1972, pp. 239–253.

11 Watt, K. E. F., *Ecology and Resource Management: A Quantitative Approach*, McGraw-Hill, New York, 1968.

12 Smith, F. E., "Analysis of Ecosystems," in *Analysis of Temperate Forest Ecosystems*, Springer-Verlag, New York, 1970, pp. 7–18.

13 Leigh, E. C., "Making Ecology an Applied Science," *Science*, Vol. 160, 1968, pp. 1326–1327.

14 Minter, K. W., "Standing Crop and Community Structure of Plankton in Oil Refinery Effluent Holding Ponds," PhD thesis, Oklahoma State Univ., Stillwater, 1964.

15 Tubb, R. A., "Population Dynamics of Herbivorous Insects in a Series of Oil Refinery Effluent Holding-Ponds," PhD thesis, Oklahoma State Univ., 1963.

16 Ewing, M. S., "Structure of Littoral Insect Communities in a Limiting Environment, Oil Refinery Effluent Holding Ponds," MS thesis, Oklahoma State Univ., 1964.

17 King, C. E., and Paulik, G. J., "Dynamic Models and Simulation of Ecological Systems," *Jour. Theor. Biol.*, Vol. 16, 1967, pp. 251–267.

18 Copeland, B. J., "Oxygen Relationships in Oil Refinery Effluent Holding Ponds," PhD thesis, Oklahoma State Univ., Stillwater, 1963.

19 Sebesta, H. R., *DYSIMP Simulation Program*, Oklahoma State University, Stillwater, Jan. 1971.

20 Martin, G. D., "Optimal Control of an Oil Refinery Waste Treatment Facility: A Total Ecosystem Approach," MS thesis, Oklahoma State University, Stillwater, 1973.

21 Copeland, B. J., "Primary Productivity in Oil Refinery Effluent Holding Ponds," MS thesis, Oklahoma State Univ., Stillwater, 1961.

22 Copeland, B. J., and Dorris, T. C., "Community Metabolism in Ecosystems Receiving Oil Refinery Effluents," *Limnol. Oceanogr.*, Vol. 9, 1964, pp. 431–447.

23 Davis, R. K., *The Range of Choice in Water Management: A Study of Dissolved Oxygen in the Potomac Estuary*, Johns Hopkins Press, Baltimore, 1968.

24 Davis, C. H., "Engineering Considerations in Design of Waste Treatment Facilities," *ASME Publ.* 72-PID-12, 1972.

25 Wilde, D. J., *Optimum Seeking Methods*, Prentice-Hall Englewood Cliffs, N. J., 1964.

26 Roberts, L. E., "The Plankton, Populations of an Oil Refinery Effluent Holding Pond System," MS thesis, Oklahoma State Univ., 1966.

27 *Water Quality Standards for the State of Oklahoma*, Oklahoma State Water Resources Boards, Publ. 20, 1968.

28 Beyers, R. J., "The Metabolism of Twelve Aquatic Laboratory Micro-Ecosystems," *Ecol. Monogr.*, Vol. 33, 1963, pp. 281–306.

29 Eley, R. L., "Physiochemical Limnology and Community Metabolism of Keystone Reservoir," PhD thesis, Oklahoma State Univ., Stillwater, 1970.

30 Hutchinson, G. E., *A Treatise in Limnology*, Vol. I, Wiley, New York, 1957.

31 Jenkins, R. W., "The Standing Crop of Fish in Oklahoma Ponds," *Proc. Okla. Acad. Sci.*, Vol. 38, 1957, pp. 157–172.

32 *Effluent Water Quality and Control*, Annual Report of Ponca City Refinery, Continental Oil Co., 1962.

A MODEL OF CALCIUM-CYCLING IN AN EAST TENNESSEE *LIRIODENDRON* FOREST: MODEL STRUCTURE, PARAMETERS AND FREQUENCY RESPONSE ANALYSIS[1]

H. H. Shugart, Jr., D. E. Reichle, N. T. Edwards, and J. R. Kercher[2]
*Environmental Sciences Division, Oak Ridge National Laboratory,[3]
Oak Ridge, Tennessee 37830 USA*

Introduction

Since 1962 a *Liriodendron tulipifera* forest stand located on the U.S. Energy Research and Development Administration's Oak Ridge [Tennessee] Reservation has served as a focal point for a number of ecological studies. Initially these studies were related to a large scale ^{137}Cs tracer experiment (Auerbach et al. 1964) initiated in May 1962. Over the years these studies provided a source of information (Francis and Tamura 1973) and a research philosophy (Olson 1968a) such that the *Liriodendron* forest was a logical site for a great variety of ecological research (e.g., Olson 1965, Reichle and Crossley 1965, Witherspoon and Brown 1965, Waller and Olson 1967, Olson 1968, Witkamp and Barzansky 1968, Witkamp 1969, Moulder and Reichle 1972, Edwards and Sollins 1973). Drawing on the large set of data on element kinetics associated with the *Liriodendron* forest, we constructed a model of the ecological transport and cycling of calcium in this ecosystem. The model was conceptualized and parameterized to represent the present operation of the forest (to the extent that it is known and over

greater-than-annual time scales). Using this model as a surrogate for the *Liriodendron* forest, we then investigated the consequences of attempting to maintain the forest's Ca cycle at some desired state by monitoring some ecosystem component and then using this information to adjust the input to the system. The question of ecosystem-monitoring/ecosystem-control is germane both to the management of ecosystem element cycles (e.g., toxic metals, radioisotopes) and to the determination of basic ecosystem properties (e.g., system stability, ecosystem responses to varying environments).

Model Structure

The structure of the calcium model (Fig. 1) is a general conceptualization of a deciduous forest stand of reasonable homogeneity, with four vegetative layers (canopy, subcanopy, shrub, herbaceous layers), and with saturated soil. Uptake of Ca by the plants is represented by transpirational flow between the plant tissue (e.g., canopy leaves) and the soil solution in the rooting zone. Although these transpirational flows move through the xylem in boles and branches, the transportation delay is considered to be instantaneous relative to the time scales considered in the present model. Chemical equilibria of Ca ion content between soil water and other soil compartments are assumed to be reasonably approximated by first-order kinetics in both directions with

[1] Manuscript received 27 November 1974; accepted 24 August 1975.
[2] Present address: Bio-Medical Division, Lawrence Livermore Laboratory, P.O. Box 808, Livermore, California 94550 USA.
[3] Operated by Union Carbide Corporation for the U.S. Energy Research and Development Administration.

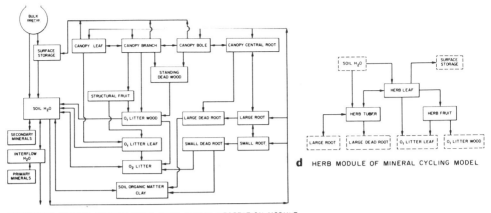

a TERRESTRIAL MINERAL CYCLING MODEL WITH CANOPY VEGETATION MODULE

d HERB MODULE OF MINERAL CYCLING MODEL

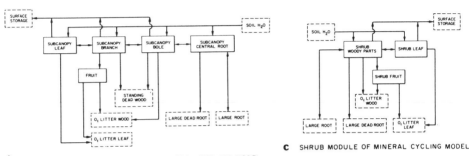

b SUBCANOPY VEGETATION MODULE OF MINERAL CYCLING MODEL

c SHRUB MODULE OF MINERAL CYCLING MODEL

Fig. 1. General structure of calcium model for *Liriodendron* forest. (*a*) Canopy and soil portions of model. (*b*) Subcanopy. (*c*) Shrub layer. (*d*) Herbaceous layer. Boxes represent state variables and arrows represent materials transfers.

nonoscillatory equilibria whose ratio can be expressed as a K_d value. The model is considered a reasonable approximation of the longer than annual time scale transfers of Ca in the present operating state of the *Liriodendron* forest, but is not appropriate for representing the subannual (seasonal) changes in Ca allocation. Studies over the past 12 yr in the *Liriodendron* forest using radioactive tracers have not demonstrated time responses that are counterindicative of the structural assumptions above. This is not to say that these time responses are not also appropriate to some set of structurally-different models as well.

The general mathematical properties and conditions for validity of this class of ecological models has been discussed most recently by Mulholland and Keener (1974). In the present study we are interested in the stability of a feedback system that monitors some component of the forest and the potential application of these monitored data to continuously alter the ecosystem's input. We focus our attention toward representing the system (the *Liriodendron* forest and a hypothetical control system) at the state in which the forest is presently operating. This corresponds directly to analyzing for stability under Lyapunov's (1893) first stability theorem by considering a linearization of the system in a small region near the system's operating point. Specifically a system is considered stable if the effects of some relatively small perturbation on the system vanish with time. A general explanation of this sort of stability definition in an ecological context is developed by Lewontin (1969). A general rationale for linearization of ecological systems may be found in Patten (1971, 1972) and in May (1973). Consistent with this approach to the calcium-cycle stability problem is the use of ordinary linear differential equations to represent the *Liriodendron* forest calcium cycle. In matrix notation the form of the model is:

$$x' = Ax + g$$

TABLE 1. The matrix, A, for the *Liriodendron* forest calcium model. Typical element, a_{ij}, is in year^{-1}

Typical element, a_{ij}, is in year^{-1}

$i \backslash j$	Compartment Name	1	2	3	4	5	6	7	8	9	10	11	12	13	14	15
1	Surface storage	-.160E2	.140E0	.869E-3		.219E-3		.140E0	.863E-3		.277E-3		.139E-3	.442E-3		.140E0
2	Canopy leaf		-.111E1	.491E-1												
3	Canopy branch		.280E-1	-.128E0												
4	Canopy fruit			.164E-1	-.100E10	.100E-1										
5	Canopy bole			.408E-1		-.181E-1	.100E-1									
6	Canopy central root					.328E-2	-.172E-1									
7	Subcanopy leaf							-.132E1	.686E-1							
8	Subcanopy branch							.190E-1	-.215E0		.100E-1					
9	Subcanopy fruit								.406E-1	-.100E1						
10	Subcanopy bole								.253E-1		-.630E-1	.100E-1				
11	Subcanopy central root										.165E-2	-.192E0				
12	Shrub leaf												-.125E1	.121E1		
13	Shrub fruit												.100E-1	-.196E1	.503E0	
14	Shrub woody parts													.111E-2	-.100E1	
15	Herb leaf															-.114E1
16	Herb tuber															.385E-2
17	Herb fruit															.385E-3
18	Standing dead wood			.499E-2					.503E-1		.511E-1					
19	O$_1$ litter wood			.158E-1	.722E0				.296E-1							
20	O$_1$ litter leaf		.945E0		.278E0			.116E1					.110E1			
21	O$_2$ litter															
22	Large dead root						.458E-2							.100E1	.500E0	
23	Small dead root											.510E-1				
24	Large root						.263E-2			.502E0		.131E0		.841E0		
25	Small root									.500E0						
26	Soil H$_2$O	.160E2														
27	Soil organic matter/clay															
28	Interflow H$_2$O															
29	Secondary minerals															

$i \backslash j$	Compartment Name	16	17	18	19	20	21	22	23	24	25	26	27	28	29
1	Surface storage														
2	Canopy leaf											.576E1			
3	Canopy branch											.129E1			
4	Canopy fruit														
5	Canopy bole											.328E0			
6	Canopy central root											.103E0			
7	Subcanopy leaf											.676E0			
8	Subcanopy branch											.195E0			
9	Subcanopy fruit														
10	Subcanopy bole											.141E0			
11	Subcanopy central root									.100E-1		.231E-1			
12	Shrub leaf											.794E-1			
13	Shrub fruit											.783E-1			
14	Shrub woody parts									.100E-1					
15	Herb leaf	.510E0										.288E0			
16	Herb tuber	-.104E1									.100E-1	.701E-1			
17	Herb fruit		-.100E1												
18	Standing dead wood			-.176E0					.625E0						
19	O$_1$ litter wood		.501E0	.176E0	-.756E0										
20	O$_1$ litter leaf		.500E0			-.281E1									
21	O$_2$ litter				.659E0	.252E1	-.176E1								
22	Large dead root							-.386E0		.200E-1	.100E0	.118E0			
23	Small dead root								-.625E0		.100E-1				
24	Large root									-.585E-2		.506E1			
25	Small root										-.102E1				
26	Soil H$_2$O				.972E-1	.299E0	.176E1	.386E0	.563E1	.855E-2		-.647E2	.663E-1	.100E1	
27	Soil organic matter/clay											.147E2	-.663E-1		.100E0
28	Interflow H$_2$O											.182E1		-.313E1	
29	Secondary minerals											.338E2			-.100E0

127

where x' is a vector of derivatives (g Ca $\cdot$ m^{-2} $\cdot$ yr^{-1});

    A is a matrix of rate constants with typical
    element, a_{ij}, dimensioned in yr^{-1};

    x is a vector of state variables (g Ca $\cdot$ m^{-2}); and

    g is a vector of inputs to the system (forcing
    functions) (g Ca $\cdot$ m^{-2} $\cdot$ yr^{-1}).

The actual structure of the model (Fig. 1) and the state variables have been developed to coincide with extant data on the *Liriodendron* forest and with identifiable functional parts of the forest.

MODEL PARAMETERS

The priority for use of values to parameterize the model were (in decreasing order): (1) Parameters estimated from data collected in the *Liriodendron* forest; (2) Parameters estimated from data collected from ecologically similar sites on the U.S. Energy Research and Development Administration's Oak Ridge Reservation; (3) Suitable parameter estimates from the literature; and (4) Parameters estimated by assumption or analogy (e.g., estimating turnover rates assuming conservation of mass; using ^{90}Sr as a biological analog of Ca to estimate rate constants). In most instances, direct measures of state variables (e.g., standing crop of Ca in boles) and transfers (e.g., litterfall) were possible from on-site data and supporting references are cited. Sometimes it was necessary to calculate fluxes from mass balance relationships (i.e., uptake of Ca by roots equals Ca stored in increment plus that returned through leaching, litterfall, mortality, etc.). In certain areas, such as the physiology of nutrition (e.g., allocation of minerals among tissue components) and soil chemistry (e.g., residence times of reaction products before subsequent plant uptake), assumptions were necessary to partition mass flows and their temporal characteristics between ecosystem compartments. The rate constants for the *Liriodendron* forest Ca budget (Table 1) are provided in standard matrix form. System inputs are documented below. Weight in the sections below refers to dry weight.

Atmospheric source terms and transfers

Total atmospheric Ca input to the forest during 1971 and 1972 was 10.51 kg $\cdot$ ha^{-1} $\cdot$ yr^{-1}, apportioned into 4.46 kg $\cdot$ ha^{-1} $\cdot$ yr^{-1} dry fallout and 6.05 kg $\cdot$ ha^{-1} $\cdot$ yr^{-1} in wetfall (precipitation) (Auerbach et al. 1972). Total Ca reaching the forest floor averaged 23.24 kg $\cdot$ ha^{-1} $\cdot$ yr^{-1}, with the difference between this value and bulk precipitation of 12.73 kg $\cdot$ ha^{-1} $\cdot$ yr^{-1} due to transfers from surface storage and plant leaching. Stemflow accounted for 0.4 kg $\cdot$ ha^{-1} $\cdot$ yr^{-1} of this Ca input (we assumed 50% of this value was lost from branches and 50% from

bole), since the remainder of 12.33 kg $\cdot$ ha^{-1} $\cdot$ yr^{-1} was due to leaching of foliage (Auerbach et al. 1972). We assumed that 10% of direct canopy input of Ca went straight to the soil (90% interception), and that surface storage turnover of Ca in the canopy was equal to the number of rains per year greater than the 2.3 cm storage capacity of canopy. This averaged 16 events per year with most of the rains occurring during the growing season (N. T. Edwards, *personal observations*). We allocated precipitation input of Ca by using rainwater concentration of Ca and partitioning rain input according to canopy interception and direct throughfall to the ground. Data for the 1971–1972 growing season show that, for precipitation events greater than the storage capacity of the canopy (2.3 cm), $\approx$ 5% of incident precipitation reached the forest floor. This is a conservative estimate and we used a value of 10% throughfall to compensate for the more open canopy during early spring. Independent measurements (Hutchinson 1975) using photographic analysis of canopy closure, gave an average for 12 separate measurements during the growing season of 5% canopy opening.

Plant uptake

Annual primary production of foliage accounted for 71.75 kg Ca $\cdot$ ha^{-1} $\cdot$ yr^{-1}; this increment plus the 12.33 kg $\cdot$ ha^{-1} $\cdot$ yr^{-1} leached amounted to annual plant uptake of 84.08 kg Ca $\cdot$ ha^{-1} $\cdot$ yr^{-1} (Reichle and Edwards 1973). We assumed that 10% of annual leaf uptake (8.41 kg $\cdot$ ha^{-1} $\cdot$ yr^{-1}) came from branch storage.[4] Therefore, root uptake transfer via the transpirational stream accounted for 75.37 kg Ca $\cdot$ ha^{-1} $\cdot$ yr^{-1} to leaves. Branch, bole, and root increment of 20.9 kg Ca $\cdot$ ha^{-1} $\cdot$ yr^{-1}, which represents the remaining component of plant uptake, was calculated by multiplying mean growing season concentrations of Ca (Reichle and Edwards 1973) times biomass of each component (Reichle et al. 1973a). Based upon ^{45}Ca tracer studies in the field (Thomas 1969), we used a value of 10 yr for the time required for Ca ratios between plant parts to re-equilibrate within individual trees after a perturbation.

Soil solution and available calcium

Total Ca in the soil is 8.13 × 10^3 kg/ha; 4.346 × 10^3 kg Ca/ha are unavailable. Total soil Ca was

---

[4] Based on ^{45}Ca cycling data for dogwood trees (Thomas 1969) and specific activity ratios of ^{45}Ca/Ca in leaves and branches one year after inoculation (Thomas 1970), we calculate $\sim$ 19% of leaf Ca to be derived from branch translocation. Since Thomas' experiment did not facilitate complete isotopic dilution of the ^{45}Ca tag into support tissues, we utilize a more conservative estimate of 10% annual leaf uptake of Ca from branch storage.

determined by HF perchloric acid digestion; exchangeable Ca was extracted by 1 N NH_4Ac at pH 7. Elemental analyses were made by atomic absorption spectrometry. Data are summarized for a 75-cm depth soil profile. Calcium concentrations in soil solution (determined by soil access tubes) ranged from 3.1 to 9.9 μg Ca/milliliter and averaged 4.97 (N. T. Edwards, *personal observations*). Water constitutes 30% of soil weight (we assume this to be a reasonable constant, since the soil is well-drained and moisture potentials in this forest do not exceed −5 bars during the growing season (R. K. McConathy, *personal communication*). The average pool of Ca in the soil solution was calculated as 11.24 kg/ha using a bulk density of 1.16 g/cm^3 (Henderson et al. 1971) for the 65-cm deep rooting zone. Using site climatological data (B. Hutchison and D. Matt, *personal communication*) and an energy balance—aerodynamic plant-soil-atmosphere water balance program, PROSPER (Goldstein and Mankin 1972), we estimated losses from the ecosystem to be 48 cm $H_2O \cdot m^{-2} \cdot yr^{-1}$ or about 27.4 kg Ca $\cdot ha^{-1} \cdot yr^{-1}$. We defined an arbitrary soil H_2O compartment called interflow water. This water is below the rooting zone, subject to mixing with soil water, and is the compartment from which H_2O (and Ca) is transported from the forest. We assumed that the volume of interflow water equaled that of the soil solution. Bedrock weathering of Ca was calculated from total ecosystem mass balance (atmospheric input + ecosystem storage − ground water output) to approximate an annual flux of 16.9 kg $\cdot ha^{-1} \cdot yr^{-1}$. Transfers of available Ca between soil water and soil humus were approximated by assuming that these two pools would approach dynamic equilibria at a known velocity. This velocity is assumed to be on the order of 0.1–0.01 yr^{-1} (0.01 yr^{-1} is used in the model based on sorption-desorption experiments by C. Francis [*personal communication*]).

Standing crop pools of calcium

Total leaf uptake of Ca is 84.08 kg $\cdot ha^{-1} \cdot yr^{-1}$, with 71.75 kg Ca annually distributed into standing crop (mean growing season concn. 1.45% Ca). With branch biomass of 2.9 $\times 10^4$ kg/ha (Reichle et al. 1973*a*, Sollins et al. 1973) and a Ca content of 0.8%, branches contain 232 kg Ca/ha. Total Ca in canopy bole wood (including stump and central root) was 309 kg/ha (Reichle and Edwards 1973). Using an empirically-based allometric function (Sollins et al. 1973), we apportioned 75.3% (233 kg/ha) to bole and 76 kg/ha to stump and central root. We assumed roots of all sizes are near equilibrium (death = growth). Standing crop of Ca in herbivores (0.004 kg/ha) was considered negligible and ignored in this version of the model, although fluxes of Ca

(1.07 kg $\cdot ha^{-1} \cdot yr^{-1}$) due to insect grazing (Reichle et al. 1973*b*) are included in the throughfall values reported earlier. Calcium incorporated in standing dead trees is 4.8 kg/ha (Reichle and Edwards 1973).

Understory biomass estimates were based upon total understory values (Sollins et al. 1973) and allocated to different structural components (leaves, branches, and central root) using allometric ratios (Whittaker and Woodwell 1968). Understory leaf Ca (1.48% [Day 1973]) was 7.3 kg/ha. Branch values were 13.9 Ca/ha (2.50% Ca in bark and 0.23% in wood [Thomas 1969]). Boles accounted for 35.2 kg Ca/ha and central root 5.8 kg/ha; Thomas (1969) reported 0.55% Ca for roots > 2 cm diam. Calcium in fruits (30 kg dry wt/ha and 1.88% Ca) was 5.64 $\times 10^{-1}$ kg/ha.

The shrub stratum is composed predominantly of *Hydrangea, Cornus,* and *Cercis* seedlings and *Smilax* with a mean annual biomass of 167.56 kg dry wt/ha. Using a stem:leaf ratio of 1 and Ca content of 1.08% (Gerloff et al. 1964), we estimated 0.904 kg Ca/ha each for both leaf and woody components of shrubs. We assumed that the proportion of Ca in fruit in the shrub layer is the same (1.5% of biomass) as in the overstory, and that 50% of fruit is transferred to O_1 wood and 50% of O_1 leaf litter annually. We used the same 10% allocation of Ca from branch to leaves as for overstory vegetation. A turnover rate of 1.16 yr^{-1} was calculated for leaves of shrubs.

Over 80% of total aboveground herbaceous biomass (482 kg dry wt/ha) is composed of *Hydrangea* and *Polystichum* (Cristofolini 1970). Additionally, 24% of this biomass is root tissue (150 kg dry wt/ha). We used 0.98% Ca (Gerloff et al. 1964) to arrive at a total herbaceous Ca pool of 5.9 kg Ca/ha. We used a turnover time of 1 yr for aboveground herbaceous and 10 yr (maximum) for belowground components based on Taylor's estimate for mayapple (Taylor 1974) and Christmas fern (F. G. Taylor *personal communication*). We assumed that tubers may lose up to 50% of weight and Ca (0.75 kg/ha) in initiating new leaves.

Belowground root biomass is 9.5 $\times 10^3$ kg/ha for large roots (> 0.5 cm diam) and 7.6 $\times 10^3$ kg/ha for small roots (< 0.5 cm diam) (Harris et al. 1974); with respective Ca concentrations of 0.8% and 0.7% (Cox 1973), total root Ca is 128 kg Ca/ha. Twenty percent of small root standing crop (10.4 kg Ca/ha) is dead (T. Cox, *personal communication*). We assumed large root mortality to be the same as for branches. A turnover rate of 0.33 yr^{-1} for large roots and 1.25 yr^{-1} for small roots was used (W. F. Harris, *personal communication*). We assumed that 10% of the Ca in small dead roots goes to O_2 litter and the remainder goes directly to soil organic matter.

Production and standing crop of detritus

Annual overstory litterfall contributes 71.75 kg Ca/ha in leaves, 3.52 kg Ca/ha in reproductive parts and 1.25 kg Ca/ha in branches (88.3% from overstory) (Reichle and Edwards 1973). Bole death accounts for 1,000 kg dry wt$\cdot$ha^{-1}$\cdot$yr^{-1} or 5.48 kg Ca$\cdot$ha^{-1}$\cdot$yr^{-1} to the forest floor (Reichle and Edwards 1973). We assumed that the stump death rate equaled bole death (with stump equal to 24.7% of bole biomass); stump death was calculated to contribute 1.35 kg Ca$\cdot$ha^{-1}$\cdot$yr^{-1} back to soil. Twigs contributed 40 kg dry wt$\cdot$ha^{-1}$\cdot$yr^{-1} (Sollins et al. 1973) or about 0.412 kg Ca$\cdot$ha^{-1}$\cdot$yr^{-1} as litter. Reproductive parts have a turnover time of 1 yr. We assumed that 80% of fruit is wood-like and goes to O_1 wood litter and 20% goes to O_1 leaf litter.

Understory twig fall of 330 kg dry wt$\cdot$ha^{-1}$\cdot$yr^{-1} (Sollins et al. 1973) contributes an additional 3.40 kg Ca$\cdot$ha^{-1}$\cdot$yr^{-1}. Understory bole and branch death contribute 1.80 and 0.70 kg Ca$\cdot$ha^{-1}$\cdot$yr^{-1}, respectively (Thomas 1969). The leaf turnover time (for leaves on the tree) of 1.16 yr^{-1} and the mean standing crop of 7.3 kg Ca/ha in understory leaves gives an annual input from leaves to O_1 litter of 8.468 kg Ca$\cdot$ha^{-1}$\cdot$yr^{-1}.

Mean annual litter biomass is 4,750 kg dry wt/ha or 71.80 kg Ca/ha (Reichle and Edwards 1973). Litter Ca was apportioned as 17.82 kg Ca/ha in O_1 wood and 36.32 kg Ca in O_1 leaf debris. The litter had 17.66 kg Ca/ha. Calcium concentration of leaf litter ranged from 0.93%–2.36% in O_1 litter and from 1.35%–2.16% in O_2 litter. We used organic weight loss (Reichle et al. 1973a) as estimated from gas exchange (Edwards and Sollins 1973) of 0.54 yr^{-1} to estimate Ca turnover. The rate of Ca leaching from small twigs was assumed to be similar to that of leaf litter.

Frequency Response Analysis

The general methodology of frequency response analysis centers around observing a system's response to a sinusoidal (sine wave) input at a given frequency and using this information to determine system stability and other total system characteristics. The techniques have potential use for ecologists in that the analytical results are graphical and easy to compare between systems (Child and Shugart 1972, Waide et al. 1974); the techniques are not predicated on the existence of some mathematical model, and are traditionally used on real systems (in this case we are simply using a mathematical model as an ecosystem surrogate); the techniques are easily applied to certain classes of time-varying systems (as well as to time invariate systems), which is an attribute not shared by a variety of matrix-analytical

techniques such as the Routh-Hurwitz stability criteria (DiStephano et al. 1967).

Frequency response characteristics of the model were determined using an available computer code (Kerlin and Lucius 1966). The importance of knowing the frequency response characteristics of an ecosystem is twofold. First, for any linear system, the system response to any arbitrary input can be determined from the knowledge of the system frequency response (i.e., the output of the system given inputs in the form of sine waves of varying frequencies). Second, since ecosystem perturbations could conceivably be periodic (e.g., annual stresses due to periodic input of some industrial byproduct of a seasonally produced material; longer-term periodic inputs of materials due to effects of economic cycles on production), the frequency response of the system is of interest in its own right.

For a given frequency, ω, the system's response is quantified by determining (Milsum 1966):

1) $MR = A_o/A_I$, where MR is the magnitude ratio, A_I is the amplitude of an input sinusoid, and A_o is the amplitude of the response to the sinusoidal input measured as variation in some system component. MR is frequently measured in dB $= 20 \cdot \log_{10} (MR)$, where dB are decibels.

2) ϕ (in degrees), where ϕ is the phase shift (i.e., the lag between the peak of an input sinusoid and the peak of the component response to that input. MR and ϕ vary as functions of ω depending on the internal nature of the system.

Results

Since there are 29 state variables in the calcium model, it is possible to analyze the properties of 29 different feedback systems that might be applied to regulate the forest. Each of these 29 structurally-distinct systems (each is distinct in that it monitors a different point) can differ as to the nature of the feedback-control system. We have inspected negative-feedback systems with gains of -1, -10, -100, and $-1,000$ for each of the 29 control systems for a total of 116 analyses. There are an infinite number of such analyses that might be performed and we have selected these 116 cases as interesting and representative cases. The gain of the feedback is a measure of the responsiveness of the control system to the monitored information (e.g., a gain of -1 indicates that if the transfer of calcium from monitoring point is 1 g/time below the expected flux then the input is increased 1 g/time in response; for a gain of -10 if the transfer from monitoring point is 0.1 g/time low then input is augmented 1 g/time, etc.).

The first question to ask about any such control system concerns its stability. Potentially any negative-feedback system can have instabilities at certain frequencies of perturbations, and knowledge of these ranges is important in assessing the suitability of a given control system. We have used Nyquist analysis to determine the stability of the various feedback systems used on the forest ecosystem calcium model. In Nyquist analysis, one plots the MR of the system's response as the radius and ϕ as functions of ω. The stability of the system can then be determined by the relation of this plot (the "Nyquist path") to the point $(1, -180°)$. The general details of Nyquist techniques for system stability analysis are considered classic control theory (DiStephano et al. 1967). All of the feedback systems added to the natural Ca cycle were found to be stable. An example case is shown in Fig. 2 and discussed below. The system considered in the example monitors loss of Ca from the system (the output from interflow water) and feeds this information back to control the input (with a gain of -1). The system is a realizable system (Fig. 2a) but the main interest is in the properties of the ecosystem that are manifest in this context. Much information can be inferred from Nyquist analysis of such a system (Fig. 2b):

1) The control system is stable. The trajectory of the system's response to increasing frequencies (ω, Fig. 2b) does not pass through or enclose (pass to the right of) the instability point $(1, -180°)$.

2) The system has high relative stability. Amplification and time delays in the control system, with a reasonable range, do not cause the system frequency trajectory to pass through or enclose the instability point. The amount of amplification in the control system needed to produce an unstable system response is quantified by the gain margin. The gain margin is defined as the degree to which the system MR would have to be multiplied to cause the Nyquist path of the system to pass through the instability point. Similarly the amount of time delay needed to produce an unstable system response (the phase margin) is quantified by number of degrees, that if added to the system ϕ, would cause the Nyquist path to pass through the instability point. Together the phase and gain margins provide measures of the relative stability of the system.

3) The system-feedback features respond over perturbations frequencies of 50–200 yr. This is evidenced by the ranges of frequencies that have MR values large enough to be easily noted on the system Nyquist path (Fig. 2b). Such frequencies of perturbations would be most likely to produce observable system responses.

For the Ca model (and to the extent that it represents the real system), emergent system properties due to the internal feedbacks in the system do not become important unless one is considering what

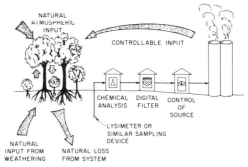

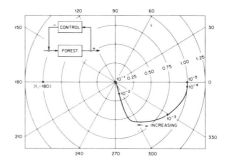

(a) EXAMPLE OF ECOSYSTEM CALCIUM CONTROL SYSTEM.

(b) NYQUIST STABILITY PLOT OF ECOSYSTEM CALCIUM CONTROL SYSTEM.

FIG. 2. Nyquist analysis of example feedback control system. (a) Example of the actual physical nature of a control system monitoring interflow water. (b) Nyquist plot of example system. Monitoring interflow water and altering input with a gain of -1.

from a management standpoint would be considered long-term (50–200 yr) perturbations. Perturbations on the system at a more frequent level (short- and medium-term perturbations) should be assessed according to effects on component parts (e.g., toxic effects of a given element on herbivores) of the ecosystem, as the system-feedback attributes do not emerge at this time scale. Thus, predicted on the representativeness the Ca model and relative to a specific set of objectives, we have determined the appropriate time scales for considering system perturbations and for considering direct effects of perturbations on ecosystem components.

The controlability of the negative-feedback system can be examined by plotting the ratio of the system response with and without the monitoring-feedback control system. Response of the system under feedback control was determined using a code developed for this application (Kercher and Funderlic 1975). Such a graph has been developed for a tropical forest magnesium cycling model (Child and Shugart 1972). Primarily, this method is used to determine over what range of frequency of perturbations a hypo-

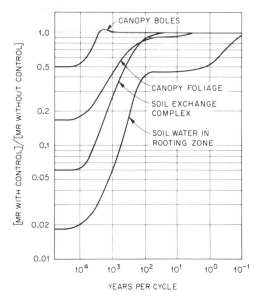

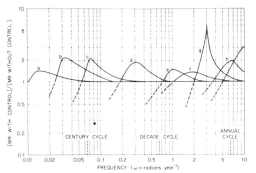

FIG. 3. Deviation curves for 4 example control systems (with a gain −1). Soil water in the rooting zone system has the most desirable control features in that it has the smallest ratio for all frequencies of perturbation. A value of 1 on the abcissa of a deviation curve indicates that a control system attached to the ecosystem will have no effect in reducing system fluctuations; a value of > 1.0 indicates the system will amplify fluctuations; < 1.0, the system will reduce fluctuations.

FIG. 4. Deviation curves for negative feedback systems that amplify input variations in 100–1 yr frequencies. Deviation curve plots the ratio of [MR with feedback]/[MR without feedback] vs. frequency (ω = radians/year). MR is the component amplitude/input amplitude. Both axes are logarithmic. Feedback control systems plotted are: (a) monitoring Ca content in canopy boles, altering input with a gain of −10; (b) monitoring Ca content in canopy central root, input with a gain of −100; (c) canopy central root, gain = −1,000; (d) standing dead wood, gain = −1,000; (e) O_1 wood, gain = −1,000; (f) O_2, gain = −10; (g) O_2, gain = −100; (h) small dead root, gain = −100; and (i) O_2 leaf, gain = −100.

thetical control system increases or decreases the effect of a perturbation. Typically such a curve has three regions (Milsum 1966)—a low-frequency region in which the control system actually causes the ecosystem to oscillate from a perturbation to a greater degree than if there were no feedback-control system, a medium-frequency region in which the control system reduces system oscillation, and a high-frequency range in which the control system has little or no effect on the system behavior. Engineers typically attempt to design systems such that the medium-frequency region corresponds to the expected frequency of perturbations. In an engineering context this may frequently be done by altering parts of the system toward this goal, but such system alteration is not a viable option in the case of natural ecosystem. The alternative is to modify the monitoring point in the system to find a feedback system with a control range that is appropriate to the frequency of perturbations. In the case of the 29-compartment Ca model there are 29 possible monitoring points. Each of the feedback-control systems associated with the 29 monitoring points would have different properties; some of the systems being more appropriate than others for a given ob-

jective. To determine which feedback-control system serves to best reduce deviations in the ecosystem Ca content caused by perturbations or stochastic variations in the input, we constructed deviation curves (e.g., Fig. 3) for each of the 116 control systems. The best monitoring point for control systems with a given gain was always the soil water and the properties of this feedback system were improved with increased gain.

Although the soil-water monitoring point seems to be the best monitoring point for the general class of feedback systems explored here, there are amplifying input fluctuations occurring at frequencies with a period of 1 year to 1 century (a range over which the Ca ecosystem model is expected to be representative of the *Liriodendron* forest). Examples of such systems are shown in Fig. 4. Of the 116 feedback control systems examined ≈ 20% had the tendency to amplify fluctuations at some frequency.

DISCUSSION

For the past decade, ecologists have had considerable interest in the diversity and stability of ecosystems, e.g., the 1969 Brookhaven Symposium (Woodwell and Smith 1969). More recently, several authors (May 1973, Waide and Webster 1975) have discussed both community and ecosystem stability, and have found that the diversity-stability relationship is neither as straightforward nor as axiomatically applicable as it was thought to be only a few years ago. Most ecosystem stability studies have focused on the stability properties of natural ecosystems relative to some perturbation. The perturbations in these cases are frequently considered as unpredictable (Slobodkin and Saunders 1969) changes in the ecosystem's typical environment. The present study focuses on the stability properties of a system consisting of a natural ecosystem in the sense we have used the term above plus both an additional information flow from some ecosystem component to a controllable input source and a transfer of Ca from this source to the ecosystem. This system is not strongly perturbed but is considered relative to its responses to input fluctuations.

The analysis of the *Liriodendron* forest Ca model indicates that the forest Ca cycle (as represented by the model surrogate) is stable, and further that it is structurally stable (Lewontin 1969, May 1973) relative to the addition of 29 additional system pathways (e.g., Fig. 2). The former result is largely a consequence of the consistent application of laws of conservation of mass in forming the equations, the diagonally-dominated nature of the system rate matrix, A, and the large slow-responding elemental compartments that seem to be universally present in state-variable conceptualizations of ecosystems. This former result is not inconsistent with observations of a natural ecosystem operating in its natural environmental conditions. Indeed, some authors (Waide and Webster 1975) have expressed doubt that any system exists that is unstable under the conditions described above.

The structural stability of the Ca model is not a result that can be expected of all element cycle models (even if they are of the same general form). [We are using structural stability in a rather loose sense in that we are considering an ecosystem model that changes stability properties, e.g., from stable to unstable, given the addition of an intuitively small degree of negative feedback to be structurally unstable. A more formal use of structural stability would be to consider stability properties of the model with slight variations in the system equations.] Also, the Ca model, if certain components are monitored, can amplify certain input variations by as much as sixfold (Fig. 4). Such properties are not unique to the Ca model of the *Liriodendron* forest but are due to the internal lags expected in any ecosystem. The important result in this case is that these highly undesirable properties can occur in cases where the ecosystem component chosen as a monitoring point is logical, the feedback system is not unreasonable, and at reasonable frequencies. This is a cautionary result that forewarns of the potential dangers of arbitrarily choosing components of ecosystems to

monitor without a thorough knowledge of ecosystem internal dynamics. A second, more positive, result is the indication that this knowledge of internal dynamics may allow more effective monitoring. In the case of the Ca model, the apparent best monitoring point was the Ca content of soil H_2O in the rooting zone (Fig. 3). The monitoring points that produced undesirable or ineffective results are usually buffered by one or more compartments from a direct input. The optimality of the soil H_2O as a monitoring point and the identity of the undesirable monitoring points are directly related to the model parameters and the model structure. It is our opinion that these results are much more sensitive to the manner in which the model is conceptualized (viz the model structure) than any errors in parameter estimation. Specifically, the general results in this case are strongly dependent upon our utilization of soil-water state variables, which were functional components often omitted from earlier ecosystem compartment models. Ecologists using the methods we have outlined should take cognizance of the importance of determining appropriate model structures for given scientific objectives in their interpretation of model-analytical results.

ACKNOWLEDGMENTS

We thank R. K. McConathy, F. G. Taylor, M. Shanks, W. F. Harris, G. S. Henderson, C. W. Francis, B. Hutchison, and D. Matt for making unpublished data collected from the *Liriodendron* forest available to us. M. Witkamp, J. S. Olson, R. V. O'Neill, and J. B. Mankin provided helpful discussions on various aspects of the manuscript. Research was supported by the Eastern Deciduous Forest Biome Project, US-IBP, funded by the National Science Foundation under Interagency Agreement AG-199, BMS69-01147 A09 with the Energy Research and Development Administration—Oak Ridge National Laboratory. This is contribution no. 230 from the EDFB-IBP, and publication no. 767, Environmental Sciences Division, ORNL.

LITERATURE CITED

Auerbach, S. I., J. S. Olson, and H. D. Waller. 1964. Landscape investigations using cesium-137. Nature **201**:761–764.

Auerbach, S. I., D. J. Nelson, and E. G. Struxness. 1972. Environmental Sciences Division annual progress report. ORNL-4848. (Oak Ridge National Laboratory, Oak Ridge, Tenn.). 127 p.

Child, G. I., and H. H. Shugart. 1972. Frequency response analysis of magnesium cycling in a tropical forest system, p. 103–134. *In* B. C. Patten [ed.] Systems analysis and simulation in ecology, Vol. II. Academic Press, New York. 526 p.

Cox, T. 1973. Production, mortality and nutrient cycling in root systems of *Liriodendron* seedlings. Ph.D. Thesis. Univ. Tennessee, Knoxville. 289 p.

Cristofolini, G. 1970. Biomassa e produttivita dello strato erbaceo di un ecosistema forestale. G. Bot. Ital. **104**:1–34.

Day, F. P., Jr. 1973. Primary production and nutrient

pools in the vegetation on a southern Appalachian watershed. Ph.D. Thesis. Univ. Georgia, Athens. 133 p.

DiStefano, J. J., A. R. Stubberud, and I. J. Williams. 1967. Feedback and control systems. Schaum, New York. 369 p.

Edwards, N. T., and P. Sallins. 1973. Continuous measurement of carbon dioxide evolution from partitioned forest floor components. Ecology 54:406–412.

Francis, C. W., and T. Tamura. 1973. Cesium-137 soil inventory of a tagged *Liriodendron* forest 1962 and 1969, p. 140–149. *In* D. J. Nelson [ed.] Radionuclides in ecosystems. Proc. of the Third National Symposium on Radioecology. CONF-710501. (US AEC, Oak Ridge, Tenn.). 1268 p.

Gerloff, G. C., P. D. Moore, and J. T. Curtis. 1964. Mineral content of native plants of Wisconsin. Univ. Wis. Agric. Exp. Stn. Res. Pap. 14. 28 p.

Goldstein, R. A., and J. B. Mankin. 1972. PROSPER: A model of atmosphere-soil-plant water flow, p. 1176–1181. *In* Proc. 1972 Summer Computer Simulation Conference, ACM, IEEE, SHARE, SCI, San Diego, Calif. California State Univ., San Diego, Center for Regional Environmental Studies, San Diego.

Harris, W. F., R. S. Kinerson, Jr., and N. T. Edwards. 1974. Comparison of the belowground biomass of natural deciduous forests and loblolly pine plantations. *In* J. K. Marshall [ed.] Proceedings of the Belowground Ecosystem Symposium (*in press*).

Henderson, G. S., N. T. Edwards, C. W. Francis, D. E. Reichle, M. H. Shanks, and P. Sollins. 1971. Physical and chemical properties of Emory Silt Loam in the yellow poplar experimental area, p. 85–91. *In* S. I. Auerbach and D. J. Nelson [ed.] Ecological Sciences Division annual progress report. ORNL-4759. (Oak Ridge National Laboratory, Oak Ridge, Tenn.). 188 p.

Hutchison, B. A. 1975. Photographic assessment of deciduous forest radiation regimes. ATDL Contribution No. 75/3. (National Oceanic and Atmospheric Administration, Atmospheric Turbulence and Diffusion Laboratory, Oak Ridge, Tenn.). 164 p.

Kercher, J. R., and R. E. Funderlic. 1975. Mathematical methods of analysis of linear models: A documentation of the Linear Ecosystem Analysis Package (LEAP). EDFB-IBP 75-7 (Oak Ridge National Laboratory, Oak Ridge, Tenn.) (*in press*).

Kerlin, T. W., and J. L. Lucius. 1966. The SFR-3 code. A Fortran program for calculating the frequency response of a multivariable system and its sensitivity to parameter changes. ORNL-TM-1571. (Oak Ridge National Laboratory, Oak Ridge, Tenn.). 81 p.

Lewontin, R. C. 1969. The meaning of stability, p. 13–24. *In* G. M. Woodwell and H. H. Smith [ed.] Diversity and stability in ecological systems. Brookhaven Symposium in Biology No. 22. U.S. Dep. Commerce, National Bureau of Standards, Springfield, Va. 264 p.

Lyapunov, A. M. 1893. Probleme general de la stabilite du mouvement. Annals of mathematics study, No. 17. [1949 American ed. of 1907 French translation.] Princeton Univ. Press, Princeton, N.J.

May, R. M. 1973. Stability and complexity in model ecosystems. Princeton Univ. Press, Princeton, N.J. 235 p.

Milsum, J. H. 1966. Biological control systems analysis. McGraw-Hill, New York. 486 p.

Moulder, B. C., and D. E. Reichle. 1972. Significance of spider predation in the energy dynamics of forest floor arthropod communities. Ecol. Monogr. 42: 473–498.

Mulholland, R. J., and M. S. Keener. 1974. Analysis of linear compartment models for ecosystems. J. Theor. Biol. 44:105–116.

Olson, J. S. 1965. Equations for cesium transfer in a *Liriodendron* forest. Health Phys. 11:1385–1392.

———. 1968a. Use of tracer techniques for the study of biogeochemical cycles, p. 271–288. *In* F. Eckhardt [ed.] Functioning of terrestrial ecosystems at the primary production level. UNESCO, Paris.

———. 1968b. Distribution and radiocesium transfers of roots in a tagged mesophytic Appalachian forest in Tennessee, p. 133–139. *In* M. S. Ghilarov [ed.] Methods of productivity studies in root systems and rhizosphere organisms. Nauka Publ. House, Leningrad.

Patten, B. C., ed. 1971. Systems analysis and simulation in ecology, vol. I. Academic Press, New York. 607 p.

———. 1972. Systems analysis and simulation in ecology, vol. II. Academic Press, New York. 526 p.

Reichle, D. E., and D. A. Crossley, Jr. 1965. Radiocesium dispersion in a cryptozoan food web. Health Phys. 11:1375–1384.

Reichle, D. E., B. E. Dinger, N. T. Edwards, W. F. Harris, and P. Sollins. 1973a. Carbon flow and storage in a forest ecosystem, p. 345–365. *In* G. M. Woodwell and E. V. Pecan [ed.] Carbon and the biosphere. Brookhaven Symposium in Biology No. 24. U.S. Dep. Commerce, National Bureau of Standards, Springfield, Va. 392 p.

Reichle, D. E., and N. T. Edwards. 1973. IBP-Eastern Deciduous Forest Biome—Oak Ridge Site, p. 151–202. *In* Modeling forest ecosystems. Report of international woodlands workshops, U.S. International Biological Program/PT Section. EDFB-IBP-73-7. (Oak Ridge National Laboratory, Oak Ridge, Tenn.). 339 p.

Reichle, D. E., R. A. Goldstein, R. I. Van Hook, and G. J. Dodson. 1973b. Analysis of insect consumption in a forest canopy. Ecology 54:1076–1084.

Slobodkin, L. B., and H. L. Sanders. 1969. On the contribution of environmental predictability to species diversity, p. 82–95. *In* G. M. Woodwell and H. H. Smith [ed.] Diversity and stability in ecological systems. Brookhaven Symposium in Biology No. 22. U.S. Dep. Commerce, National Bureau of Standards, Springfield, Va. 264 p.

Sollins, P., D. E. Reichle, and J. S. Olson. 1973. Organic matter budget and model for a southern Appalachian *Liriodendron* forest. EDFB-IBP-73-2. (Oak Ridge National Laboratory, Oak Ridge, Tenn.). 150 p.

Taylor, F. G. 1974. Phenodynamics of production in a mesic deciduous forest, p. 237–254. *In* H. Lieth and F. Sterns [ed.] Phenology and seasonality modeling. Ecological Studies 8. Springer-Verlag, New York. 444 p.

Thomas, W. A. 1969. Accumulation and cycling of calcium by dogwood trees. Ecol. Monogr. 39:101–120.

———. 1970. Retention of calcium-45 by dogwood trees. Plant Physiol. 45:510–511.

Waide, J. B., J. E. Krebs, S. P. Clarkson, and E. M. Setzler. 1974. A linear systems analysis of the calcium cycle in a forested watershed ecosystem, p. 261–345. *In* R. Rosen and F. M. Snell [ed.] Progress in theoretical biology. Academic Press, New York. 349 p.

Waide, J. B., and J. R. Webster. 1975. Engineering systems analysis: Applicability to ecosystems. *In* B. C. Patten [ed.] Systems analysis and simulation in ecology, Vol. IV. Academic Press, New York. (*in press*).

Waller, H. D., and J. S. Oilson. 1967. Prompt transfers of cesium-137 to the soils of a tagged *Liriodendron* forest. Ecology **48**:15–25.

Whittaker, R. H., and G. M. Woodwell. 1968. Dimension and production relations of trees and shrubs in the Brookhaven Forest, New York. J. Ecol. **56**:1–25.

Witherspoon, J. P., and G. N. Brown. 1965. Translocation of cesium-137 from parent trees to seedlings of *Liriodendron tulipifera*. Bot. Gaz. **125**:181–185.

Witkamp, M. 1969. Cycles of temperature and carbon dioxide evolution from litter and soil. Ecology **50**:922–924.

Witkamp, M., and B. Barzansky. 1968. Microbial immobilization of ^{137}Cs in forest litter. Oikos **19**:392–395.

Woodwell, G. M., and H. H. Smith, ed. 1969. Diversity and stability in ecological systems. Brookhaven Symposium in Biology No. 22. U.S. Dep. Commerce, National Bureau of Standards, Springfield, Va. 264 p.

135

11

Reprinted from pages 1–27 of *Mineral Cycling in Southeastern Ecosystems*,
F. G. Howell, J. B. Gentry, and M. H. Smith, eds., ERDA Symp. Ser. (CONF-740513),
Springfield, Va.: TIC and ERDA, 1974, 898pp.

NUTRIENT RECYCLING AND THE STABILITY OF ECOSYSTEMS

JACKSON R. WEBSTER, JACK B. WAIDE, and BERNARD C. PATTEN
Department of Zoology and Institute of Ecology, University of Georgia,
Athens, Georgia

ABSTRACT

A theoretical perspective on ecosystems is elaborated which relates alternative strategies of stability to observable and measurable attributes of ecosystems. Arguments are presented for viewing nutrient cycling as positive feedback. Any resultant tendency for unlimited growth is resisted by (1) finiteness of resources, (2) kinetic limitations on resource mobilization, and (3) processes of nutrient regeneration. Ecosystem structure, a static inertia defined by the mass of biotic and abiotic components, is opposed by dynamic dissipative forces related to metabolism and erosion. Balance between these two factors (structural mass and dissipative force) guarantees the asymptotic stability of ecosystems. Attention is thus focused on two aspects of relative stability: resistance and resilience. Resistance, the ability of an ecosystem to resist displacement, results from the accumulated structure of the ecosystem. Resilience, the ability of an ecosystem to return to a reference state once displaced, reflects dissipative forces inherent in the ecosystem. A linear ecosystem model that embodies these concepts is discussed, and four relative stability indexes are derived. Random matrices, subject to mass-conservation limitations, and hypothetical ecosystem models, constructed according to a characterization of alternative properties of nutrient cycles, are analyzed to examine relationships between the relative stability indexes and specific properties of nutrient cycles.

Resistance is shown to be related to large storage, long turnover times, and large amounts of recycling. Resilience reflects rapid turnover and recycling rates. Thus resistance and resilience are inverse concepts. Factors that determine what balance between resistance and resilience an ecosystem exhibits are considered, including the degree and frequency of environmental fluctuation and the limitations placed on resource mobilization. The contribution of turnover rates of ecosystem components to the balance between resistance and resilience is also examined, involving consideration of (1) the population concepts of r and K selection, (2) the contribution of early successional species to ecosystem stability, and (3) the relation of herbivory to nutrient regeneration. The theory put forth in this paper is seen as a rigorous, operational approach to ecosystems which is testable by both observation and experimental analysis.

A dialectical point of departure for studying ecosystems is provided by the antithetical processes of biological growth and decay. At the cellular level, balance between the opposing forces of anabolism and catabolism determines both structure and reaction kinetics. Anabolic and catabolic phenomena similarly operate at the ecosystem level but are less well understood. On the one hand are the mobilization of energy and nutrient resources into organic configurations and the accretion of biomass; on the other are dissipative forces tending to erode whatever biotic structures have been realized, returning the system toward physicochemical equilibrium while regenerating assimilated nutrients.

Morowitz (1966) postulated that energy dissipation is sufficient to cause associated material cycles. Such a postulate is fundamental since in the materially closed biosphere, maintenance of life requires nutrient regeneration. For most natural ecosystems, recycling rates limit primary production and so regulate, at the source, biotic energy flows. A positive-feedback loop is thus inherent in the structure of every ecosystem: energy flow produces nutrient cycles, which lead to greater energy flow. Any tendency for unlimited growth is resisted by (1) finiteness of the resource base, (2) kinetic requirements of resource mobilization, and (3) restorative processes of nutrient regeneration.

Thus biotic growth tendencies are bounded by resource availability as well as by limitations on resource assimilation. The dialectical viewpoint outlined above must account for these facts. The biotic structure of ecosystems results from the tendency of living organisms to acquire resources, as limited by the requirements of resource mobilization. Acting to erode structure are dissipative forces that tend to degrade both organic and inorganic configurations. Degradation of biotic structure is related to metabolic processes of living organisms. Decay of abiotic structure relates both to the biotic decomposition of minerals and to the purely abiotic processes of weathering and erosion. Hence, on the one hand is the structure of the ecosystem, a static inertia defined by the mass of biotic and abiotic components. On the other hand is the dissipative force tending to erode this structure, a dynamic force defined by metabolism and erosion. At the ecosystem level these two factors (structural mass and dissipative force) are not necessarily antithetical. Both contribute, in different ways, to the stability of ecosystems.

A recurrent theme in ecological literature is that ecosystem stability is related to nutrient-cycling characteristics. E. P. Odum (1969) suggested that the closing of nutrient cycles through ecosystem development contributes to increased stability. Pomeroy (1970) related the stability of several ecosystem types to elemental standing crops and turnover times, biomass, and productivity. Jordan, Kline, and Sasscer (1972) examined ecosystem stability in relation to models of forest nutrient cycles. Hutchinson (1948a, 1948b), H. T. Odum (1971), Child and Shugart (1972), and Waide et al. (1974) also suggested causal links between nutrient cycling and ecosystem stability. These arguments were largely intuitive or heuristic, however, and did not seek the basis for causal relationships in specific properties of ecosystem nutrient cycles. In this paper we investigate relations between observable characteristics of nutrient cycles and system-level concepts of stability.

STABILITY CONCEPTS AND DEFINITIONS

Absolute Stability

Liapunov (1892) provided the basis of stability theory. Let $x(t)$ be a vector of n time-dependent state variables, with $\| x(t) \|$ a norm such as

$$\| x(t) \| = \sum_{i=1}^{n} |x_i(t)| \qquad (i = 1, 2, \ldots, n)$$

An equilibrium state $x^0 (\dot{x} = 0$ when $x = x^0)$ is said to be stable in the sense of Liapunov if for every initial time t_0 and every $\epsilon > 0$ there exists $\delta > 0$ such that, if $\| x(t_0) - x^0 \| < \delta$, then $\| x(t) - x^0 \| < \epsilon$ for all $t > t_0$. In other words, a system is stable if, following displacement from equilibrium, its subsequent behavior is restricted to a bounded region of state space. A stronger stability concept involves return to equilibrium following initial displacement. An equilibrium state x^0 is said to be asymptotically stable (1) if it is stable in the sense of Liapunov and (2) if for any t_0 there exists $\alpha > 0$ such that, if $\| x(t_0) - x^0 \| < \alpha$, then $x(t) \to x^0$ as $t \to \infty$.

Holling (1973) suggested that such classical stability concepts are little more than theoretical curiosities in ecology. We suggest instead that natural ecosystems are asymptotically stable (Child and Shugart, 1972; Waide et al., 1974; Patten, 1974; Waide and Webster, 1975). A dynamic balance between the maintenance and dissipation of structure produces nonzero ecosystem states that are stable. Around this nominal (unperturbed, reference) trajectory exist basins or domains of attraction (Lewontin, 1970a; Holling, 1973) within which ecosystem displacements from nominal behavior are followed by return to the original condition. The relevant question for ecologists' attention is not "Are ecosystems stable?" but rather, "How stable?" Ecologists' concern should thus be focused on relative rather than absolute stability and on the mechanisms by which differing levels of relative stability are achieved.

Relative Stability

Attempts to measure the relative stability of ecosystems have met with limited success (e.g., MacArthur, 1955; Patten and Witkamp, 1967) because relative stability is not well defined mathematically or ecologically. Relative stability concerns the nature of an ecosystem's response to small displacement from a nominal trajectory. Two aspects of this response may be identified (Patten and Witkamp, 1967; Child and Shugart, 1972; Holling, 1973; Marks,

1974). The first aspect concerns the resistance of an ecosystem to displacement. An ecosystem that is easily displaced has low resistance, whereas one that is difficult to displace is highly resistant and is, in this sense, very stable. The second aspect of relative stability concerns return to the reference state, or resilience.* An ecosystem that returns to its original condition rapidly and directly following displacement is more resilient, more stable in this sense, than one that responds slowly or with oscillation.

Thus, given that an ecosystem is asymptotically stable, two aspects of its relative stability are (1) immovability, or resistance, which determines extent of displacement, and (2) recoverability, or resilience, which reflects rate of recovery to the original condition. This view of ecosystems identifies two alternatives for persistence. Resistance to displacement results from the formation and maintenance of large biotic and abiotic structures. Resilience following displacement reflects inherent tendencies for the dissipation of such structure, but, because it is related to ecosystem metabolism, it also reflects rates with which structure is reformed following its destruction. In the closed biogeochemical cycles of the biosphere, the observable structural and functional attributes of ecosystems are determined by the realized balance between factors favoring resistance and resilience. Nutrient cycling, a fundamental process inherent in ecosystems, thereby becomes a central issue in the consideration of mechanisms of macroscopic relative stability.

NUTRIENT CYCLING AND FEEDBACK

The use of flow diagrams to represent conservative energy and material flows in ecosystems has partly confused the concepts of input, output, and feedback. Input is any exogenous signal† that impinges on a system. Output is any endogenous attribute of a system transmitted as signal flow to an observer. Output generation is exclusively the province of the system, while output selection is the prerogative of the observer. Often output is equated with the state of the system, where state provides the necessary and sufficient information for a determinate mapping from input to output (Zadeh and Desoer, 1963).

Feedback exists in a system if any of its inputs are determined by its state. If the measure of state is directly related to such inputs, the feedback is positive; if the two are inversely related, the feedback is negative.

---

*Holling (1973) used *resilience* to denote what we term *resistance*, and *stability* for our *resilience*. Our use of resistance and resilience is consistent with common and accepted English usage (*Webster's New World Dictionary of the American Language*, Second College Edition, 1972, The World Publishing Company, New York).

†"Signal" denotes an observable and measurable flow of conserved (energy or matter) or unconserved (information) quantities.

A flow diagram of an ecosystem or ecosystem component (Fig. 1) shows inflow of material or energy which is processed by the system, resulting in outflow. Inflows and outflows are conserved. In a control diagram of this system (Fig. 1), output has been equated to state. Inflow and outflow both constitute possible inputs to the system and may be subject to feedback control. Inflow and outflow are still conserved, but no such conservation restriction applies to input and output.

It can be argued that, if feedback control of outflow exists in ecosystems, it must be negative and therefore stabilizing. That is, ecosystem component losses are regulated by density-dependent mechanisms. These losses of conserved

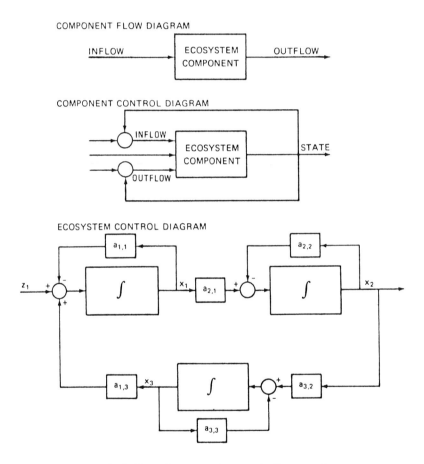

Fig. 1 Generalized flow diagram and control diagram of an ecosystem component and a control diagram of a three-component ecosystem model. Circles indicate summing junctions. Rectangles are storage (integrative) elements. z_i is an input; x_i is the state of the ith component; and $a_{i,j}$ is the rate coefficient for transfer from x_j to x_i.

quantities must be offset by inflows to maintain nonzero states. At the organism and population levels, positive-feedback mechanisms operate to promote inflow and are therefore potentially destabilizing (Milsum, 1968). Mobilization of resources is the essence of life processes (Smith, 1972); however, many density-dependent mechanisms exist which regulate inflow in a negative-feedback sense (Whittaker and Woodwell, 1972). Further, a macroscopic perspective leads to the conclusion that ecosystems and their components are ultimately resource limited (Hairston, Smith, and Slobodkin, 1960; Wiegert and Owen, 1971; Patten et al., 1974; Waide and Webster, 1975; Webster and Waide, 1975). Under unperturbed conditions ecosystems are maximally expanded within the resource hyperspace to the point of kinetic limitation of material transfers as set by the physicochemical environment (Blackburn, 1973). Thus inflow is limited by matter-recycling kinetics that ensure boundedness. Bounded inflow and negative-feedback control of outflow coupled with the first law of thermodynamics (mass conservation) form the basis of our argument for nonzero ecosystem states that are stable.

These ideas lead to a representation of ecosystems (Fig. 1) as sets of interacting components, each regulated by a negative-feedback loop related to its dissipative (i.e., turnover) character. Material recycling is displayed as feedback involving multiple system components. Because material flow is involved, recycling must be interpreted as positive feedback (H. T. Odum, 1971). This point emphasizes a fundamental difference between feedback in a control diagram and material recycling in a flow diagram. In the control diagram control is mediated by nonconservative information flows, whereas in the flow diagram control among components is exerted only through material or energy flows that must be conserved. Feedback mechanisms are not explicit in flow diagrams but must, nevertheless, be incorporated into any mathematical model of the system.

Thus a systems theoretic interpretation of nutrient cycling as feedback leads to the general conclusions already elaborated: (1) biotic tendency for unlimited growth is bounded by the first law of thermodynamics (mass conservation), as mediated through material-recycling kinetics and (2) negative-feedback decay to abiotic physicochemical equilibrium, if material and energy inflows are removed, is assured by the dissipative character of ecosystems and the second law of thermodynamics. The first conclusion guarantees Liapunov stability. The two conclusions together are sufficient to establish the stability of nonzero ecosystem trajectories (Patten, 1974).

MEASURES OF RELATIVE STABILITY

The General Linear Ecosystem Model

The dynamics of conserved quantities in an ecosystem with n components can be described mathematically as

$$\dot{x}_i - \text{inflow} - \text{outflow} \qquad (i = 1, 2, \ldots, n) \qquad (1)$$

Inflow can emanate from outside the ecosystem (z_i) or from other system components ($F_{i,j}$, $j = 1, 2, \ldots, n; j \neq i$). Outflow may pass to other components ($F_{j,i}$) or out of the system ($F_{0,i}$). Hence Eq. 1 may be reformulated in compartmental form as

$$\dot{x}_i = (z_i + \sum_{\substack{j=1 \\ j \neq i}}^{n} F_{i,j}) - (F_{0,i} + \sum_{\substack{j=1 \\ j \neq i}}^{n} F_{j,i}) \qquad (i = 1, 2, \ldots, n) \qquad (2)$$

Material transfers within the ecosystem represent inflows to some components and outflows from others. On the basis of the arguments given above and elsewhere (Patten et al., 1974; Webster and Waide, 1975), these internal flows, as well as outflows from the system, can be modeled as donor-based according to the equation

$$F_{i,j} = a_{i,j} x_j \qquad (3)$$

If we define component turnover rates as

$$a_{i,i} = - \sum_{\substack{j=1 \\ j \neq i}}^{n} a_{j,i} - a_{0,i} \qquad (i = 1, 2, \ldots, n) \qquad (4)$$

Eq. 2 becomes

$$\dot{x}_i = z_i + \sum_{j=1}^{n} a_{i,j} x_j \qquad (i = 1, 2, \ldots, n) \qquad (5)$$

Because all x_i and $F_{i,j}$ represent material or energy, they must be nonnegative, which ensures that

$$a_{i,j} \geqslant 0 \qquad (i \neq j) \qquad (6)$$

Equation 5 can be expressed in matrix form as

$$\dot{x} = Ax + z \qquad (7)$$

where x is the state vector, z is the input vector, and A is a matrix of (possibly time dependent) rate coefficients defined by Eq. 3. The mathematical constraints defined in Eqs. 4 to 6 are sufficient to guarantee the asymptotic stability of this model (Hearon, 1953, 1963). In addition, the model is sufficient for simulating nominal and small displacement dynamics of ecosystems (e.g., Olson, 1963; Patten, 1972; Patten et al., 1974). Implicit within the model structure defined by Eqs. 1 to 7 are both accumulative and dissipative tendencies; thus this model is useful for examining macroscopic questions of ecosystem relative stability.

nth-Order Measures

The system defined by Eq. 7 is an nth-order system, being composed of n first-order equations. Relative stability indexes can be derived for this system. Specifically, the characteristic roots or eigenvalues of the system defined by Eq. 7, denoted λ_k (k = 1, 2, . . . , n), can be found by solving the matrix equation

$$\det (\lambda I - A) = 0 \qquad (8)$$

where det denotes the determinant of the indicated matrix, and I is the n X n identity matrix. The solution to Eq. 7 can be expressed in terms of these characteristic roots, where each eigenvalue defines a particular mode of system behavior, as

$$x = \sum_{k=1}^{n} c_k b_k e^{\lambda_k t} + p \qquad (9)$$

where c_k is a constant, b_k is the eigenvector associated with the eigenvalue λ_k, and p is a particular solution to Eq. 7 determined by z.

Clearly, if any $\lambda_k > 0$, the system will grow exponentially. According to a theorem attributed to Liapunov and Poincaré (Bellman, 1968), a system is asymptotically stable if all the characteristic roots have negative real parts.

Two relative stability measures may be derived from these n eigenvalues. The first is the critical root, defined as the characteristic root with the smallest absolute value (Funderlic and Heath, 1971). Given that the system is asymptotically stable, the critical root is the one most likely to become positive. Hence this index indicates the system's margin of stability. This critical root is the smallest turnover rate (the longest time constant) in the system. Thus the system does not recover fully from displacement until this slowest component of the transient response decays away. Second, the trace of the matrix A (the sum of the diagonal elements) relates to the response time following perturbation (Makridakis and Weintraub, 1971b). Since the sum of the main diagonal elements of A equals the sum of the eigenvalues, we have used the mean root, defined as the mean value of the n eigenvalues, as an equivalent measure of response time. The mean root reflects the time required for most of the system, or for some hypothetical mean component of the system, to recover following displacement.

Second-Order Measures

Extensive experience in control-systems engineering has demonstrated the utility of approximating higher order linear systems as second order for analytical purposes (DiStefano, Stubberud, and Williams, 1967; Shinners, 1972). Child and Shugart (1972) provided a rationale for implementing such an approach in studying ecosystem behavior and applied it to an analysis of magnesium cycling in a tropical forest. Waide et al. (1974) used this approach in analyzing a model of calcium dynamics in a temperate forest. Hubbell (1973a, b) demonstrated the benefits of a frequency-domain analysis of second-order population models.

In this approach the behavior of an nth-order system of the form of Eq. 7 is approximated as second order with the equation

$$\ddot{y} + 2\zeta\omega_n\dot{y} + \omega_n^2 y = \omega_n^2 z \tag{10}$$

where ζ is the damping ratio and ω_n is the undamped natural frequency (DiStefano et al., 1967). The characteristic roots of this equation are given by

$$\lambda_1,\lambda_2 = -\zeta\omega_n \pm \omega_n (\zeta^2 - 1)^{\frac{1}{2}} \tag{11}$$

The roots of this second-order approximation represent the apparent roots of the original nth-order system. That is, these two eigenvalues, as well as the natural frequency, represent weighted mean roots of the higher order system. They capture most of the information contained in the nth-order trajectories. The weighting function that determines these second-order parameters from the n original eigenvalues is related to the magnitude of the eigenvector components of the nth-order system (Eq. 9).

From Eq. 11, if $\zeta = 1$, the system is said to be critically damped, the system responds rapidly and without oscillation following displacement, and λ_1, $\lambda_2 = -\omega_n$. If $\zeta > 1$, the system is overdamped, the response of the system is slower than that of a critically damped system, though still nonoscillatory, and the eigenvalues are real and unequal. If $\zeta < 1$, the system in underdamped, and the roots are complex and are given by

$$\lambda_1,\lambda_2 = -\zeta\omega_n \pm j\omega_n (1 - \zeta^2)^{\frac{1}{2}} \tag{12}$$

where $j = (-1)^{\frac{1}{2}}$. The response of such a system to displacement, though initially more rapid than a critically damped system, is oscillatory. If $\zeta = 0$, the roots are imaginary, and ω_n is the radian frequency of oscillation. If $\zeta < 0$, the eigenvalues have positive real parts, and the system is unstable.

Given that the system under study is asymptotically stable (i.e., $\zeta > 0$), the two parameters ω_n and ζ may be used as measures of relative stability. The natural frequency ω_n measures (inversely) the resistance of the system to displacement. A system with a large natural frequency is especially susceptible to disturbance, whereas a system with a small natural frequency strongly resists displacement. Similarly the magnitude of the damping ratio ζ indicates the rate of system response following displacement, the resilience of the system. If the system is overdamped, the return to steady state is monotonic but slow. If underdamped, the system responds in an oscillatory fashion. A critically damped system exhibits the most rapid response possible without oscillation and thus has maximum resilience.

In this paper we investigate relationships between specific properties of ecosystem nutrient cycles and discuss the four above-mentioned relative stability indexes: critical root, mean root, natural frequency, and damping ratio. We take two approaches. The first is a stochastic approach, using Monte Carlo techniques. In the second approach we construct hypothetical ecosystem models based on a characterization of alternative properties of nutrient cycles and investigate the relative stability of these models. We also provide further ecological understanding of the four relative stability indexes and extend the basis for their implementation. Attention is restricted to time-invariant systems for heuristic purposes.

STOCHASTIC APPROACH

Construction and analysis of random matrices was used successfully to further understanding of general system properties and to investigate effects of specific system characteristics (e.g., connectivity) on such system-level properties as stability (Ashby, 1952; Gardner and Ashby, 1970; Makridakis and Weintraub, 1971a, b; May, 1972, 1973; Makridakis and. Faucheux, 1973; Waide and Webster, 1975; Webster and Waide, 1975). We initially followed such an approach to establish general relationships among relative stability indexes and system properties, focusing especially on the amount of recycling.

Methods

In constructing random matrices, off-diagonal elements $a_{i,j}$, $i \neq j$, of the A matrix (Eq. 7) were chosen from a specified statistical distribution (e.g., uniform on [0,1]). Rates of nutrient loss to the environment ($a_{0,i}$) were chosen from the same distribution and main diagonal elements calculated according to Eq. 4. For some analyses, off-diagonal elements were defined as nonzero according to a specified probability of connectivity. Only a single input z_1 was used for all analyses.

Following matrix construction, eigenvalues were calculated (Westley and Watts, 1970), and the critical root and mean root were determined. We also calculated an index of recycling (I) as the summed flows represented by the upper triangle divided by the input. That is, the ratio of nutrients recycled to nutrient input from the environment is

$$I = \frac{\left(\sum_{i=1}^{n-1} \sum_{j=i+1}^{n} F_{i,j} \right)}{z_1} \tag{13}$$

The synthetic division algorithm of Ba Hli (1971) was used to estimate the values of the natural frequency and damping ratio. A unit step input was applied to each randomly constructed matrix to generate the required discrete input—output time series. Synthetic division yielded the coefficients of a general second-order transfer function, which were equated with coefficients of the specific transfer function of Eq. 10, allowing estimation of the natural frequency and damping ratio (Hill, 1973).

The above process was repeated 50 or 100 times for each type of matrix constructed. The resulting sets of values were subjected to linear regression analysis to determine the presence of significant relationships among calculated variables. To ensure that results were not biased by methods of matrix construction, we analyzed a variety of matrices of three sizes (n = 4, 6, 10). In various experiments, matrix elements were sampled from uniform distributions of different ranges and from normal distributions with various means and variances. We tried a wide range of upper and lower triangle connectivity, and selected several different outputs for use in the synthetic division. In some cases modifications were made to obtain a pyramid-type structure of compartmental standing crops. We also examined results of increased input and recycling.

Results

The following trends were generally observed across the range of matrices analyzed. Increases in the amount of recycling relative to input led to increases in the critical root (moved closer to zero), decreases in the mean root (moved farther from zero), and decreases in the natural frequency. Also, larger critical and mean roots were both associated with smaller natural frequencies.

Trends in the damping ratio initially appeared to be variable. In some cases ζ tended to decrease with increasing recycling, critical root, and mean root. In other cases ζ showed the opposite behavior. Closer inspection revealed that, in the first case, all systems were underdamped, whereas in the second case they were overdamped. Thus, when the quantity $|1 - \zeta|$ was considered, the results were unambiguous: $|1 - \zeta|$ increased with increasing values of recycling, critical root, and mean root.

DETERMINISTIC APPROACH

Our second approach to investigating relationships between material recycling and ecosystem stability involved construction and analysis of hypothetical ecosystem models. Two basic assumptions are inherent in these analyses: (1) ecosystems are units of selection and evolve from systems of lower selective value to ones of higher selective value (we are not invoking any superorganism concept; this evolution is accomplished through species coevolution) (Slobodkin, 1964; Darnell, 1970; Lewontin, 1970; Dunbar, 1972; Whittaker and Woodwell, 1972; Blackburn, 1973); (2) those ecosystems with highest selective value are ones which optimize utilization of essential resources. Exceptions to the selection for ecosystems geared to efficient resource utilization would exist where resources were extremely abundant or where the system as a whole was operating under other environmental stress (Odum, 1967; Waide et al., 1974). An example might be a stream which receives large allochthonous inputs of detritus and which is strongly influenced by current action. In other ecosystems selective value involves efficient conservation and recycling of essential nutrients.

We suggest that three factors are involved in nutrient utilization in ecosystems: (1) the presence or absence of large abiotic nutrient reserves, (2) the degree of localization of nutrients within the biota, and (3) the turnover rate of the actively recycling pool of nutrients. Figure 2 schematically depicts these factors. In this figure a specific ecosystem type is associated with a given combination of factors. This conceptual scheme is clearly idealized since there exists a great range of each of these distinct types of ecosystems. However, this scheme is consistent with current ecological theory and represents a convenient method of examining relationships between nutrient cycling and stability.

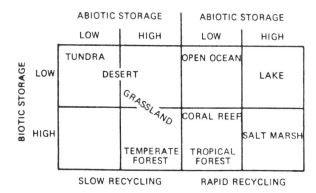

Fig. 2 Alternative properties of nutrient cycles. Shown in each box is an idealized ecosystem type that seems to exhibit the indicated combination of properties.

Methods

To facilitate quantitative comparisons among these various idealized ecosystems, we constructed a general model of nutrient cycling (Fig. 3). In this diagram the food base (x_1) may be either primary producers or detritus. Consumers (x_2) are organisms that feed directly on the food source. The $F_{3,1}$ is either death or mechanical breakdown of the food base to detritus (x_3). In an ecosystem with internal primary production, detritus is essentially dead primary producers (litter). In detritus-based systems this component is fine particulate organic matter. Decomposers (x_4) are those organisms which feed directly or indirectly on detritus. Available nutrients (x_5) are directly available for use in primary production. Nutrients in reserve (x_6) are not available but are tied up in sediments, primary minerals, clay complexes, or other refractory materials (e.g., humics). However, they may become available through transfer to x_5. Inflows and outflows occur primarily through the available nutrient pool.

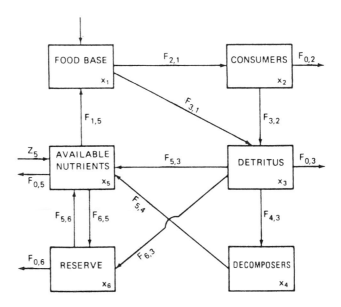

Fig. 3 General nutrient-flow model of an ecosystem. x_i is the size of the ith compartment; z_i is inflow to compartment x_i; $F_{i,j}$ is the flow from x_j to x_i; and $F_{0,j}$ is the outflow to the environment from x_j.

We have quantified this general model for seven of the ecosystem types shown in Fig. 2 (Table 1). We also applied this model to an idealized stream, which typifies an ecosystem without large abiotic reserves, with low biotic localization of nutrients, with little or no recycling, and with large nutrient throughflows. Standing-crop values were normalized to an available nutrient pool of 100 units. All transfers were per year. The values given in Table 1 are relative estimates that reflect differences among the idealized ecosystems, rather than exact, absolute estimates of nutrient transfers and standing crops. A variety of sources was consulted for each ecosystem type (Table 1). However, gaps and inconsistencies existed which were filled from general references or qualitative considerations. Each system was assumed to be at steady state.

From these numbers we derived several indexes which reflect structural characteristics of the eight ecosystems and which quantify the concepts of abiotic storage, biotic storage, and recycling (Table 2). Both the turnover time of the reserve ($T_6 = 1/|a_{6,6}|$) and the proportion of nutrients localized in the two abiotic pools $[(x_5 + x_6)/\Sigma x]$ are indexes of abiotic storage. Reserve turnover varies from slow in forests to fast in oceans and streams. The proportion of total nutrients in abiotic compartments is highest in temperate forests and lakes and lowest in tundra.

Biotic storage, given by the turnover time of biotic compartments $[(x_1 + x_2 + x_3 + x_4)/F_{1,5}]$, is higher in terrestrial ecosystems and lower in aquatic ecosystems.

We calculated two indexes of recycling. The turnover rate of the detritus pool ($F_{1,5}/x_3$) is higher in aquatic systems and generally lower in terrestrial ecosystems, except for tropical forests where there is a rapid turnover of detritus. The ratio of recycling to input ($F_{1,5}/\Sigma z$), as used in stochastic analyses, is approximately the inverse of the other recycling index. However, since systems with larger biotic pools typically recycle more nutrients than do systems with smaller biotic standing crops, this index partially confounds storage and recycling. This index ranges from 500 for grasslands to 0 for streams.

Two other useful indexes are the ratios of total standing crop to recycling material ($\Sigma x/F_{1,5}$) and total standing crop to total inflow ($\Sigma x/\Sigma z$). Both indexes estimate system turnover time. Longest turnover times occur in temperate forests and grasslands, whereas there is rapid turnover in stream and ocean ecosystems.

The specific values given in Table 1 have obvious deficiencies. Each idealized ecosystem represents a wide spectrum of actual ecosystems differing in many important characteristics. Similarly the kinetics of specific nutrients within a given ecosystem differ, quantitatively and qualitatively. In quantifying the general model shown in Fig. 3, we have attempted to suppress such specific details and to focus instead on the alternative properties of nutrient cycles depicted in Fig. 2. Our emphasis is thus on macroscopic properties of ecosystems rather than on specific differences between systems or nutrients. Comparison of the structural indexes (Table 2) with Fig. 2 reveals that the chosen values agree well with the idealized conceptualization.

Results

The eight models were analyzed in the same fashion as described previously, providing values for critical root, mean root, natural frequency, and damping ratio (Table 3). Both critical root and natural frequency were smallest (in absolute value) for the temperate forest and grassland models and largest for the stream model and tended to be smaller (in absolute value) for the four terrestrial ecosystem models. All values of damping were greater than 1, indicating all eight ecosystem models to be overdamped. The smallest value was obtained for the ocean, the largest for the stream. No clear separation between terrestrial and aquatic ecosystems was obvious.

The relative stability indexes were then compared with the structural indexes given in Table 2, using least-squares regression. Correlation coefficients are shown in Table 4. Both critical and mean roots were directly related to the turnover time of the reserve nutrient pool T_6, whereas the natural frequency exhibited an inverse relationship. For longer turnover times, critical and mean roots were nearer zero, and the natural frequency was smaller.

Regressions against the proportion of nutrients in the two abiotic pools were not significant. However, when terrestrial and aquatic ecosystems were considered separately, a trend was evident. Increased abiotic storage or slower abiotic turnover produced critical and mean roots nearer zero and smaller natural frequencies.

All four stability indexes were related to recycling. A greater recycling rate ($F_{1,5}/x_3$) or a smaller ratio of recycling to input ($F_{1,5}/\Sigma z$) resulted in roots farther from zero, a larger natural frequency, and greater damping.

Both critical and mean roots, as well as natural frequency, were significantly related to system turnover ($\Sigma x/F_{1,5}$). All four indexes were correlated with turnover as related to system input ($\Sigma x/\Sigma z$). In general, the slower the system turnover rate (the greater the turnover time), the nearer the critical and mean roots were to zero, the smaller the natural frequency, and the smaller the damping ratio.

The results clearly indicate that increased storage and turnover times (abiotic, biotic, or total), as well as increased amounts of recycling, lead to critical and mean roots nearer zero and to smaller natural frequencies. Increased recycling and turnover rates (of biotic or abiotic elements, or their sum), on the other hand, lead to critical and mean roots that are farther from zero and to larger natural frequencies. Relationships involving the damping ratio are less clear. However, if we ignore the stream, which has no recycling ($F_{1,5} = 0$) and for which the second-order approximation may not be accurate owing to dominance by the extremely large nutrient inflow, other trends become apparent (Table 4). Although correlations are not as large as for the other stability indexes, damping generally tended to be directly related to storage or turnover times but inversely related to recycling or turnover rates. Thus damping and natural frequency typically showed opposite behavior relative to the structural indexes considered.

DISCUSSION

The preceding arguments were presented for the asymptotic stability of ecosystems. This stability is guaranteed by limitations on resource mobilization and by the dissipative character of ecosystems. Resistance, the ability of an

TABLE 1

SUMMARY OF RELATIVE VALUES USED IN QUANTIFYING THE GENERAL
NUTRIENT CYCLING MODEL (FIG. 3) IN THE EIGHT IDEALIZED ECOSYSTEMS
INVESTIGATED (FIG. 2)

Parameter[*]	Tundra[a]	Grassland[b]	Temperate forest[c]	Tropical forest[d]	Ocean[e]	Lake[f]	Salt marsh[g]	Stream[h]
x_1	200	500	100	500	10	10	1,000	500
x_2	15	50	0.5	2.5	10	1	25	50
x_3	200	1,000	25	5	10	25	1,000	10
x_4	20	100	1	1	0.5	25	100	20
x_5	100	100	100	100	100	100	100	100
x_6	100	1,000	5,000	1,500	50	2,000	50,000	1,000
z_1	0	0	0	0	0	0	0	1,000
z_5	1	1	1	1	110	100	75	100,000
$F_{2,1}$	20	100	1	5	500	20	100	200
$F_{3,1}$	30	400	5	46	545	180	900	800
$F_{3,2}$	20	100	1	5	500	20	100	190
$F_{4,3}$	50	480	5.5	49.9	50	180	500	300
$F_{5,3}$	0	10	0.4	1	900	10	400	600
$F_{5,4}$	50	480	5.5	49.9	50	180	500	300
$F_{5,6}$	1	10	0.6	1.1	10	20	1,000	100
$F_{6,3}$	0	10	0.1	0.1	50	10	50	0
$F_{6,5}$	1	0	0.5	1	20	10	950	100
$F_{1,5}$	50	500	6	51	1,045	200	1,000	0
$F_{0,2}$	0	0	0	0	0	0	0	10
$F_{0,3}$	0	0	0	0	45	0	50	90
$F_{0,5}$	1	1	1	1	5	100	25	100,900
$F_{0,6}$	0	0	0	0	60	0	0	0

[*]x_i represents the size of the *i*th compartment; z_i is the input to x_i; $F_{i,j}$ is the flow of nutrients from x_j to x_i; and $F_{0,j}$ represents nutrient loss to the environment from x_j. All values are normalized against x_5, which was set to 100 units/unit area for each system. References consulted in deriving these values are listed below.

[a]Rodin and Bazilevich, 1967; Schultz, 1969.

[b]Reuss, 1971; Rodin and Bazilevich, 1967; Sims and Singh, 1971.

[c]Bormann and Likens, 1970; Likens and Bormann, 1972; Rodin and Bazilevich, 1967.

[d]Child and Shugart, 1972; McGinnis et al., 1969; Rodin and Bazilevich, 1967.

[e]Brylinsky, 1972; E. P. Odum, 1971; Riley, 1972.

[f]Juday, 1940; Likens and Bormann, 1972; Lindeman, 1941, 1942; Williams, 1971.

[g]E. P. Odum, 1971; Pomeroy et al., 1969; Teal, 1962; Wiegert et al., 1974.

[h]Boling et al., 1974; Cummins, 1971; Woodall, 1972.

Additional general references consulted include Collier et al., 1973; Golley, 1972; Pomeroy, 1970; Wiegert and Evans, 1964.

TABLE 2

INDEXES SUMMARIZING VARIOUS STRUCTURAL CHARACTERISTICS OF THE EIGHT HYPOTHETICAL ECOSYSTEMS AND DIFFERENTIATING AMONG THE PROPERTIES OF NUTRIENT CYCLES SHOWN IN FIG. 2*

System	Abiotic storage T_6	Abiotic storage $\dfrac{x_5 + x_6}{\Sigma x}$	Biotic storage $\dfrac{x_1 + x_2 + x_3 + x_4}{F_{1,5}}$	Recycling $\dfrac{F_{1,5}}{x_3}$	Recycling $\dfrac{F_{1,5}}{\Sigma z}$	System turnover $\dfrac{\Sigma x}{F_{1,5}}$	System turnover $\dfrac{\Sigma x}{\Sigma z}$
Tundra	100	0.31	8.7	0.25	50	12.7	635
Grassland	1,000	0.86	3.3	0.5	500	23.5	11,750
Temperate forest	8,333	0.97	21.1	0.24	6	870.93	5,226
Tropical forest	1,364	0.76	9.97	10.2	51	41.34	2,108
Ocean	0.714	0.83	0.029	104.5	9.5	0.173	1.6†
Lake	100	0.97	0.305	8	2	10.80	21.6
Salt marsh	50	0.96	2.12	1	13.3	52.22	696
Stream	10	0.65	∞	0	0	∞	0.017
			(0.58)†	(99)†	(0.0098)†	(1.70)†	

*x_i is the size of the ith compartment; z_i is the input to x_i; $F_{i,j}$ is the flow of nutrients from x_j to x_i; $F_{0,j}$ is the nutrient loss to the environment from x_j; and T_6 is the time constant of x_6.

†Since $F_{1,5} = 0$ for the stream, the indicated index was recalculated using the total loss from x_3 instead of $F_{1,5}$.

TABLE 3

RESULTS OF RELATIVE STABILITY ANALYSIS OF
NUTRIENT-CYCLING MODELS FOR EIGHT
HYPOTHETICAL ECOSYSTEMS

System	Critical root	Mean root	Natural frequency	Damping ratio
Temperate forest	−0.0001	−1.312	0.000227	1.2174
Grassland	−0.0001	−2.218	0.000228	1.1794
Tropical forest	−0.0003	−10.456	0.001039	1.2585
Salt marsh	−0.0013	−5.128	0.003898	1.1852
Tundra	−0.0015	−0.810	0.004413	1.1840
Lake	−0.0083	−9.718	0.02924	1.2954
Stream	−0.0999	−188.350	6.2947	1.4700
Ocean	−0.7678	−61.85	1.8478	1.1404

ecosystem to resist perturbation, results from the accumulated structure of the ecosystem. Resilience, the ability of an ecosystem to return to a nominal trajectory once displaced, reflects dissipative forces inherent in the ecosystem. These concepts were shown to be implicit within the linear donor-based model formulation of Eqs. 1 to 7, from which four relative stability indexes were derived: Critical root measures the system's margin of stability. Mean root is an index of system response time. Natural frequency measures resistance to displacement, and damping ratio measures resilience following displacement. Randomly constructed matrices (subject to the restriction of mass conservation; Eqs. 4 and 6) and hypothetical ecosystem models were analyzed to examine relationships between relative stability and specific properties of nutrient cycles.

Results of the stochastic analyses indicated that an increase in the amount of recycling relative to input resulted in a decreased margin of stability, faster mean response time, greater resistance, and less resilience. Analyses of the hypothetical ecosystem models revealed similar relationships among stability measures. Greater amounts of recycling were correlated with a smaller margin of stability, slower mean response time (not consistent with stochastic results), greater resistance, and less resilience (ignoring the stream value). Deterministic results also revealed that increased storage and turnover times resulted in exactly the same relationships as described for the amount of recycling. Increases in both recycling and turnover rates produced opposite results, however, leading to a larger stability margin, faster response time, smaller resistance, and greater resilience.

The inconsistent correlations between amount of recycling and mean response time can be explained. In the stochastic analyses, increases in recycling

TABLE 4

CORRELATION COEFFICIENTS FOR RELATIONSHIPS BETWEEN RELATIVE STABILITY MEASURES AND INDEXES OF STRUCTURAL PROPERTIES*

	Structural indexes													
	Abiotic storage				Biotic storage		Recycling				System turnover			
	T_6		$\dfrac{x_5 + x_6}{\Sigma x}$		$\dfrac{x_1 + x_2 + x_3 + x_4}{F_{1,5}}$		$\dfrac{F_{1,5}}{x_3}$		$\dfrac{F_{1,5}}{\Sigma z}$		$\dfrac{\Sigma x}{F_{1,5}}$		$\dfrac{\Sigma x}{\Sigma z}$	
Critical root	1	0.99	1	0.10†	4	0.89	3	−0.89	4	0.59	4	0.90	4	0.90
Mean root	4	0.70	1	0.24	4	0.72	4	−0.97	2	0.85	4	0.72	2.4	0.88
Natural frequency	4	−0.89	1	0.22†	4	−0.81	3	0.91	2	−0.85	4	−0.85	4	−0.98
Damping ratio	3	0.36	1	0.14	1	0.29	2	0.49	2	−0.85	1	−0.10	2	−0.66
	(3	0.58‡)	(1	0.25)	(4	0.26)	(3	−0.50)	(2	−0.38)	(4	0.42)	(1	−0.16)

*Each indicated variable pair was tested for (1) linear; (2) log–log relationships. The model with the largest correlation is reported and indicated to the left of the correlation coefficient. Levels of significance are 0.666 (5%) and 0.798 (1%). x_i is the size of the ith compartment; z_i is the input to x_i; $F_{i,j}$ is the flow of the nutrients from x_i to x_j; $F_{0,j}$ represents the nutrient loss to the environment from x_i; and T_6 is the time constant of x_6.

†These relationships were greatly improved by considering terrestrial and aquatic ecosystems separately. In each case the correlation coefficient was 0.99 (model 4). The relationship was positive for critical root and negative for natural frequency damping factor.

‡Values in parentheses represent correlations and model numbers, if the stream system is not considered (ξ only).

coefficients forced increases in turnover rates of donor compartments ($|a_{i,i}|$, Eq. 4). Since randomly constructed matrices exhibited a narrow range of coefficient values, a change in any one turnover rate was reflected in the mean response time. The deterministic models exhibited a much wider range in values of transfer coefficients (several orders of magnitude), so that larger turnover rates of x_i did not correspond to longer mean response times. The opposite relationship, in fact, existed. Those systems with large amounts of recycling also had large storage and hence mean roots near zero. Indeed, the presence of rate coefficients that range over several orders of magnitude is one important characteristic of ecosystems that differentiates them from randomly organized systems.

Table 3 shows that the eight hypothetical ecosystems, ordered from least to most resistant (largest to smallest ω_n), were stream, ocean, lake, tundra, salt marsh, tropical forest, grassland, and temperate forest. The four terrestrial ecosystem models were, on the whole, much more resistant than the four aquatic models. Analyses did not reveal such a clear separation of ecosystems with high and low resilience, nor did the eight systems differ as much with respect to the resilience aspect of relative stability as they did in relation to resistance. From least to most resilient (largest to smallest ζ), the ecosystems were stream, lake, tropical forest, temperate forest, salt marsh, tundra, grassland, and ocean. This factor is tied to system characteristics (such as recycling) which do not differ strictly between aquatic and terrestrial ecosystems. Although several of the aquatic models were more resilient than most terrestrial ones, the lake model showed one of the smallest resilience values, probably related to slow turnover of the large abiotic storage pool. These results should be interpreted cautiously, in light of the data used in this analysis. Certainly the order-of-magnitude differences in the natural frequencies would seem to reflect real differences in the idealized ecosystems. The differences in damping ratios are apparently much smaller. However, these differences actually reflect large differences in the time dynamics of the ecosystem types because ζ appears as an exponent in the time-domain solutions (Eqs. 9 and 11).

These results agree well with previous analyses. Pomeroy (1970) related ecosystem stability to the presence or absence of abiotic reserves, system turnover rate, and predictability of the physical environment. Specifically, he noted that ecosystems with low abiotic storage and rapid recycling (tropical forests and coral reefs) are slow to recover following disturbance. Consistent with this observation, Table 3 shows the tropical forest to have one of the lowest resilience values. Also, the relative rankings of ecosystems in terms of stability given by Pomeroy correspond closely to rankings depicted in Table 4. Jordan et al. (1971) also showed an inverse relationship between recovery time following displacement and the amount of nutrient recycling relative to input. Comparisons between tropical and temperate forests in this study also agree with the analyses of Child and Shugart (1972) and Waide et al. (1974).

155

Inverse Relationships Between Resistance and Resilience

Taken together our results indicate an inverse relationship between resistance and resilience. Those factors which tend to increase resistance decrease resilience, and those factors which increase resilience decrease resistance. In addition, those systems which are highly resistant have low resilience, and vice versa. Thus ecosystem evolution would seem to involve a compromise or balance between resistance and resilience. In some situations, selection has favored ecosystems with large storage and a large amount of recycling, factors that contribute to ecosystem persistence by increasing resistance to displacement. Other ecosystems in other environments have low storage and rapid recycling and persist by responding rapidly following disturbance. The relationship is not an exact inverse, however. Results show, for example, the tropical forest to be both less resistant and less resilient than either the temperate forest or grassland. Also, the grassland model is next to the most stable in terms of both resistance and resilience, and the stream is least stable in both regards. Still, the notion of a functional balance between ecosystem properties favoring resistance or resilience is substantiated.

Environmental conditions that favor ecosystem resistance or resilience must be considered. In general, those environments in which resources are scarce or which place severe physicochemical limitations on resource mobilization will not favor the accumulation of large biotic stores of nutrients. Systems that recycle nutrients rapidly, and hence are highly resilient, should be favored in such environments. However, kinetic limitations on resource assimilation could be so severe as to produce systems that are neither resistant nor resilient, as streams seem to be. On the other hand, environments in which resources are available and which place less severe limitations on resource mobilization should favor the development of ecosystems that accumulate large nutrient reserves that turn over slowly and hence are relatively more resistant. Such considerations in part explain the separation between aquatic and terrestrial ecosystems in terms of resistance. With the exception of coral reefs, aquatic systems are generally limited in their ability to retain and recycle essential resources (Pomeroy, 1970; Riley, 1972). Such systems are typically more resilient, and less resistant, than terrestrial systems.

Also, as emphasized by Holling (1973), the balance between resistance and resilience is strongly influenced by the types of environmental fluctuations commonly encountered by an ecosystem. For example, results suggest that the hypothetical ocean is the least resistant ecosystem next to the stream. It is not reasonable to expect selection for maximum resistance of such an ecosystem since the environment typically encountered by oceanic ecosystems is buffered (by the surrounding water mass) compared to that impinging upon a temperate forest, the most resistant ecosystem considered. Similar buffering is attained in terrestrial ecosystems through large biotic storage.

As a corollary to these two last points, the kinds of environmental fluctuations an ecosystem "sees," and hence to which it responds, depend upon the degree of resistance or resilience it exhibits. A system will filter out or attenuate inputs with a frequency greater than its natural frequency but will pass and hence react to inputs with a lower frequency. Thus analyses indicate that terrestrial ecosystems are, on the average, currently responding to lower frequency environmental signals than are aquatic ecosystems. From the opposite perspective, we could perhaps argue that higher frequency inputs may be more damaging to terrestrial ecosystems and that selection has thus favored large, slowly recycling biotic structures that attenuate such persistent, potentially destabilizing inputs. Thus the degree of resistance or resilience a given ecosystem exhibits is determined by the types and frequencies of environmental fluctuations commonly encountered by the system, as well as by the environmental limitations on resource mobilization which the system experiences.

Contribution of Component Turnover Rates to Stability

It was suggested above that one of the factors which characterizes ecosystems is the presence of a large range in values of transfer rate coefficients and turnover rates, typically over several orders of magnitude. Each component turnover rate contributes to the resultant balance between resistance and resilience for a given ecosystem.

The concept of r and K selection define alternative evolutionary strategies at the population level (Pianka, 1970, 1972). These ideas may be reformulated in an ecosystem context by considering r selected species to be ones that have rapid turnover and low storage, thereby contributing to ecosystem resilience, whereas K specialists exhibit slow turnover and high storage, and thus contribute to resistance. Hence the degree of resistance or resilience observed in a given ecosystem results from the relative proportions of K and r selected components, respectively. This treatment does not seek to destroy the original meaning of these ideas but rather to suggest their implications for behavior at the ecosystem level.

During succession, ecosystems progress from stages that are relatively more resilient to ones that are relatively more resistant. Although differing degrees of environmental limitation and fluctuation will produce different balances between resilience and resistance, all developmental processes involve some amount of biomass accretion and nutrient storage. However, even at steady state a large variation in turnover rates of component populations is still present. It is the presence of such a variety of adaptations of component populations in steady-state ecosystems which ensures their ability to respond following disturbance and hence which confers the property of resilience on ecosystems. For example, pin cherry is an early successional woody plant common in northeastern deciduous forests, which ensures their rapid return to steady-state function following major perturbation (Marks and Bormann, 1972; Marks, 1974). Black locust seems to play a similar role in forest ecosystems in the southern Appalachians. Yet neither species is anything more than a minor component of steady-state ecosystems in either locality. Clearly, their persistence within these ecosystems represents a system-level adaptation for resilience which is not explained by considering dominant steady-state components alone. Similar examples could be cited for other ecosystem types.

The role of component turnover rates in regulating ecosystem stability is also emphasized by a consideration of the contribution of primary consumers to ecosystem stability. Primary biophages are generally viewed as being able to regulate their rate of resource supply and hence the ability of a specific ecosystem to accumulate biomass and store nutrients (Odum, 1962; Wiegert and Owen, 1971). Where environments favor ecosystem resistance, selection would thus seem to lead to mechanisms that suppress primary consumption, allelochemically, structurally, and via predators and parasites. However, in situations where ecosystem resilience is favored, mechanisms for reducing primary consumption would not necessarily be advantageous. Indeed, in such systems herbivory would seem to be a major mechanism of nutrient regeneration and recycling (Johannes, 1968; Pomeroy, 1970). Comparison of resilience values for the eight hypothetical ecosystems investigated with estimates of the amount of primary production passing through primary biophages (Wiegert and Evans, 1967; Wiegert and Owen, 1971; Golley, 1972) reveals a direct relationship between these two parameters, with those ecosystem types in which primary consumption is higher typically being more resilient. Such a relationship between herbivory and nutrient regeneration requires further experimental verification, especially in terrestrial ecosystems.

SUMMARY

The theoretical perspective embodied in this paper represents an attempt to account for alternatives for persistence at the ecosystem level and at the same time to relate ecosystem response to specific observable and measurable attributes of ecosystems. The argument that ecosystems are asymptotically stable focuses attention on the critical area of relative stability. It clearly identifies two aspects of ecosystem relative stability, resistance and resilience. Resistance is related to the formation and maintenance of persistent ecosystem structure. Resilience results from the tendencies inherent in ecosystems for the erosion of such structures. Thus this perspective offers to integrate various areas of ecological theory into a unified picture of ecosystem structure and function. Further research should help to establish the validity of these ideas. However, at present, they seem to represent a rigorous, operational approach to ecosystem theory which is testable by both observation and experimental analysis.

ACKNOWLEDGMENTS

Research was supported in part by the Coweeta site of the Eastern Deciduous Forest Biome, U. S. International Biological Program, funded by the National Science Foundation under Interagency Agreement AG-199, 40-193-69 with the Energy Research and Development Administration—Oak Ridge National Laboratory. This is paper No. 20, University of Georgia *Contributions in Ecosystems Ecology*. We appreciate the helpful comments of R. V. O'Neill and R. G. Wiegert.

REFERENCES

Ashby, W. R., 1952, *Design for a Brain*, Chapman & Hall, Ltd., London.

Ba Illi, F., 1971, A Time Domain Approach, in *Aspects of Network and System Theory*, pp. 313-325, R. E. Kalman and N. DeClaris (Eds.), Holt, Rinehart, and Winston, Inc., New York.

Bellman, R. E., 1968, *Some Vistas of Modern Mathematics: Dynamic Programming, Invariant Imbedding, and the Mathematical Biosciences*, University of Kentucky Press, Lexington.

Blackburn, T. R., 1973, Information and the Ecology of Scholars, *Science*, **181**: 1141-1146.

Boling, R. H., Jr., R. C. Peterson, and K. W. Cummins, 1974, Ecosystem Modeling for Small Woodland Streams, in *Systems Analysis and Simulation in Ecology*, Vol. 3, pp. 183-204, B. C. Patten (Ed.), Academic Press, Inc., New York.

Bormann, F. H., and G. E. Likens, 1970, The Nutrient Cycles of an Ecosystem, *Sci. Amer.*, **220**: 92-101.

Brylinsky, M., 1972, Steady-State Sensitivity Analysis of Energy Flow in a Marine Ecosystem, in *Systems Analysis and Simulation in Ecology*, Vol. 2, pp. 81-101, B. C. Patten (Ed.), Academic Press, Inc., New York.

Child, G. I., and H. H. Shugart, Jr., 1972, Frequency Response Analysis of Magnesium Cycling in a Tropical Forest Ecosystem, in *Systems Analysis and Simulation in Ecology*, Vol. 2, pp. 103-135, B. C. Patten (Ed.), Academic Press, Inc., New York.

Collier, B. D., G. W. Cox, A. W. Johnson, and P. C. Miller, 1973, *Dynamic Ecology*, Prentice-Hall, Inc., Englewood Cliffs, N. J.

Cummins, K. W., 1971, *Predicting Variations in Energy Flow Through a Semicontrolled Lotic Ecosystem*, Michigan State University, Institute of Water Research, Technical Report 19, pp. 1—21.

Darnell, R. M., 1970, Evolution and the Ecosystem, *Amer. Zool.*, **10**: 9-16.

DiStefano, J. J., A. K. Stubberud, and I. J. Williams, 1967, *Feedback and Control Systems*, McGraw-Hill Book Company, New York.

Dunbar, M. J., 1972, The Ecosystem as a Unit of Natural Selection, *Trans. Conn. Acad. Arts Sci.*, **44**: 111-130.

Funderlic, R. F., and M. T. Heath, 1971, Linear Compartmental Analysis of Ecosystems, USAEC Report ORNL-IBP-71-4, Oak Ridge National Laboratory.

Gardner, M. R., and W. R. Ashby, 1970, Connectedness of Large Dynamic (Cybernetic) Systems: Critical Value of Stability, *Nature*, **228**: 784.

Golley, F. B., 1972, Energy Flux in Ecosystems, in *Ecosystem Structure and Function*, pp. 69-90, J. A. Wiens (Ed.), Oregon State University Press, Corvallis.

Hairston, N. G. F. E. Smith, and L. B. Slobodkin, 1960, Community Structure, Population Control, and Competition, *Amer. Natur.*, **94**: 421-425.

Hearon, J. Z., 1963, Theorems on Linear Systems, *Ann. N. Y. Acad. Sci.*, **108**: 36-68.

——, 1953, The Kinetics of Linear Systems with Special Reference to Periodic Reactions, *Bull. Math. Biophys.*, **15**: 121-141.

Hill, J., IV, 1973, Component Description and Analysis of Environmental Systems, M.S. Thesis, Utah State University, Logan.

Holling, C. S., 1973, Resilience and Stability of Ecological Systems, *Annu. Rev. Ecol. Syst.*, **4**: 1-24.

Hubbell, S. P., 1973a, Populations and Simple Food Webs as Energy Filters, I. One-Species Systems, *Amer. Natur.*, **107**: 94-121.

——, 1973b, *ibid*, II. Two-Species Systems, *Amer. Natur.*, **107**: 122-151.

Hutchinson, G. E., 1948a, Circular Causal Systems in Ecology, *Ann. N. Y. Acad. Sci.*, **50**: 221-246.

——, 1948b, On Living in the Biosphere, *Sci. Mon.*, **67**: 393-397.

Johannes, R. E., 1968, Nutrient Regeneration in Lakes and Oceans, in *Advances in Microbiology of the Sea*, pp. 203-213, M. R. Droop and E. J. F. Wood (Eds.), Academic Press, New York.

Jordan, F., J. R. Kline, and D. S. Sasscer, 1972, Relative Stability of Mineral Cycles in Forest Ecosystems, *Amer. Natur.*, **106**: 237-253.

Juday, C., 1940, The Annual Energy Budget of an Inland Lake, *Ecology*, **21**: 438-450.

Lewontin, R. C., 1970a, The Meaning of Stability, in Diversity and Stability in Ecological Systems, Symposia in Biology, No. **22**, USAEC Report BNL-50175, Brookhaven National Laboratory.

——, 1970b, The Units of Selection, *Annu. Rev. Ecol. Syst.*, **1**: 1-18.

Liapunov, M. A., 1892, *Problème Générale de la Stabilité du Mouvement*, Kharkov. Reprinted as *Annals of Mathematical Study*, No. 17, Princeton University Press, Princeton, N. J.

Likens, G. E., and F. H. Bormann, 1972, Nutrient Cycling in Ecosystems, in *Ecosystem Structure and Function*, pp. 25-67, J. A. Wiens (Ed.), Oregon State University Press, Corvallis.

Lindeman, R. L., 1942, The Trophic-Dynamic Aspect of Ecology, *Ecology*, **23**: 399-418.

——, 1941, Seasonal Food-Cycle Dynamics in a Senescent Lake, *Amer. Midland Natur.*, **26**: 636-673.

MacArthur, R. H., 1955, Fluctuations of Animal Populations, and a Measure of Community Stability, *Ecology*, **36**: 533-536.

Makridakis, S., and C. Faucheux, 1973, Stability Properties of General Systems, *Gen. Sys.*, **18**: 3-12.

——, and E. R. Weintraub, 1971a, On the Synthesis of General Systems, Part I, The Probability of Stability, *Gen. Sys.*, **16**: 43-50.

——, and E. R. Weintraub, 1971b, On the Synthesis of General Systems, Part II, Optimal System Size, *Gen. Sys.*, **16**: 51-54.

Marks, P. L., 1974, The Role of Pin Cherry (*Prunus pennsylvanica* L.) in the Maintenance of Stability in Northern Hardwood Ecosystem, *Ecol. Monogr.*, **44**: 73-88.

——, and F. H. Bormann, 1972, Revegetation Following Forest Cutting: Mechanisms for Return to Steady-State Nutrient Cycling, *Science*, **176**: 914-915.

May, R. M., 1973, *Stability and Complexity in Model Ecosystems*, Princeton University Press, Princeton, N. J.

——, 1972, Will a Large Complex System be Stable?, *Nature*, **238**: 413-414.

McGinnis, J. T., F. B. Golley, R. G. Clements, G. I. Child, and M. J. Duever, 1969, Elemental and Hydrologic Budgets of the Panamanian Tropical Moist Forest, *BioScience*, **19**: 697-702.

Milsum, J. H., 1968, Mathematical Introduction to General System Dynamics, in *Positive Feedback*, pp. 23-65, J. H. Milsum (Ed.), Pergamon Press, New York.

Morowitz, H. J., 1966, Physical Background of Cycles in Biological Systems, *J. Theor. Biol.*, **13**: 60-62.

Odum, E. P., 1971, *Fundamentals of Ecology*, W. B. Saunders Co., Philadelphia.

——, 1969, The Strategy of Ecosystem Development, *Science*, **164**: 262-270.

——, 1962, Relationships Between Structure and Function in the Ecosystem, *Jap. J. Ecol.*, **12**: 108-118.

Odum, H. T., 1971, *Environment, Power and Society*, Wiley-Interscience, Inc., New York.

——, 1967, Work Circuits and System Stress, in *Primary Productivity and Mineral Cycling in Natural Ecosystems*, pp. 81-138, H. E. Young (Ed.), University of Maine Press, Orono, Maine.

Olson, J. S., 1963, Analog Computer Models for Movement of Nuclides Through Ecosystems, in *Radioecology*, pp. 121-125, V. Shultz and A. W. Klements, Jr. (Eds.), Van Nostrand Reinhold Company, Cincinnati, and American Institute of Biological Sciences, Washington.

Patten, B. C., 1974, The Zero State and Ecosystem Stability, in *Proceedings of the First International Congress of Ecology*, Supplement, The Hague, Sept. 8—14, 1974, Centre for Agricultural Publishing and Documentation, Wageningen.

——, 1972, A Simulation of the Shortgrass Prairie Ecosystem, *Simulation*, **19**: 177-186.

——, D. A. Egloff, T. H. Richardson, and 38 Additional Coauthors, 1974, A Total Ecosystem Model for a Cove in Lake Texoma, Texas—Oklahoma, in *Systems Analysis and Simulation in Ecology*, Vol. 3, pp. 205-421, B. C. Patten (Ed.), Academic Press, Inc., New York.

——, and M. Witkamp, 1967, Systems Analysis of ^{134}Cs Kinetics in Terrestrial Microcosms, *Ecology*, **48**: 813-824.

Pianka, E. R., 1972, r and K Selection or b and d Selection? *Amer. Natur.*, **106**: 581-588.

——, 1970, On r and K Selection, *Amer. Natur.*, **102**: 592-597.

Pomeroy, L. R., 1970, The Strategy of Mineral Cycling, *Annu. Rev. Ecol. Syst.*, **1**: 171-190.

——, R. E. Johannes, E. P. Odum, and B. Roffman, 1969, The Phosphorus and Zinc Cycles and Productivity of a Salt Marsh, in Symposium on Radioecology, pp. 412-419, Proceedings of the Second National Symposium, May 15—17, 1967, D. J. Nelson and F. C. Evans (Eds.), Ann Arbor, Mich., USAEC Report CONF-670503.

Reuss, J. O., 1971, Soils of the Grassland Biome Sites, in *Preliminary Analysis of Structure and Function in Grasslands*, pp. 35-39, N. R. French (Ed.), Colorado State University, Fort Collins.

Riley, G. A., 1972, Patterns of Production in Marine Ecosystems, in *Ecosystem Structure and Function*, pp. 91-112, J. A. Wiens (Ed.), Oregon State University Press, Corvallis.

Rodin, L. E., and M. I. Bazilevich, 1967, *Production and Mineral Cycling in Terrestrial Vegetation*, Oliver & Boyd, Edinburgh.

Schultz, A. M., 1969, A Study of an Ecosystem: The Arctic Tundra, in *The Ecosystem Concept in Natural Resource Management*, pp. 77-93, G. M. VanDyne (Ed.), Academic Press, Inc., New York.

Shinners, S. M., 1972, *Modern Control System Theory and Application*, Addison-Wesley Publishing Company, Inc., Reading, Mass.

Sims, P. L., and J. S. Singh, 1971, Herbage Dynamics and Net Primary Production in Certain Ungrazed and Grazed Grasslands in North America, in *Preliminary Analysis of Structure and Function in Grasslands*, pp. 59-124, N. R. French (Ed.), Colorado State University, Fort Collins.

Slobodkin, L. B., 1964, The Strategy of Evolution, *Amer. Sci.*, **52**: 342-357.

Smith, F. E., 1972, Spatial Heterogeneity, Stability, and Diversity in Ecosystems, *Trans. Conn. Acad. Arts Sci.*, **44**: 309-335.

Teal, J. M., 1962, Energy Flow in the Salt Marsh Ecosystem of Georgia, *Ecology*, **43**: 614-624.

Waide, J. B., J. E. Krebs, S. P. Clarkson, and E. M. Setzler, 1974, A Linear Systems Analysis of the Calcium Cycle in a Forested Watershed Ecosystem, in *Progress in Theoretical Biology*, Vol. 3, pp. 261-345, R. Rosen and F. M. Snell (Eds.), Academic Press, Inc., New York.

——, and J. R. Webster, 1975, Engineering Systems Analysis: Applicability to Ecosystems, in *Systems Analysis and Simulation in Ecology*, Vol. 4, B. C. Patten (Ed.), Academic Press, Inc., New York (in press).

Webster, J. R., and J. B. Waide, 1975, Complexity, Resource Limitation, and the Stability of Ecosystems, in preparation.

Westley, G. W., and J. A. Watts, 1970, The Computing Technology Center Numerical Analysis Library, Technical Report No. CTC-39, Oak Ridge National Laboratory.

Whittaker, R. H., and G. M. Woodwell, 1972, Evolution of Natural Communities, in *Ecosystem Structure and Function*, pp. 137-159, J. A. Wiens (Ed.), Oregon State University Press, Corvallis.

Wiegert, R. G., R. R. Christian, J. L. Gallagher, J. R. Hall, R. D. H. Jones, and R. L. Wetzel, 1974, A Preliminary Ecosystem Model of Coastal Georgia *Spartina* Marsh, in *Recent Advances in Estuarine Research*, J. Costlow (Ed.) (in press).

——, and F. C. Evans, 1967, Investigations of Secondary Productivity in Grasslands, in *Secondary Productivity of Terrestrial Ecosystems*, pp. 499-518, K. Petrusewicz (Ed.), Polish Academy of Science, Warsaw.

——, and F. C. Evans, 1964, Primary Production and the Disappearance of Dead Vegetation on an Old Field in S. E. Michigan, *Ecology*, **45**: 49-63.

——, and D. F. Owen, 1971, Trophic Structure, Available Resources, and Population Density in Terrestrial vs. Aquatic Ecosystems, *J. Theor. Biol.*, **30**: 69-81.

Williams, R. B., 1971, Computer Simulation of Energy Flow in Cedar Bog Lake, Minnesota, Based on the Classical Studies of Lindeman, in *Systems Analysis and Simulation in Ecology*, Vol. 1, pp. 543-582, B. C. Patten (Ed.), Academic Press, Inc., New York.

Woodall, W. R., Jr., 1972, Nutrient Pathways in Small Mountain Streams, Ph.D. Dissertation, University of Georgia, Athens.

Zadeh, L. A., and C. A. Desoer, 1963, *Linear System Theory: The State Space Approach*, McGraw-Hill Book Company, New York.

Part II

MAJOR DEVELOPMENTS IN
SYSTEMS ECOLOGY

Major expansions of the field of systems ecology have occurred during the present decade. By 1970, the desirability of formulating ecological models was firmly established and a number of research groups were beginning to make contributions.

The divergent and concurrent lines of development make it difficult to organize the material or choose specific problem areas. One series of important papers (Clymer and Bledsoe 1970; Patten 1971; Bledsoe and Jameson 1970) presented detailed descriptions of how to construct an ecological model. Other papers (O'Neill 1970a; Gustafson and Innis 1973; Bledsoe 1970; Bledsoe and Olson 1970; Sollins 1971; Brennan et al. 1970) discussed methods for implementing models on the digital computer. These papers were influential in determining the techniques that are still utilized in ecological modeling. However, since the papers are addressed to professional modelers and are of little general interest, they have been omitted from this volume.

Other major developments considered philosophical problems associated with model development. A number of papers concerned the relative merits of linear versus nonlinear models (Patten 1975; Bledsoe 1975), stochastic versus deterministic models (O'Neill 1970b; Tiwari and Hobbie 1976) and continuous-time versus discrete-time models (Innis 1972). An excellent review of these and other technical questions has been provided by Wiegert (1975).

Major modeling developments have also occurred in fields closely related to ecology, such as water resource management (Thomann et al. 1970; Chen and Orlob 1972; Crawford and Linsley 1966; Huff et al. 1973). Agronomic modeling also developed rapidly during this period (Lemon et al. 1971; DeMichele and Sharp 1971; Hesketh et al. 1972). While these modeling efforts may not be properly designated as developments in systems ecology itself, the advances they represent have influenced ecological models.

Within the field of ecology, advances were made in the modeling of populations, communities, and specific physiological processes. Based on the early research of Monsi and Saeki (1953), models have been developed for primary production processes ranging from single leaves (Goldstein and Mankin 1972; Sinclair 1972) to entire plant communities (Miller 1972). In like manner, models have considered trophic interactions (DeAngelis et al. 1975), decomposition (Bunnell 1972), mineral cycling (Shugart et al. 1976), and evolution (Conrad and Pattee 1970).

From the multiple lines of development that might be considered, we have chosen to emphasize three. First, the diversity/stability hypothesis was considered in an important series of papers that can be considered as a unit. Second, several papers have attempted to develop theory by viewing the ecosystem in an holistic manner. The third line of development, and by far the most important from the viewpoint of total effort, was the development of large, complex models of total ecosystems. Each of these three lines of development are represented in the sections that follow.

REFERENCES

Bledsoe, L. J. 1970. *ODE, Numerical Analysis for Ordinary Differential Equations*, Tech. Rep. 46, Grassland Biome. Natural Resource Ecology Laboratory, Colorado State Univ., Fort Collins, 42 pp.

Bledsoe, L. J. 1975. Linear and Nonlinear Approaches for Ecosystem Dynamic Modeling. In Patten, B. C. (ed.), *Systems Analysis and Simulation in Ecology*, vol. 4. Academic Press, New York (in press).

Bledsoe, L. J., and D. A. Jameson. 1970. Model structure for a grassland ecosystem, pp. 410–35. In Dix, R. L., and R. G. Beidleman (eds.), *The Grassland Ecosystem—A Preliminary Synthesis*, Range Sci. Ser. No. 2. Range Sci. Dep., Colorado State Univ., Fort Collins, 437 pp.

Bledsoe, L. J., and J. S. Olson. 1970. *COMSYS1—A Stepwise Compartmental Simulation Program*, ORNL-2413. Oak Ridge National Laboratory, Oak Ridge, Tenn.

Brennan, R. D., C. T. deWit, W. A. Williams, and E. V. Quattrin. 1970. The utility of a digital simulation language for ecological modeling. *Oecologia* 4:113–32.

Bunnell, F. L. 1972. Modeling decomposition and nutrient flux, pp. 116–20. *Proc. 1972 Tundra Biome Symp.* Lake Wilderness Center, Univ. of Washington, Seattle.

Chen, C. W., and G. T. Orlob. 1972. *Ecologic Simulation for Aquatic Environments*, Final Report for the Office of Water Resources Research. Washington, D.C., 156 pp.

Clymer, A. B., and L. J. Bledsoe. 1970. A guide to the mathematical modeling of an ecosystem, pp. 175–99. In Wright, R. G., and G. M. Van Dyne (eds.), *Simulation and Analysis of Dynamics of a Semi-desert Grassland*, Range Sci. Ser. No. 6. Range Sci. Dep., Colorado State Univ., Fort Collins, 297 pp.

Conrad, M., and H. H. Pattee. 1970. Evolution experiments with an artificial ecosystem. *J. Theor. Biol.* **28**:393–409.

Crawford, N. H., and R. L. Linsley. 1966. *Digital Simulation in Hydrology—Stanford Watershed Model IV*, Tech. Rep. 39. Civil Eng., Stanford Univ., Stanford, CA.

DeAngelis, D. L., R. A. Goldstein, and R. V. O'Neill. 1975. A model for trophic interaction. *Ecology* **56**:881–92.

DeMichele, D. W., and P. Sharpe. 1971. *A Morphological and Physiological Model of the Stomata*, EDFB Memo Rep. 71-27. Oak Ridge National Laboratory, Oak Ridge, Tenn., 15 pp.

Goldstein, R. A., and J. B. Mankin. 1972. *Space-Time Considerations in Modeling the Development of Vegetation*, EDFB Memo Rep. 71-138, Oak Ridge National Laboratory, Oak Ridge, Tenn., 11 pp.

Gustafson, J. D., and G. S. Innis. 1973. *SIMCOMP (Verson 2.1) Users Manual and Maintenance Document*, Tech. Rep. 217, Grassland Biome. Natural Resource Ecology Laboratory, Colorado State Univ., Fort Collins, 96 pp.

Hesketh, J. D., H. Hellmers, and B. R. Strain. 1972. *Carbon Dioxide Fertilization as a Tool in Developing Logic for Plant Growth Models*, EDFB Memo Rep. 72–2. Oak Ridge National Laboratory, Oak Ridge, Tenn., 16 pp.

Huff, D. D., J. F. Koonce, W. R. Ivarson, P. R. Weiler, E. H. Dettmann, and R. F. Harris. 1973. Simulation of urban runoff nutrient loading and biotic response of a shallow eutrophic lake, pp. 33–55. In Middlebrooks, E. J., D. H. Falkenborg, and T. E. Maloney (eds.), *Modeling the Eutrophication Process*. Utah Water Res. Lab., College of Engr., Utah State Univ., Logan, 228 pp.

Innis, G. S. 1972. *ELM: A Grassland Ecosystem Model*, Preprint No. 35, Grassland Biome. Natural Resource Ecology Laboratory, Colorado State Univ., Fort Collins.

Lemon, E. R., D. W. Stewart, and R. W. Shawcroft. 1971. The sun's work in a corn field. *Science* **174**:371–78.

Miller, P. C. 1972. Bioclimate, leaf temperature, and primary production in red mangrove canopies in south Florida. *Ecology* **53**:22–45.

Monsi, M. and T. Saeki. 1953. Über den Lichtfaktor in den Pflanzengesellschaften und seine Bedeutung für die Stoffproduktion. *Jap. J. Bot.* **14**:22–52.

O'Neill, R. V. 1970a. *An Introduction to the Numerical Solution of Differential Equations in Ecosystems Models*, ORNL–IBP–70–4. Oak Ridge National Laboratory, Oak Ridge, Tenn., 29 pp.

O'Neill, R. V. 1970b. A stochastic model of energy flow in predator compartments of an ecosystem. In Patil, G. P. (ed.), *Int. Symp. Stat. Ecol.* Penn. State Univ. Press, University Park.

Patten, B. C. 1971. A primer for ecological modeling and simulation with analog and digital computers, pp. 3–121. In Patten, B. C. (ed.), *Systems Analysis and Simulation in Ecology*, vol. 1. Academic Press, New York.

Patten, B. C. 1975. Ecosystem linearization: An evolutionary design problem. *Am. Nat.* **109**:529–39.

Shugart, H. H., D. E. Reichle, N. T. Edwards, and J. R. Kercher. 1976. A model of calcium cycling in an east Tennessee LIRIODENDRON forest: Model structure, parameters, and frequency response analysis. *Ecology* **57**: 99–109.

Sinclair, T. R. 1972. *A Leaf Photosynthesis Submodel for Use in General Growth Models*, EDFB Memo Rep. 72–14. Oak Ridge National Laboratory, Oak Ridge, Tenn., 14 pp.

Sollins, P. 1971. *CSS, a Computer Program for Modeling Ecological Systems*, ORNL–IBP–71–5. Oak Ridge National Laboratory, Oak Ridge, Tenn., 96 pp.

Thomann, R. V., D. J. O'Connor, and D. M. DiToro. 1970. Modeling of the nitrogen and algal cycles in estuaries. *Fifth Int. Water Pollut. Res. Conf.* San Francisco, Calif.

Tiwari, J. L., and J. E. Hobbie. 1976. Random differential equations as models of ecosystems: Monte Carlo simulation approach. *Math. Bios.* **28**: 25–44.

Wiegert, R. G. 1975. Simulation models of ecosystems. *Ann. Rev. Ecol. Syst.* **6**:311–38.

Editors' Comments
on Papers 12 Through 15

THE DIVERSITY/STABILITY PROBLEM

In 1958, R. Margalef wrote a book, entitled *Perspectives in Ecological Theory*, that provided some exciting conjectures about ecosystem properties. Margalef was already well known for his studies of communities in fresh-water and oceanic environments. The combination of his extensive data and his fresh insight into the meanings of these data generated considerable interest in the work of this Spanish ecologist. Margalef theorized that there should be a relationship between diversity (as measured by an information-theoretic index derived from the work of Weiner (1948) and ecosystem stability (the ability of an ecosystem to either resist change or to return to its normal operating state after some perturbation).

For a time it was assumed that the diversity-stability relationship was a natural law that applied axiomatically to ecosystems. A significant literature developed on the subject (Van Voris 1976) and it became an important subject for both experimental and theoretical studies. But a series of empirical studies (Watt 1964; Paine 1966; Zaret and Paine 1973; Hairston et al. 1968; Hurd et al. 1971; Hurd and Wolf 1974) failed to demonstrate the expected relationship.

In the series of papers reprinted below, it is hypothesized that the more interconnections among components of a system, the lower the probability that the system will be stable. In the context of the development of systems ecology, this series of papers provides an excellent example of how mathematical treatment of a theoretical problem can proceed toward a resolution.

REFERENCES

Hairston, N. G., J. D. Allan, R. K. Colwell, D. J. Futuyma, J. Howell, M. D. Lubin, J. Mathias, and J. H. Vandermeer. 1968. The relationship between species diversity and stability: An experimental approach with protozoa and bacteria. *Ecology* **49**:1091–101.

Hurd, L. E., M. V. Mellinger, L. L. Wolf, and S. J. McNaughton. 1971. Stability and diversity at three trophic levels in terrestrial successional ecosystems. *Science* **173**:1134–36.

Hurd, L. E., and L. L. Wolf. 1974. Stability in relation to nutrient enrichment in arthropod consumers of old field successional ecosystems. *Ecol. Monogr.* **44**:465–82.

Margalef, R. 1968. *Perspectives in Ecological Theory.* University of Chicago Press, Chicago, 111 pp.

Paine, R. T. 1966. Food web complexity and species diversity. *Am. Nat.* **100**: 65–75.

Van Voris, P. 1976. *Ecological Stability: An Ecological Perspective,* ORNL/TM-5517. Oak Ridge National Laboratory, Oak Ridge, Tenn., 36 pp.

Watt, K. E. F. 1964. Comments on fluctuations of animal populations and measures of community stability. *Can. Entomol.* **96**:1434–42.

Wiener, N. 1948. *Cybernetics.* John Wiley & Sons, New York.

Zaret, T. M., and R. T. Paine. 1973. Species introduction in a tropical lake. *Science* **182**:777–83.

12

Reprinted from *Nature* **228**:784 (1970)

CONNECTANCE OF LARGE DYNAMIC (CYBERNETIC) SYSTEMS: CRITICAL VALUES FOR STABILITY

Mark R. Gardner and W. Ross Ashby

MANY systems being studied today are dynamic, large and complex: traffic at an airport with 100 planes, slum areas with 10^4 persons or the human brain with 10^{10} neurones. In such systems, stability is of central importance, for instability usually appears as a self-generating catastrophe. Unfortunately, present theoretical knowledge of stability in large systems is meagre: the work described here was intended to add to it.

Most of these large systems. often biological or social, are grossly non-linear, which increases the difficulties associated with them. Here we consider linear systems merely as a first step towards a more general treatment.

We have attempted to answer: What is the chance that a large system will be stable? If a large system is assembled (connected) at random, or has grown haphazardly, should we expect it to be stable or unstable? And how does the expectation change as n, the number of variables, tends to infinity?

Monte Carlo-type evidence[1,2] had suggested that the probability of stability decreased rapidly as n was increased, in some cases perhaps as fast as 2^{-n}, an exponentially-fast vanishing of the chance that the system will be stable. This result, however, was for systems that were fully connected, where every variable had an immediate effect on every other variable. While this case is obviously important in theory, it is not the case in most large systems in real life: not every person in a slum has an immediate effect on every other person, and not every cell in the brain directly affects every other cell. The amount of connectedness ("connectance") is often far below 100 per cent. We have studied how such incomplete connectance affects the probability of a system's stability.

Let the linear system's state be represented by the vector $\mathbf{x}$ ($= <x_1, \ldots, x_n>$, where each x_i is a variable, a function of time), and its changes in time by the matrix equation

$$\mathbf{x} = A\mathbf{x}$$

To "join the variables at random" is to give the elements in A values taken from some specified distribution. "Non-connexion from x_i to x_j" corresponds to giving the element a_{ji} the value zero. Thus, if the specified distribution has a peak at zero, sampling from it will give the equivalent of a dynamic system with many non-connexions. The connectance, C, of the system can then be conveniently defined as the percentage of non-zero values in the distribution with 99 per cent zeros, and if $n = 1,000$, then each line of the equation would contain about ten non-zero coefficients, corresponding to a system in which each variable is directly affected by about ten other variables.

Because our work was essentially exploratory, we used the distribution in which the non-zero elements were distributed evenly between -1.0 and $+1.0$. The elements in the main diagonal, corresponding to the intrinsic stabilities of the parts, were all negative, distributed evenly between -1.0 and -0.1. Thus each sampled value of A corresponded to a system of individually stable parts, connected so that each part was affected directly by about 7 per cent of the other parts.

On a digital computer, a value for n was given and a value for C. Random numbers appropriately distributed were then sampled to provide a matrix A. Hurwitz's criterion was applied to test whether the real parts of

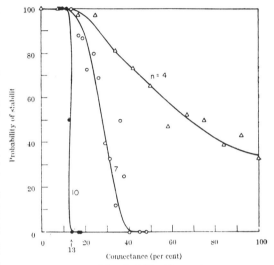

Fig. 1. Variation of stability with connectance.

A's latent roots were all negative (the stable case) and the result recorded. Further samples, giving further As, allowed the probability of stability (P) to be estimated. The probability was then re-estimated for another value of C. and so on, until the variation of P with C became clear.

The results showed the feature that we wish to report here. As the system was made larger, a new simplicity appeared. Fig. 1 shows a selection of the results, enough to illustrate the principal fact.

When $n = 4$, the probability that the system would be stable depended on C in a somewhat complex curve (which could perhaps be predicted exactly). But as n increases, the curve changes shape rapidly towards a step-function, so that even when n is only 10, the shape might be so regarded, at least for some practical purposes. Thus, even at $n = 10$, questions of stability can be answered simply by asking whether the connectance is above or below 13 per cent: 2 per cent deviation either way being sufficient to convert the answer from "almost certainly stable" to "almost certainly unstable".

The matter is being investigated further, but it may be of general interest to notice that this work suggests that all large complex dynamic systems may be expected to show the property of being stable up to a critical level of connectance, and then, as the connectance increases, to go suddenly unstable.

This work was supported in part by the US Office of Scientific Research.

[1] Ashby, W. R., *Design for a Brain* (Chapman and Hall, London, 1952).
[2] Gardner, M. R., *Critical Degenerateness in Linear Systems*, Tech. Rep. No. 5.8 (Biological Computer Laboratory, University of Illinois, Urbana, Illinois 61801, 1968).

PARADOX OF ENRICHMENT: DESTABILIZATION OF EXPLOITATION ECOSYSTEMS IN ECOLOGICAL TIME

Michael L. Rosenzweig

State University of New York, Albany

Schemes for increasing primary productivity by enriching an ecosystem's energy or nutrient flow are much in evidence today and are probably a reflection of the increasing demands of the world's population. Such schemes may end in catastrophe.

In 1963, Huffaker, Shea, and Herman (*1*) reported destabilization of a stable exploitation ecosystem which resulted in the extinction of both the exploiter (an acarophagous mite) and its victim (an herbivorous mite). They produced this result by trebling the herbivore's food density. By using a variety of realistic models, I predict that instability should often be the result of nutritional enrichment in two-species interactions.

Rosenzweig and MacArthur (*2*) showed that exploitation (or predator-prey) ecosystems do not necessarily exhibit any oscillations. Furthermore, even if there are oscillations, they do not last under ordinary circumstances. If the exploiter is quite proficient at reproducing in the presence of few of its victims, then the ecosystem does not persist. If, however, the victims are relatively proficient at escape or their exploiters have a relatively poor reproductive efficiency or digestive efficiency, then the system will persist in ecological time (*3*).

The dividing line between persistent and explosive systems is definable from a general graph of exploitation (*2*). The victim's density V is plotted against P, the exploiter's density. The collection of graph points at which $dV/dt = 0$ is called the victim's isocline. The collection of points at which $dP/dt = 0$ is called the exploiter's isocline. Any point of intersection between the two isoclines is an ecosystem equilibrium, but not all such equilibria will result in a steady state. The usual form of the prey isocline is a hump (*4*). If the equilibrium is at a point on the left side of the hump, the predator is too proficient and the system will ordinarily not persist.

If equilibrium is at a point on the right-hand (downslope) side of the hump, the system will persist. Thus, the hump's peak is over a critical value of V, V^*. If the equilibrium value of V is larger than V^*, the system is safe. If not, it is in danger of extinction.

If the exploiters do not actually interfere with each other directly—if they never battle over the same individual victim or engage in cannibalism or territorial defense—then the P isocline is a simple vertical line ($V = J$). The position of this line is fully determined by the phenotypes of the exploiter and its victim. It does not change with nutrient flow or energy supply.

To discover the effect of enriching a system, one needs to find how V^* changes as enrichment proceeds. If enrichment increases V^*, then it is jeopardizing the system, because eventually V^* will be made greater than J.

Briefly, the method is this. Set $dV/dt = 0$ and solve for P. This is the algebraic equation for the V isocline. Take $\partial P/\partial V$. The value of V that satisfies $\partial P/\partial V = 0$ is V^*. If K represents the

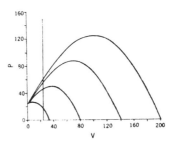

Fig. 1. Isoclines at four levels of productivity ($K = 34$, 80, 140, and 200) when model 4 is used. Symbol V is victim's density; P is exploiter's. The curved lines are V isoclines, which peak over higher values of V at higher K. The vertical line, $V = 25$, is the P isocline. From the slopes of the V isoclines at the points where they intersect the P isocline, one expects only the $K = 34$ system to have a steady state.

standing crop of V where $P = 0$, then K must be directly proportional to the flow rates of limiting nutrients. Thus, enrichment implies K increase. The final step, then, is to obtain $\partial V^*/\partial K$. Since this is always positive, enrichment leads toward system instability.

Each analytical model used is the difference between inherent rate of increase of V and the number of V that die (or are not born) owing to the activities of P.

Assume that each V must receive a quota of nutrients at a rate Q in order just to replace itself. Nutrients flow to the lone individual V at a rate R. Thus, this individual reproduces at a net rate $r (R - Q)$. As V increases, however, each individual V is subjected to intraspecific competition. One may assume that an individual V's effective "feeding" rate diminishes with increasing V^a, where a is the victim's competition constant, $1 \geqslant a > 0$ (*5*). Thus the per capita reproductive rate is $r (RV^{-a} - Q)$ and the inherent rate is $rV (RV^{-a} - Q)$. From the parentheses one obtains $K = (R/Q)^{1/a}$. Hence, to increase K, R must be increased (a and Q are constant), and $\partial V^*/\partial K$ has the same sign as $\partial V^*/\partial R$ (*6*).

In addition to the above model, I have used the traditional Pearl-Verhulst logistic $rV (1 - V/K)$ and the Gompertz $rV (\ln K - \ln V)$.

Kill rate models. Lotka and Volterra's approach was to treat the two populations like molecules kVP [see (*7*) and references therein]. This has been shown to be inadequate. Gause (*7*) obtained a reasonable fit to a kill rate curve by taking the square root of V. We can generalize this procedure by taking V to the gth power $0 < g \leqslant 1$. Thus, a kill rate model is kPV^g.

Another model is based on observations (*5*, *8*) that one lone exploiter will attack $k(1 - e^{-cV})$ victims in a fixed amount of time. Since the exploiters compete only by reducing each other's food supply, P exploiters will kill $kP(1 - e^{-cV})$.

Thus models for dV/dt include

$$dV/dt = rV(RV^{-a} - Q) - kP(1 - e^{-cV}) \tag{1a}$$

$$dV/dt = rV(1 - V/K) - kPV^g \tag{2a}$$

$$dV/dt = rV(RV^{-a} - Q) - kPV^g \tag{3a}$$

$$dV/dt = rV(1 - V/K) - kP(1 - e^{-cV}) \tag{4a}$$

$$dV/dt = rV(\ln K - \ln V) - kPV^g \tag{5a}$$

$$dV/dt = rV(\ln K - \ln V) - kP(1 - e^{-cV}) \tag{6a}$$

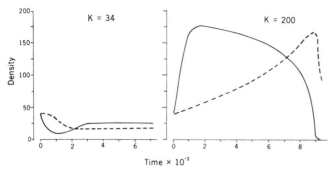

Fig. 2. Iteration of the model 4 exploitation at $K = 34$ and $K = 200$. Solid curve is V; dashed curved is P. Enrichment causes the simulated extinction of both species. The exploiter equation used was $dP/dt = AkP \ (e^{-cJ} - e^{-cV})$. Time units are in calculator cycles (10).

In view of the lack of convincing tests of any of the models as a general case for all systems, I have analyzed all six.

The first step in each analysis is omitted here: solution of each equation for P when $dV/dt = 0$. For example, Eq. 4a becomes

$$P = \frac{rV(1 - V/K)}{k(1 - e^{-cV})}$$

This set of equations is the set of V isoclines.

Next we obtain $\partial P/\partial V$ and determine the conditions under which this will be zero. These are the V^* equations:

$$R = \frac{Q(V^*)^a \ (e^{cV^*} - 1 - cV^*)}{(e^{cV^*} - 1) \ (1 - a) - cV} \quad (1b)$$

$$K = \frac{(2 - g)}{(1 - g)} V^* \quad (2b)$$

$$R = \frac{Q(V^*)^a \ (1 - g)}{(1 - a - g)} \quad (3b)$$

$$K = V^* \frac{(2e^{cV^*} - cV^* - 2)}{(e^{cV^*} - cV^* - 1)} \quad (4b)$$

$$\ln K = \ln V^* + 1/(1 - g) \quad (5b)$$

$$\ln K = \ln V^* + 1 + \frac{cV^*}{e^{cV^*} - 1 - cV^*} \quad (6b)$$

The final step requires a small explanaton. We need the sign of $\partial V^*/\partial K$ or $\partial V^*/\partial R$. Often the equation systems are easily solved for K or $\ln K$, but not V^*. However, $\partial V^*/\partial K$ is positive if and only if $\partial K/\partial V^*$ is. And $\partial K/\partial V^*$ is positive if and only if $\partial \ln K/\partial V^*$ is. Hence, we can readily proceed with these latter two partial derivatives.

Three are positive for any set of values of the constants:

$$\frac{\partial K}{\partial V^*} = (2 - g)/(1 - g) \quad (2c)$$

$$\frac{\partial K}{\partial V^*} = \frac{2(e^{cV} - 1)(e^{cV} - 1 - cV - c^2V^2/2)}{(e^{cV} - 1 - cV)^2} \quad (4c)$$

$$\frac{\partial \ln K}{\partial V^*} = \frac{1}{V^*} \quad (5c)$$

Equation 4c is always positive because the MacLaurin series for e^{cV} is $1 + cV + c^2V^2/2 + c^3V^3/6 + \cdots$ (see Figs. 1 and 2).

The other three cases are not quite so readily handled. Equation 3b does not always have a positive solution for V^*. In fact V^* is negative if and only if $(1 - a - g)$ is also negative. The V isocline of Eq. 3a is humpless for such values of $(a + g)$. Values this great imply intense intraspecific producer competition and also a relatively low tendency for exploiters to become hungry or satiated and to modify their behavior accordingly. In such a system there is no tendency for extinction regardless of productivity.

However, if $(a + g)$ is less than 1, there is a positive V^* and

$$\frac{\partial R}{\partial V^*} = \frac{aQ(1 - g) \ (V^*)^{a-1}}{(1 - a - g)} \quad (3c)$$

Clearly Eq. 3c is always positive if V^* is biologically real. Hence, in model 3, if there is any threat of system extinction, it is increased by enrichment.

Models 1 and 6 are similar and most complex. It turns out that Eq. 6b is satisfied by two values of V. One is V^*. Another is a very small value of V

that occurs over a trough in the V isocline. Thus, there is ambiguity in the following:

$$\frac{\partial \ln K}{\partial V^*} = \frac{(e^{cV^*} - 1) \ (e^{cV^*} - 1 - cV^* - c^2V^{*2})}{(e^{cV^*} - cV^* - 1)^2} \quad (6c)$$

This equation, set to zero, holds for both V^* and the V under the trough. The unstable equilibrium values of V are those between V (trough) and V^*. Model 6c is positive for V^* and negative for V (trough) (9). Hence, as enrichment proceeds, the range of unstable V is increasing at both ends. Therefore again, enrichment unambiguously tends to weaken the steady state. Model 1 has the same characteristics (9).

Until we are confident that the conclusions based on these systems do not apply to natural ecosystems, we must remain aware of the danger in setting enrichment as a human goal.

References and Notes

1. C. B. Huffaker, K. P. Shea, S. G. Herman, *Hilgardia* **34** (9), 305 (1963).
2. M. L. Rosenzweig and R. H. MacArthur, *Amer. Natur.* **97**, 209 (1963).
3. Throughout, phenotypes of both species are held constant, ignoring evolution. Current hopes for a high magnitude increase in productivity would probably deny natural selection the time to act.
4. E. J. Maly, *Ecology* **50**, 59 (1969); M. L. Rosenzweig, *Amer. Natur.* **103**, 81 (1969). The isocline equations of this report are further evidence for the hump, though one is "humpless" in some cases and two others have, in addition, an (artifactual?) extreme left-hand rise to $+\infty$.
5. K. E. F. Watt, *Can. Entomol.* **91**, 129 (1959).
6. This model is sigmoidal. Its point of inflection is at $V = (1 - a)^{1/a} K$. Thus it depends on the intensity of intraspecific competion. By L'Hospital's rule, as a approaches 0, $(1 - a)^{1/a}$ or exp in $(1 - a)/a$ approaches $1/e$. Thus, if intraspecific competition is slight, this equation closely resembles the Gompertz curve. In fact, I doubt if any $a < 0.5$ is readily distinguishable from the hump. However, the Pearl-Verhulst yields an inflection at $K/2$, which is always greater than that predicted from this model.
7. G. F. Gause, *The Struggle for Existence* (Williams & Wilkins, Baltimore, 1934).
8. V. S. Ivlev, *Experimental Ecology of the Feeding of Fishes* (Yale Univ. Press, New Haven, 1961).
9. A proof is on file and available to those who desire it.
10. Iteration was performed with the use of difference equations on a Wang programmable calculator. I thank Dr. Fred Walz for its use.
11. I thank Drs. E. Leigh, III, S. Levin, D. McNaught, L. Segel, and M. Slatkin for valuable comments; and J. Riebesell, whose careful work was helpful in making mathematical errors scarce, if not entirely absent. Supported by the National Science Foundation and the Research Foundation of the State University of New York.

26 October 1970

14

Reprinted from *Nature* **238**:413–414 (1972)

WILL A LARGE COMPLEX SYSTEM BE STABLE?

Robert M. May

Gardner and Ashby[1] have suggested that large complex systems which are assembled (connected) at random may be expected to be stable up to a certain critical level of connectance, and then, as this increases, to suddenly become unstable. Their conclusions were based on the trend of computer studies of systems with 4, 7 and 10 variables.

Here I complement Gardner and Ashby's work with an analytical investigation of such systems in the limit when the number of variables is large. The sharp transition from stability to instability which was the essential feature of their paper is confirmed, and I go further to see how this critical transition point scales with the number of variables n in the system, and with the average connectance C and interaction magnitude α between the various variables. The object is to clarify the relation between stability and complexity in ecological systems with many interacting species, and some conclusions bearing on this question are drawn from the model. But, just as in Gardner and Ashby's work, the formal development of the problem is a general one, and thus applies to the wide range of contexts spelled out by these authors.

Specifically, consider a system with n variables (in an ecological application these are the populations of the n interacting species) which in general may obey some quite nonlinear set of first-order differential equations. The stability of the possible equilibrium or time-independent configurations of such a system may be studied by Taylor-expanding in the neighbourhood of the equilibrium point, so that the stability of the possible equilibrium is characterized by the equation

$$dx/dt = Ax \qquad (1)$$

Here in an ecological context $\mathbf{x}$ is the $n \times 1$ column vector of the disturbed populations x_j, and the $n \times n$ interaction matrix $\mathbf{A}$ has elements a_{jk} which characterize the effect of species k on species j near equilibrium[2,3]. A diagram of the trophic web immediately determines which a_{jk} are zero (no web link), and the type of interaction determines the sign and magnitude of a_{jk}.

Following Gardner and Ashby, suppose that each of the n species would by itself have a density dependent or otherwise stabilized form, so that if disturbed from equilibrium it would return with some characteristic damping time. To set a time-scale, these damping times are all chosen to be unity: $a_{jj} = -1$. Next the interactions are "switched on", and it is assumed that each such interaction element is equally likely to be positive or negative, having an absolute magnitude chosen from some statistical distribution. That is, each of these matrix elements is assigned from a distribution of random numbers, and this distribution has mean value zero and mean square value α. (For a fuller account of such a formulation, see refs. 2 and 3.) α may be thought of as expressing the average interaction "strength", which average is for simplicity common to all interactions. In short,

$$A = B - I \qquad (2)$$

where **B** is a random matrix, and **I** the unit matrix. Thus we have an unbounded ensemble of models, one for each specific choice of the interaction matrix elements drawn individually from the random number distribution.

It is important to note that randomness only enters in the initial choice of the coefficients a_{jk}, which then define a particular model. Once the dice have been rolled to get a specific system, the subsequent analysis is purely deterministic.

The system (1) is stable if, and only if, all the eigenvalues of **A** have negative real parts. For a specified system size n and average interaction strength α, it may be asked what is the probability $P(n,\alpha)$ that a particular matrix drawn from the ensemble will correspond to a stable system. For large n, analytic techniques developed for treating large random matrices may be used to show[*] that such a matrix will be almost certainly stable ($P \rightarrow 1$) if

$$\alpha < (n)^{-1/2} \qquad (3)$$

and almost certainly unstable ($P \rightarrow 0$) if

$$\alpha > (n)^{-1/2} \qquad (4)$$

The transition from stability to instability as α increases from the regime (3) into the regime (4) is very sharp for $n \gg 1$; indeed the relative width of the transition region scales as $n^{-2/3}$.

---

[*] From equation (2) it is obvious that the eigenvalues of **A** are λ-1, where λ are those of **B**. The "semi-circle law" distribution for the eigenvalues of a particular random matrix ensemble was first obtained by Wigner[4], and subsequently generalized by him to a very wide class of random matrices whose elements all have the same mean square value[5]. Although the matrix **B** does not in general possess the hermiticity property required for most of these results to be directly applicable, the present results for the largest eigenvalue and its neighbourhood can be obtained by using Wigner's[4] original style of argument on $(B)^N (B^T)^N$ where N is very large. Indirectly relevant is Mehta[5] and Ginibre[6].

Such a precise answer for any model in the ensemble in the limit $n \gg 1$ is a consequence of the familiar statistical fact that, although individual matrix elements are liable to have any value, by the time one has an $n \times n$ matrix with n^2 such statistical elements, the total system has relatively well defined properties.

Next we introduce Gardner and Ashby's connectance, C, which expresses the probability that any pair of species will interact. It is measured as the percentage of non-zero elements in the matrix, or as the ratio of actual links to topologically possible links in the trophic web. The matrix elements in **B** now either, with probability C, are drawn from the previous random number distribution, or, with probability $1 - C$, are zero. Thus each member of the ensemble of matrices A corresponds to a system of individually stable parts, connected so that each part is affected directly by a fraction C of the other parts. For large n, $\alpha^2 C$ plays the role previously played by α^2, and we find the system (1) is almost certainly stable ($P(n, \alpha, C) \to 1$) if

$$\alpha < (nC)^{-1/2} \qquad (5)$$

and almost certainly unstable ($P \to 0$) if

$$\alpha > (nC)^{-1/2} \qquad (6)$$

It is interesting to compare the analytical results with Gardner and Ashby's computer results for smallish n. (Their choice of **A** differs slightly from ours, but in essence they have the fixed value $\alpha^2 = 1/3$, and diagonal elements intrinsically -0.55 rather than -1.) Although our methods are based on the assumption that n is large, and are therefore only approximations when applied to $n = 4$, 7, 10, the two approaches in fact agree well when compared, being not more than 30% discrepant even for $n = 4$.

The central feature of the above results for large systems is the very sharp transition from stable to unstable behaviour as the complexity (as measured by the connectance and the average interaction strength) exceeds a critical value. This accords with Gardner and Ashby's conjecture.

Applied in an ecological context, this ensemble of very general mathematical models of multi-species communities, in which the population of each species would by itself be stable, displays the property that too rich a web connectance (too large a C) or too large an average interaction strength (too large an α) leads to instability. The larger the number of species, the more pronounced the effect.

Two corollaries are worth noting, although they should not be taken to have more than qualitative significance.

First, notice that two different systems of this kind, with average interaction strengths and connectances α_1, C_1 and α_2, C_2 respectively, have similar stability character if

$$\alpha_1^2 C_1 \simeq \alpha_2^2 C_2 \tag{7}$$

Roughly speaking, this suggests that within a web species which interact with many others (large C) should do so weakly (small α), and conversely those which interact strongly should do so with but a few species. This is indeed a tendency in many natural ecosystems, as noted, for example, by Margalef[7]: "From empirical evidence it seems that species that interact feebly with others do so with a great number of other species. Conversely, species with strong interactions are often part of a system with a small number of species. . . ."

A second feature of the models may be illustrated by using Gardner and Ashby's computations (which are for a particular α) to see, for example, that 12-species communities with 15% connectance have probability essentially zero of being stable, whereas if the interactions be organized into three separate 4×4 blocks of 4-species communities, each with a consequent 45% connectance, the "organized" 12-species models will be stable with probability 35%. That is, of the infinite ensemble of these particular 12-species models, essentially none of the general ones are stable, whereas 35% of those arranged into three "blocks" are stable. Such examples suggest that our model multi-species communities, for given average interaction strength and web connectance, will do better if the interactions tend to be arranged in "blocks"—again a feature observed in many natural ecosystems.

This work was sponsored by the US National Science Foundation.

[1] Gardner, M. R., and Ashby, W. R., *Nature*, **228**, 784 (1970).
[2] Margalef, R., *Perspectives in Ecological Theory* (University of Chicago, 1968).
[3] May, R. M., *Math. Biosci.*, **12**, 59 (1971)
[4] Wigner, E. P., *Proc. Fourth Canad. Math. Cong., Toronto*, 174 (1959).
[5] Mehta, M. L., *Random Matrices*, 12 (Academic Press, New York, 1967).
[6] Ginibre, J., *J. Math. Phys.*, **6**, 44 (1965).
[7] Margalef, R., *Perspectives in Ecological Theory*, 7 (University of Chicago, 1968).

Reprinted from *Ecology* 56:238–243 (1975), by permission of the publisher, Duke University Press, Durham, North Carolina

STABILITY AND CONNECTANCE IN FOOD WEB MODELS[1]

D. L. DE ANGELIS[2]

Institute for Soil Science and Forest Fertilization, University of Goettingen,
34 Goettingen-Weende, Federal Republic of Germany

INTRODUCTION

Recent reports by Gardner and Ashby (1970) and May (1972) have described the property that randomly connected systems tend to become less stable as the number of connections among the components increases. Specifically Gardner and Ashby (1970) considered a set of linear equations describing the system. In matrix notation the equations can be written

$$\dot{x} = (B - D) \cdot x. \tag{1}$$

The matrix x of the $n \times 1$ column represents the n dependent variables, the magnitudes of the deviations of the system's components from their equilibrium values. Then B is an $n\text{-}\times\text{-}n$ matrix representing interactions between pairs of components, and D is an $n\text{-}\times\text{-}n$ diagonal matrix representing self-damping forces acting on the variables. This set of linear equations may be thought of as the first order terms of a Taylor series expansion of a set of nonlinear equations about the equilibrium point of the system. Whether the system is stable about its equilibrium point is decided by the eigenvalues of equation (1). If at least one of the n eigenvalues has a positive real part, the system is unstable. This is more fully discussed in many places (e.g., Rosen 1970).

To test the stability of the above system as a function of the relative number of interactions among the components, Gardner and Ashby (1970) used a Monte Carlo technique. They allowed varying percentages of the elements, b_{ij}, of B to be nonzero. Each percentage is defined to be a particular connectance C of the system; in effect, every component of the system interacts with about C percent of the other components. The nonzero elements were as-

signed random values uniformly in the interval $-1.0 \leqslant b_{ij} \leqslant 1.0$. The elements, d_{ii}, of the self-damping diagonal matrix D were assigned random values in the range $0.1 \leqslant d_{ii} \leqslant 1.0$. Systems with no nonzero b_{ij}'s were therefore stable by virtue of the self-damping terms. From stability analysis of a large number of randomly connected systems, Gardner and Ashby (1970) were able to determine that as C increased, the probability, P, of the equilibrium point being stable always decreased. May (1972) extended these results, in the case where D is the identity matrix, to very large systems. He showed in the limit $n \rightarrow \infty$ that an abrupt transition from $P = 1$ to $P = 0$ occurs as C is increased beyond some critical value.

These results are important to ecology, especially so because they seem to contradict the traditional idea that ecosystem stability is positively related to complexity, measured here by the system's connectance, C (e.g., Odum 1971: 149). But when one tries to assess the implications of these results for actual ecosystems, for example a food web, it becomes clear that equation (1), with the above assumptions, is in some respects too general and in others not general enough. It is not general enough, for if the variable x represents the perturbed biomasses of species in a food web, allowance must be made for the fact that at least some autotrophic species, if they are in part regulated by heterotrophs, will not be subject to self-damping forces at equilibrium. Instead, their tendency would be to grow in the absence of trophic interactions. These same remarks apply to herbivore species, if these are considered the lowest trophic level in the particular model, and if the herbivores are to some extent regulated by predators at the equilibrium point. On the other hand equation (1) is too general because the assumed randomness in the values of the elements of B and D ignores the relationships that must exist among some of these elements. The model as

[1] Manuscript received January 24, 1974; accepted July 19, 1974.

[2] Present address: Environmental Sciences Division, Oak Ridge National Laboratory, Oak Ridge. TN 37830.

defined can also include systems which violate conservation of matter.

In this paper the question of stability and connectance is examined using a model appropriate to an ecological food web. It is shown that the relationship of these two system characteristics depends on the detailed nature of the interactions. There exist quite plausible cases where the probability of stability of the food web increases as the connectance is increased.

PROPERTIES OF FOOD WEBS

Food webs exhibit a number of structural characteristics determining the mathematical nature of their models. Two such characteristics are the following:

(i) If the rate of decrease of biomass x_j, due to interaction with (i.e., predation by) x_i can be represented by a function $f_{ij}(x_i, x_j)$, the corresponding rate of increase of x_i must be some function $e_{ij}(x_i, x_j)$ where $e_{ij} < f_{ij}$ for all values of x_i and x_j. This reflects the fact that assimilation efficiency is always less than unity. For simplicity it is assumed here that the assimilation efficiency is a constant, γ_{ij}, i.e.,

$$e_{ij} = \gamma_{ij} f_{ij} \qquad (\gamma_{ij} < 1). \qquad (2)$$

(ii) The food web exhibits a trophic structure. Herbivores feed on autotrophs, carnivores on herbivores, etc. This hierarchical pattern must be accounted for in any abstract model by a distinction among the variables in terms of their position in the trophic structure.

The equation representing the dynamics of the biomass of a particular species is

$$\dot{x}_i = \sum_{\substack{j=1 \\ (j \neq i)}}^{n} \gamma_{ij} f_{ij}(x_i, x_j) - \sum_{\substack{k=1 \\ (k \neq i)}}^{n} f_{ki}(x_i, x_k) + s_i(x_i), \qquad (3)$$

where the first term represents the contribution to biomass from species of lower trophic levels and the second term represents loss caused by predation from higher trophic levels. The function $s_i(x_i)$ denotes the intrinsic tendency of a species to increase or decrease in the absence of interactions with other species explicitly represented in the model.

Equation (3) is, in general, nonlinear. If the food web system has an equilibrium point, x_o, then it is possible to linearize (3) about this point to obtain a set of perturbation equations, each having the form

$$\dot{y}_i = \sum_{\substack{j=1 \\ (j \neq i)}}^{n} \gamma_{ij} \left(\frac{y_i \, \partial f_{ij}}{\partial x_i} + \frac{y_j \, \partial f_{ij}}{\partial x_j} \right)$$

$$- \sum_{\substack{k=1 \\ (k \neq i)}}^{n} \left(\frac{y_i \, \partial f_{ki}}{\partial x_i} + \frac{y_k \, \partial f_{ki}}{\partial x_k} \right) + \frac{y_i \, ds_i}{dx_i}, \qquad (4)$$

where y_i represents a small perturbation about the equilibrium point,

$$y_i = x_i - x_{io} \qquad (y_i \ll x_{io}), \qquad (5)$$

and where derivatives of f_{ij}, f_{ki}, and s_i are evaluated at the equilibrium point. An additional n-1 equations of the same from as (4) complete the mathematical description of the system near its equilibrium point.

In this report two further assumptions will be made concerning food webs. Although exceptions to these assumptions may occur, they should be generally applicable:

(iii) In the absence of interactions with heterotrophs, some autotrophs will have a tendency to increase near the equilibrium point; i.e.,

$$ds_i / dx_i > 0 \text{ (some species of lowest} \qquad (6)$$
$$\text{trophic level).}$$

Of course, this is not necessarily true for all autotrophs. Feeding by heterotrophs may limit the competitors of some autotrophs, which would therefore tend to decrease in the absence of such feeding. If herbivores are considered to constitute the lowest trophic level in the model in question, a sufficient food source for the herbivores being implicitly assumed at the equilibrium point, (6) will apply to some of these species.

On the other hand, in the absence of trophic interactions, all species of higher trophic levels should tend to decrease, since they would not receive any biomass input; i.e.,

$$ds_i / dx_i < 0 \quad \text{(higher trophic level species).} \qquad (7)$$

(iv) At the equilibrium point it is reasonable to assume that the magnitude of the interaction terms, $f_{ij}(x_i, x_j)$ would increase if either x_i or x_j were increased;

$$\partial f_{ij} / \partial x_i, \ \partial f_{ij} / \partial x_j > 0. \qquad (8)$$

The system studied is, in view of the above assumptions, quite restricted compared with that studied by Gardner and Ashby (1970). It is, however, a very general system within the framework of food web models. It is difficult to imagine a food web not satisfying (i)–(iv).

The set of equations describing the system can be written in matrix form:

$$\dot{y} = G \cdot y, \qquad (9)$$

where the elements of G are as follows:

$$g_{ij} = \begin{cases} \gamma_{ij}\partial f_{ij}/\partial x_j & \text{(if } j \text{ is the donor)} \\ -\partial f_{ji}/\partial x_j & \text{(if } i \text{ is the donor) (10)} \\ \displaystyle\sum_{\substack{k=1 \\ (k\neq i)}}^{n} \gamma_{ik}\frac{\partial f_{ik}}{\partial x_i} - \sum_{\substack{m=1 \\ (m\neq i)}}^{n} \frac{\partial f_{mi}}{\partial x_i} + \frac{ds_i}{dx_i} & (i=j). \end{cases}$$

The terms in the first summation in (10) can be non zero only when species i is in a higher trophic level than species k, and similarly, terms in the second summation can be non zero only when species m is in a higher trophic level than species i.

In Equation (10) it is seen that the diagonal elements would be, except for the presence of ds_i/dx_i, linear combinations of off-diagonal elements. This relationship is an expression of an implicitly assumed balance of biomass flows; the rate of change of perturbed biomass of one species is balanced by compensating rates of change in the other species plus the rates of loss from the system. The fact that the diagonal and off-diagonal elements are intimately related is at the heart of differences in stability-complexity behavior between the matrices of typical food web models and random matrices. These differences will first be exhibited by reference to a particular example. Then the mechanisms accounting for them will be displayed by an examination of the diagonal elements in (10).

The example studied consists of six species in the lowest trophic level, three species in the next trophic level, and one species that can feed on any of the others (Fig. 1). Such a system might, for example, consist of autotrophs, herbivores and an omnivore, or herbivores, carnivores and a carnivore–super-carnivore. The food web is assumed always to have at least those connections indicated by the solid lines in Fig. 1, each species of the lowest trophic level being fed on by one member of the next level. In its most simple level of connectance, the system may be stable or unstable, depending on the particular set of matrix elements. Complexity is increased by increasing the probability, C, that the other food web connections (indicated by dashed lines in Fig. 1) are realized.

Stability of the above system as a function of C is investigated using a Monte Carlo method, differing from Gardner and Ashby's procedure only in that random values are assigned to the derivatives of f_{ij} and s_i rather than to the matrix elements directly. For simplicity it is assumed that the assimilation efficiency is the same for all interactions; i.e., $\gamma_{ij} = \gamma$ for all values of i and j, where γ is a constant less than unity.

Case 1: Suppose, in view of conditions (i)–(iv), the derivatives of f_{ij} and s_i are allowed to assume random values uniformly in the following intervals:

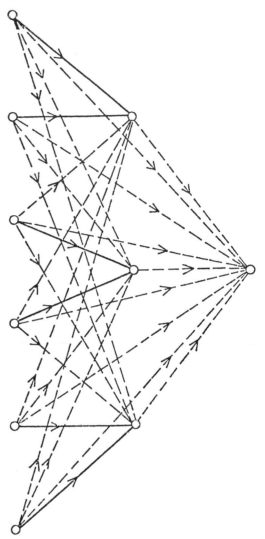

Fig. 1. Food web consisting of 10 species. Solid lines represent connections always assumed to exist; dashed lines represent potential connections, and the arrows denote the direction of biomass flow.

$$-1.0 \leqslant ds_i/dx_i \leqslant 1.0 \quad \text{(lowest trophic level),} \quad (11a)$$

$$-1.0 \leqslant ds_i/dx_i \leqslant 0.0 \quad \text{(higher trophic levels),} \quad (11b)$$

$$0.0 \leqslant \partial f_{ij}/\partial x_i, \partial f_{ij}/\partial x_j \leqslant 1.0, \quad (11c)$$

with γ being allowed to take on a series of values between zero and unity. It is found that for small values of ($\gamma \lesssim 0.2$) the probability, P, of the system being stable increases as C is increased. This effect is shown in Fig. 2. For intermediate values ($\gamma \approx 0.?$) a peaking effect is observed, P increasing for small values of C, then decreasing. The curves in Fig. (and the subsequent figures) were fitted by eye

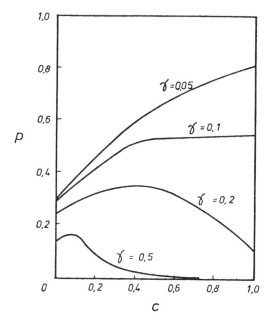

FIG. 2. Probability of system stability, P, as a function of connectance, C, for various values of assimilation efficiency.

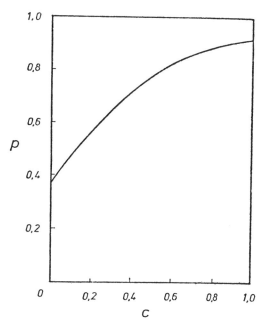

FIG. 3. Probability of system stability, P, increases with connectance, C, when there is a strong bias toward self-regulation of heterotrophic species, $R_1 = 20.0$, $\gamma = 0.6$.

the values of P, which had been computed at intervals of .05 along the C-axis. For each value of connectance, 200 random samples of matrix stability were made. According to the statistics of the binomial distribution, each value of P has a confidence interval of at most about ± 0.07 at the 95% level. Since only a qualitative picture of the relationship between P and C was sought, greater precision was not considered necessary.

Measurements have shown, for the case of grassland herbivores, for example, that assimilation efficiency (the ratio of assimilated to ingested biomass) may range from .25 to .90 (Weigert and Evans 1967). The assimilation efficiencies of carnivore species are higher on the average (Kozlovsky 1968). However the ratio γ used in this article is not, precisely speaking, the ratio of assimilated to ingested biomass, but the ratio of assimilated biomass to the amount of biomass removed from the population of prey species. If, in the act of feeding, biomass is removed from the prey species without being ingested, γ will be less than the true assimilation efficiency. It is therefore possible to imagine situations where γ will be small enough to be in the range where stability of the system is enhanced by greater complexity.

Case 2: Suppose now that the system is biased toward strong self-regulation of species in the higher trophic levels. Examples of this type of system in nature could occur where territorial or migrational

behavior act as population limiters. In these cases the self-damping forces of the higher trophic level species can be larger than the other elements of the G-matrix. Consider, as a case in point, the system,

$$-1.0 \leqslant ds_i/dx_i \leqslant 1.0 \quad \text{(lowest trophic level),} \quad (12a)$$

$$-R_1 \leqslant ds_i/dx_i \leqslant 0.0 \quad \text{(higher trophic levels),} \quad (12b)$$

$$0.0 \leqslant \partial f_{ij}/\partial x_i, \, \partial f_{ij}/\partial x_j \leqslant 1.0, \quad (12c)$$

where $R_1 > 1$. For this system it is found that connectance and stability may be related in a positive manner, if the constant R_1 is large enough. An example is shown in Fig. 3 where $R_1 = 20.0$ and $\gamma = 0.6$. For smaller values of R_1, the P-versus-C curve often exhibits a peak, the probability of stability at first increasing, then decreasing, as C increases.

Case 3: Suppose finally that the system is biased toward donor dependence. It is well known that a linear donor-dependent system is necessarily stable (Hearon 1963). Donor dependence is commonly assumed in ecosystem models largely because of this characteristic, although it is doubtless often a poor assumption. However, it may not be unrealistic to assume in many cases that trophic interactions are biased toward donor dependence. Mathematically, such a bias in an interaction between species i and j is expressed as

$$\partial f_{ij}/\partial x_j > \partial f_{ij}/\partial x_i, \qquad (13)$$

where j is the donor. A whole system may be said to be biased toward donor dependence when the relationship

$$0.0 \leqslant \partial f_{ij}/\partial x_i \leqslant R_2, \qquad (14a)$$

$$0.0 \leqslant \partial f_{ij}/\partial x_j \leqslant R_3 \qquad (14b)$$

exists, where $R_2 < R_3$ and where the index j runs over donor species and i runs over receptor species. In this regime, stability may also be positively related to connectance, more strongly as the ratio $R_3 : R_2$ increases. An example of this behavior is shown in Fig. 4 for the case where $R_3 : R_2 = 10.0$, $R_1 = 1.0$, and $\gamma = 0.6$.

ANALYSIS BY AN APPROXIMATION TECHNIQUE

The stability-versus-connectance behavior shown in figures 2, 3, and 4 simply reflects the nature of the food web model, in which the implicit balance of perturbed biomass flows constrains the diagonal elements, g_{ii}, of (10) to consist, in part, of linear combinations of off-diagonal elements. An argument by R. M. May (*pers. comm.*) makes quite transparent the effects of these constraints on system stability. In an $n+1$ species system, the value of a particular diagonal element, averaged over many random samples, is

$$\langle g_{ii} \rangle = n[\gamma \langle \partial f_{ij}/\partial x_i \rangle - \langle \partial f_{ij}/\partial x_j \rangle] + \langle ds_i/dx_i \rangle. \quad (15)$$

If $\partial f_{ij}/\partial x_j$ and $\partial f_{ij}/\partial x_i$ take random values uniformly in the intervals 0.0 to R_3 and 0.0 to R_2, respectively, and the connectance is C, then crudely,

$$\langle g_{ii} \rangle \approx (nC/2)[\gamma R_2 - R_3] + \langle ds_i/dx_i \rangle. \quad (16)$$

Clearly, when γ is small enough or R_2 is smaller than R_3, the first term on the right-hand side of (16) is negative. This negative term in the diagonal element is a stabilizing influence on the system. Since the term varies linearly with C, the stability of the system from this source increases roughly linearly with increasing connectance. On the other hand, there is a destabilizing influence from the off-diagonal elements which increases only as $C^{1/2}$ (May 1972), and can therefore be overshadowed by the influence of the diagonal terms provided they are sufficiently negative.

Although this mathematical clarification does not explain the results of Case 2, where the higher trophic level species have very large self-damping terms, these results are perhaps also understandable. A low connectance model will often have certain subsystems that are unstable but are stabilized when new connections are made to other parts of the system. Although a destabilizing tendency, increasing as $C^{1/2}$, also accompanies this increasing connectance,

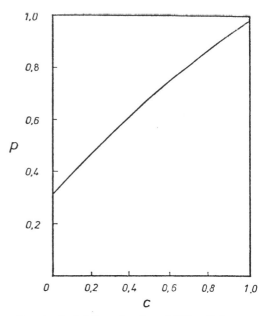

FIG. 4. Probability of system stability, P, increases with connectance, C, when there is a bias toward donor dependence; $R_1 = 1.0$, $R_3 = 10.0$, $R_2 = 1.0$, $\gamma = 0.6$.

it may not be significant if the self-damping terms of the higher trophic level species are large enough. Hence P increases with increasing values of C.

CONCLUSIONS

The above analysis of a food web model demonstrates the existence of three cases in which the increased connectance of the model is accompanied by increased stability. To summarize briefly, these are the cases where; (1) the ratio of biomass assimilated by the consumer species is a small fraction of the biomass removed from the prey species in the feeding process, (2) the higher trophic level species experience a strong self-damping force which controls their population growth, (3) there is a bias toward donor dependence in the interactions. May's approximation analysis of the diagonal matrix terms provides evidence that these conclusions are probably of general significance for food web models.

ACKNOWLEDGMENTS

This research was supported by the Deutsches Forschungsgemeinschaft, Contract No. E1/37. Computations were performed using the facilities of the Max Planck Institute for Biophysical Chemistry, Goettingen, West Germany.

LITERATURE CITED

Gardner, M. R., and W. R. Ashby. 1970. Connectance of large dynamical (cybernetic) systems: critical values of stability. Nature 228:784.

Hearon, J. Z. 1963. Theorems on linear systems. Ann. N.Y. Acad. Sci. 108:36–38.

Kozlovsky, D. G. 1968. A critical evaluation of the trophic level concept. I. Ecological efficiencies. Ecology 49:48–60.

May, R. M. 1972. What is the chance that a large complex system will be stable? Nature 237:413–414.

Odum, E. P. 1971. Fundamentals of ecology, 3rd ed. W. B. Saunders, Philadelphia. 574 p.

Rosen, R. 1970. Dynamical system theory in biology. Vol. I: Stability theory and its applications. John Wiley, New York. 302 p.

Weigert, R. G., and F. C. Evans. 1967. Investigations of secondary productivity in grasslands, p. 499. In K. Petrusewicz [ed.] Secondary productivity of terrestrial ecosystems. Panstwowe Wydawnictwo Naukove, Waszawa-Krakow. Int. Biol. Program PT; Inst. Ecol. Polish Acad. Sci.

Editors' Comments
on Papers 16, 17, and 18

16 ODUM
Ecological Potential and Analogue Circuits for the Ecosystem

17 SMITH
Analysis of Ecosystems

18 O'NEILL
Ecosystem Persistence and Heterotrophic Regulation

THE ECOSYSTEM AS AN OBJECT OF STUDY

Many systems ecology studies result in large, complex simulation models. However, a second line of development utilizes simple models to explore general, theoretical properties. In contrast with large simulation models, these models consider only a few variables and aim at elucidating overall properties. These models do not attempt to duplicate field data, but to draw out the implications contained in a specific conceptualization of the system.

For example, linear models of the flow of a material through an unperturbed ecosystem (e.g., Jordan et al. 1972) have focused on stability of the ecosystem. Other studies (Paper 10; Waide et al. 1974) have applied engineering analyses to the control of system function.

An important divergent development in systems ecology is represented by Paper 16 by Odum, which presents an energy-based method with relatively simple analog models. Individual functional components (e.g., primary producers, decomposers) are represented by different symbols and explicitly dealt with according to their function (e.g., energy production versus storage). The publication date of this paper indicates that this approach was among the earliest in the field. Possibly because the approach was well ahead of its time and possibly because of the insistence on an esoteric methodology, the analog/energy models have exerted a minimal influence on other developments in systems ecology.

The other two papers presented in this section, Paper 17 by Smith and Paper 18 by O'Neill, attempt to analyze total ecosystem dynamics through minimal mathematical models. Both studies indicate the complexity of the dynamics that are inherent in the system and emphasize counterintuitive results in which interactions within the system lead to unexpected results.

REFERENCES

Jordan, C. F., J. R. Kline, and D. S. Sasscer. 1972. Relative stability of mineral cycles in forest ecosystems. *Am. Nat.* **106**:237–53.

Waide, J. B., J. E. Krebs, S. P. Clarkson, and E. M. Setzler. 1974. A linear systems analysis of the calcium cycle in a forested watershed ecosystem. *Prog. Theor. Biol.* **3**:261–345.

16

Reprinted by permission from *Am. Sci.* **48**:1–8 (1960), journal of Sigma Xi, The Scientific Research Society of North America

ECOLOGICAL POTENTIAL AND ANALOGUE CIRCUITS FOR THE ECOSYSTEM*

By HOWARD T. ODUM

The Ecosystem and the Ecomix

A PATCH of forest is a mysterious thing, growing, repairing, competing, holding itself against dispersion, oscillating in low entropy state, getting its daily quota of free energy from the sun. It is an ecosystem.

Understanding the basic nature of the ecosystem is a principal objective of ecology. Ecosystems are phenomena such as forests, deserts, lakes, reefs, lagoons, microcosm cultures, and polluted streams. They usually contain three kinds of processes, (*a*) photosynthesis, (*b*) respiration, and (*c*) circulation. Although extremely diverse, ecosystems have some basic structures and functional processes in common. To permit quantitative comparison on similar bases, a growing number of measures of general structure and function have been devised such as photosynthetic production, community metabolism, biomass, species variety, efficiencies, storage ratios, chlorophyll per area, assimilation ratio, turnover, etc. The time is now ripe for the further synthesis of the new data into generalized theorems of the ecosystem.

In ecosystems as in many other kinds of open systems, energy is supplied in concentrated form from the outside driving a sequence of branching energy flows, maintaining complex structure, recycling materials, and finally passing out from the system in a dispersed state of high entropy. The rate adjustments are set by natural selection which constitutes a fourth law of thermodynamics applicable to those open steady states which have self-reproduction and maintenance (Odum and Pinkerton, 1955).

The flow of energy in an ecosystem is represented by energy flow diagrams like that in Figure 1 (Odum, 1956). Organisms in an eco-

* These studies were aided by a grant from the National Science Foundation NSF G3978 on Ecological Microcosms.

system in their food and energy roles participate at one of five relative positions in the flow circuit. These levels are usually called trophic levels as follows: primary photosynthetic producing plants, P; herbivore animals, H; carnivore animals, C; second order top carnivores, TC; and decomposer microorganisms and other components whose position is uncertain and for practical purposes must be lumped in a miscellaneous category pending elucidation of their exact role, D. These trophic levels are indi-

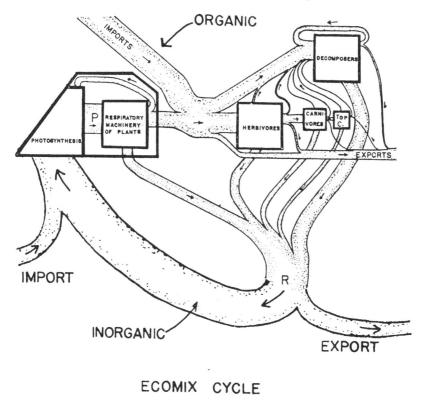

ECOMIX CYCLE

Fig. 2. Ecomix cycle diagram for an ecosystem.

cated in Figures 1, 2, and 3 by the boxes as labeled. Whereas the flow of energy is unidirectional towards dispersed, unavailable form according to the second law of thermodynamics, the materials such as carbon, nitrogen, phosphorus, and trace elements circulate in a cyclic manner being elevated into high energy combinations by plants, passing subsequently through a sequence of diminishing energy levels in the consumers. The cycle of the materials of an ecosystem can be represented by a diagram like that in Figure 2. Since the elemental ratios in the primary photosynthetic production tend to be similar to those in the respiratory-regenerator aspect of the ecosystem according to Redfield's principle

(Redfield, 1934), one may consider the cycle of the raw materials as a group. The word ecomix is used in Figure 2 to represent the particular ratio of elemental substances being synthesized into biomass and subsequently released and recirculated. For example, the ratios of some of the elements in the ecomix of a planktonic system are indicated in the overall equation for the primary producers process:

1,300,000 Cal. radiant energy + 106 CO_2 + 90 H_2O + 16 NO_3 + 1 PO_4 plus mineral elements——→ 13,000 Cal. potential energy in 3258 gm protoplasm (106 C, 180 H, 45 O, 16 N, 1 P, 815 gm mineral ash) + 154 O_2 and 1,287,000 Cal. heat energy dispersed.

This and other explanations and examples of ecosystems may be found in an ecological text (Odum and Odum, 1959).

The Ecological Analogue of Ohm's Law

The familiar Ohm's law states that the flow of electrical current, A, is proportional to the driving voltage, V, with R, the resistance, a property of the circuit.

$$A = \frac{1}{R} V \qquad \text{(Ohm's Law)} \qquad (1)$$

or, in an alternative form,

$$A = CV \qquad (2)$$

where $C = 1/R$ is the conductivity.

In terms of steady state thermodynamics, Ohm's law is a special case of the more generalized theorem that the flux, J, is proportional to the driving thermodynamic force, X, with C the conductivity (Denbigh, 1951).

$$J = CX \qquad (3)$$

Just as the product of voltage, V, and amperage, A, is power (wattage), so, in the general case, the product of thermodynamic force, X, and flux, J, is power, JX.

The ecosystem also has a flow of material under the driving influence of a thermodynamic force. The flux is the flow of food through a food chain circuit (Fig. 2) as expressed in units such as carbon per square meter of ecosystem area per unit of time. The force is some function of the concentration gradient of organic matter and biomass above and below the food circuit. A number of authors have related consumption to concentration. Jenny, Gessel, and Bingham (1949) have shown that the rate of organic decomposition by microorganisms in soil is proportional to the concentration of organic matter. A similar relationship is involved in the equation for the oxygen sag downstream from pollution outfall. Sinkoff, Geilker, and Rennerfelt (unpublished report obtainable from the Taft Environmental Health Center) use an analogue circuit for simulation of oxygen sag curves based on the principle of decomposition rate being dependent on the amount of organic matter. In comparing the ecosystem to the Ohm's law analogue, the consumption of living food as well as dead organic matter is considered to be dependent on the concentration of the food.

The validity of this application may be recognized when one breaks away from the habit of thinking that a fish or a bear catches food and thinks instead that accumulated food by its concentration practically

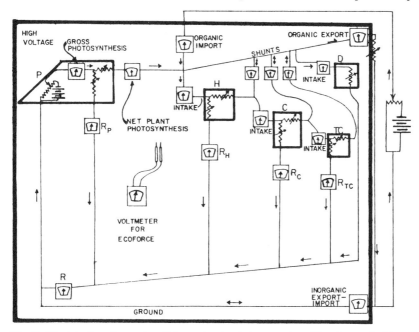

Fig. 3. Electrical analogue circuit for a steady state ecosystem like the one in Fig. 2. The flow of electrons corresponds to the flow of carbon.

forces food through the consumers. Any aggression by the fish is paid for by the food. When there are no consumers there is a state of high resistance. Usually, ecosystems rapidly develop circuits to drain reservoirs of organic free energy, often being self-organized to maintain suitable biomass structure for the purpose

Thus we may write the equation for the force and flux in the ecosystem in the form of equations 1, 2 and 3 as follows:

$$J_e = C_e X_e \tag{4}$$

where X_e is the thermodynamic force (ecoforce); J_e the ecoflux; and C_e the ecological conductivity of the food circuit. The application of equation (4) and the elucidation of the nature of the ecoforce follow in a subsequent section.

An Analogue Circuit for the Biogeochemical Cycle of the Ecosystem

Since the form of equation (4) relating ecoflux and ecoforce is the same as the form for Ohm's law, an electrical circuit can be constructed analogous to the flows of the ecomix in Figure 2. This has been done as diagrammed in Figure 3. Like the biological system in Figure 2, the electrical system in Figure 3 is an open steady state. Application of more complex analogue circuits with feedbacks, oscillations, and transient phenomena remain for the future.

In the electrical circuit of Figure 3 resistances are grouped at the locations of the producing and consuming populations. Batteries supply the concentrated energy representing the sun and the energy imported as organic matter from the outside. The various branching flows of food energy to consumers are presented with branching electrical wires. Variable resistances and switches permit the observer to set up various special situations and combinations. Milliammeters are placed in each circuit to permit rapid visual examination of the electrical flow which represents the flow of carbon and associated ecomix. As in the real ecosystem the energy is in the state of the flowing matter and is radiated from the computer as heat during passage from the high energy state of the battery to ground level. The amount of energy dispersed in any flow is readily measured by the product of the amperage and the voltage. A voltmeter with leads is available for measuring the voltage drop in any circuit adjustment.

Determination of the Ecoforce from Ecosystem Data Using the Electrical Analogue Circuit

Although approximate, fairly complete data now exist on rates of flux in circuits of real ecosystems. Data from one stable ecosystem, a fresh water stream, Silver Springs, Florida (Odum, 1957) available in the form of Figure 2, were put into the electrical analogue circuit (Figure 3) on a scale of 13.9 milliamperes per gm/M^2/year of carbon flow. The variable resistances were adjusted so that the rates of current flow were in scale with the average rates of flow of carbon estimated by various ecological means in the field. The voltage drops between the various parts of the ecosystem and the ground were then measured with the voltmeter. The results are included in Figure 4. The ecoforce, defined as a linear function of flux (equation 4), was thus measured directly from real data for an ecosystem using the electrical analogue circuit as a computation device.

Biomass Concentration, Ecoforce, and Ecopotential

As indicated previously, the flow of energy in a food chain circuit may be intuitively related to the concentration of food, just as the rates of reaction in simple chemical systems are related to the concentrations of reactants. However, the flow of energy between complex, self-reproducing entities organized within the ecosystem need not have, *a priori*, similarities to the chemical systems even though organic chemicals are involved in both.

The organic matter accumulated in the biomass of part of an ecosystem may be defined as ecopotential, E, equal to the free energy per unit carbon. This free energy, F, is the chemical free energy of the packages of biomass, prorated over the area of the ecosystem. Thus, ecopotential is a function of the concentration of biomass and organic matter.

The product of ecopotential and ecoflux has the dimensions of power.

$$EJ_e = \frac{\Delta F}{C}\frac{dC}{dt} \tag{5}$$

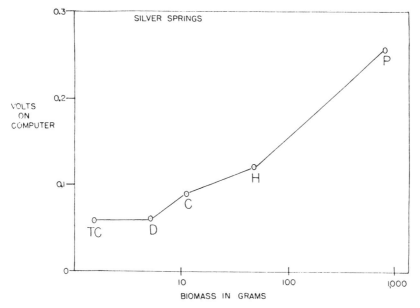

Fig. 4. Measured rates of metabolism of the organisms in the trophic levels of Silver Springs, Florida, were set in the circuit of Fig. 3 by adjusting the resistances. Then the voltages at this steady state were measured between the trophic level and ground. Since Ohm's law is a linear relationship, the voltages on the circuit are the ecoforces as defined in equation (4). The linearly defined ecoforce is some function of the organic food matter upstream from the trophic level. That is, the driving impetus to the metabolic circuits is some function of the concentration of food supply. At any point in the circuits the food supply is the standing concentration of biomass and other edible organic matter in the trophic level just above.

Measurements of the biomass concentration of organic matter in Silver Springs are available. In the graph of Fig. 4, the ecoforce is plotted as a function of the logarithm of biomass available to drive the flow. As one might predict from the basic similarity of ecological systems to chemical systems, the ecoforce is not the biomass concentration but may be a logarithmic function of it.

where E is ecopotential; J_e, ecoflux; ΔF, free energy change; C, carbon; and t, time.

Ecopotential is defined in energy units by equation (5); ecoforce was defined without specifying its dimensions except that it was linearly related to flux (equation 4). What is the relationship of ecopotential E and ecoforce X_e? In the case of Ohm's law, the potential and the force are identical. In the chemical flow systems, however, away from equilibrium, the potential is the logarithm of the force. The question arises as to the relationship of ecopotential and ecoforce in the ecosystem. Stating the question in another way, how is the flux related to the potential? Linearly? Logarithmically?

On the ordinate in Figure 4 are plotted the biomass concentrations from Silver Springs. On the abscissa are plotted the voltages from the electrical computer set for the average flux in Silver Springs. It is apparent that the relationship of biomass and voltage is not linear, but may be logarithmic. If the voltage of the computer represents the ecoforce defined linearly (equation 4) and the biomass concentration is a potential (equation 5), then ecopotential and ecoforce are not equal or linearly related. Many more such data need to be tested.

To avoid confusion it should be stated here that the concepts discussed here have nothing to do with the misnomer, biotic potential, which is not a potential in the energy sense but is a specific growth rate. Chapman's efforts (1928) to draw an analogy between Ohm's law and biotic potential are fallacious as can be recognized by dimensional analysis.

Hints About Ecosystems Derived from the Analogue Circuit

The construction and manipulation of the analogue is a powerful stimulant to the imagination concerning the behavior of ecosystems. The following are some suggestions from the analogue for experimental testing in the real ecosystems. The tests employed were made with the circuit in steady states resembling natural systems such as Silver Springs.

1. Competition exists when two circuits are in parallel.

2. Consumer animals compete with plant respiratory systems.

3. When unusual biomass and ecoforce distributions (potentials) are postulated, circuits reverse direction with food passing in unusual direction. For example, with large rates of import of organic matter, energy flows into the plants heterotrophically increasing plant respiration over its photosynthesis.

4. As sources of power, the primary producers and the import system compete.

5. If shunts exist with bacteria in important roles, a steep pyramid of metabolism develops.

6. If consumer respiration is increased, gross photosynthesis is also increased due to the lowered resistance.

7. Doubling the power supply doubles the metabolism at all levels.

8. Cutting off export increases metabolism of consumers.

9. Cutting off top carnivores does very little to the remainder of the energy flows.

10. Increasing import increases respiratory metabolism and diminishes gross photosynthesis.

11. Cutting out herbivores reduces photosynthesis and increases bacterial and plant respiration.

12. Higher trophic levels compete in part with the trophic level which it consumes.

13. A change in plant respiration has a major compensatory effect on the consumers.

14. A decrease in respiration increases the voltage (biomass concentration) upstream.

15. A short circuit is comparable to a forest fire.

ACKNOWLEDGMENTS

The author acknowledges the stimulation of theoretical discussions with Mr. Robert Beyers and Mr. Ronald Wilson of the Institute of Marine Science and with Dr. E. P. Odum, Dr. J. Olson, Dr. F. Golley, Dr. A. Smalley, and Dr. E. Kuenzler, participants in the 1959 Ecological Society Symposium on Energy Flow.

REFERENCES

CHAPMAN, R. N. 1928. The quantitative analysis of environmental factors. *Ecology, 9*, 111–122.

DENBIGH, K. G. 1951. Thermodynamics of the steady state. Methuen.

JENNY, H., S. P. GESSEL, and F. T. BINGHAM. 1959. Comparative study of decomposition rates of organic matter in temperate and tropical regions. *Soil Science, 68*, 419–432.

ODUM, E. P. with collaboration of H. T. ODUM. 1959. Fundamentals of Ecology. Saunders, Philadelphia.

ODUM, H. T. 1956. Primary production in flowing waters. *Limnolog. and Oceanogr., 1*, 102–117.

——— 1957. Trophic structure and productivity of Silver Springs, Florida. *Ecol. Monogr., 27*, 55–112.

ODUM, H. T., and R. C. PINKERTON. 1955. Time's speed regulator: the optimum efficiency for maximum power output in physical and biological systems. *Amer. Sci., 43*, 331–343.

REDFIELD, A. C. 1934. On the proportions of organic derivations in sea water and their relation to the composition of plankton, pp. 176–192 in James Johnstone Memorial Volume, Liverpool Univ. Press, 348 pp.

17

Reprinted from pages 7–18 of *Temperate Forest Ecosystems*, D. E. Reichle, ed.,
N. Y.: Springer-Verlag, 1970, 304pp.

Analysis of Ecosystems

Frederick E. Smith

The immediate goal of ecosystem analysis studies is to understand ecosystems and ecosystem processes; many of these goals are directly related to such applied problems as organic matter production or the influence of management practices on various ecosystem parameters. A beginning point for such studies must be research design. We first ask what has to be done to accomplish our goals. Oftentimes, subject matter will dictate the strategy. In this case, scientists must work together in multidisciplinary teams, study whole ecosystems (parts cannot be left out for lack of interest, personnel, or technology), share data immediately and completely, and devote a considerable effort to the process of synthesis. Systems oriented techniques with high-speed computers offer the only means by which this synthesis can be accomplished.

Criteria for Developing a Systems Model

Conceptual definitions must be converted into operational definitions lacking assumptions that may prove to be false in the real world. Conceptually the ecosystem is viewed as a functional unit with recognizable boundaries and an internal homogeneity. Operationally, we first recognize that boundaries are arbitrary and that functional unity may exist like beauty only in the eye of the beholder. We have, therefore, defined the ecosystem as everything that exists and happens within a precisely bounded region. Two sets of criteria are used for the location of the boundary. For many purposes the region must be large enough to contain a full set of ecosystem processes and their interactions; secondly, the boundary should be placed where inputs and outputs across it are most easily measured.

Once an ecosystem is defined by the location of its boundary (perimeter, roof, and floor), the next stage is to identify all its significant components. The air, land, and water can be subdivided into a number of components, and the plants and animals can be broken down to their species or to major species and groups of minor species. By lumping or splitting in various ways, the total number of components in an idealized ecosystem can vary from 5 to 50,000 the only restriction being that they must always add up to the whole ecosystem. The subdivision or fusion of components is itself one of the activities that will continue throughout the research program, and is in fact guided by the research. The ecosystem boundary also may be altered as we learn more about the system. If categories can be lumped, I think we will prefer to lump them. If they cannot be lumped, because the distinction between them is meaningful to the operation of the system, we will have to keep them separate. Hopefully, the total number of components needed to account for the significant ecosystem processes will not be more than several hundred.

The four major groups of components are the producers, the consumers, the decomposers, and the abiotic environment. So far, in the development of our programs, these account for approximately equal portions of the total cost of field and laboratory research.

Program goals require that research plans be organized to a degree unprecedented in ecology. Component projects must be compatible and, collectively, complete. The bulk of the research is conventional, however, so that most of the new emphasis is on organization. One method of formalizing the system is to reduce it to a series of tables, as follows.

Let us suppose that our system has n components, each of which can be described quantitatively according to many criteria: e.g., caloric content, carbon, phosphorus, nitrogen, and water (many more exist and are important). For each one of these, the following set of tables can be considered. Let us consider only phosphorus for the moment, remembering that additional sets of tables can be constructed for the other parameters.

The amount of phosphorus in each component (x_i) yields the top row of estimates in Table 1. The sum of this row is the total amont of phosphorus in the ecosystem.

Table 1. *The amounts (x) of phosphorus in each of the n components of an ecosystem, and the rates at which phosphorus is entering (a) and leaving (z) the system via each component. In any real system many of the rates will be virtually zero*

Component	1	2	3	.	.	.	n
Amount	x_1	x_2	x_3	.	.	.	x_n
Inflow	a_1	a_2	a_3	.	.	.	a_n
Outflow	z_1	z_2	z_3	.	.	.	z_n

Phosphorus may be entering the system, being added to some or all of the components. These rates (a_i) are the inflows, and form the second row of estimates in Table 1. Their sum is the total rate of inflow of phosphorus into the system. Phosphorus also may be leaving the system, being subtraced from some or all of the components. These rates (z_i) are the third row of estimates in Table 1, and their sum is the total rate of outflow from the system.

The location of a boundary for an ecosystem study will influence greatly how many of these rates are significant, and also how difficult they may be to estimate. One approach is to locate the boundary where the number of zero rates is maximal (where the system is least open in its connections with adjacent systems), such as the boundary between field and forest or between lake and land. It has been the custom, in fact, to draw boundaries where discontinuities are the most clear. Here inflows and outflows may be minimized. Here they will also be the least alike. That is, inflows will differ in both kind and amount from outflows.

A second approach is to locate the boundary where inflows and outflows are most similar. An example is to delimit a region of forest within a much larger forest. Since inflows and outflows will tend to be equal and opposite, their difference may have little net effect. The result, however, may be to maximize the total flow — i.e., selection of the most open system available. Finally, the significance of inflow and outflow also decreases as the system is made larger, leading to the choice of the largest system which is feasible for study.

In addition to inflow and outflow, phosphorus may be transferred from one component to another within the ecosystem. Movement through the food web is a good example. These rates can be presented as a square table of n rows and

Table 2. *Transfer rates of phosphorus among n components of an ecosystem. In any real system many of the rates will be zero*

| | | Transfer to each component | | | | | | | |
		1	2	3	4	.	.	.	n
Transfer from each component	1	—	y_{12}	y_{13}	y_{14}	.	.	.	y_{1n}
	2	y_{21}	—	y_{23}	y_{24}	.	.	.	y_{2n}
	3	y_{31}	y_{32}	—	y_{34}	.	.	.	y_{3n}
	4	y_{41}	y_{42}	y_{43}	—	.	.	.	y_{4n}
	.	.	.	.	.	.	.	.	.
	n	y_{n1}	y_{n2}	y_{n3}	y_{n4}	.	.	.	—

columns, as shown in Table 2. Each row represents losses from a particular component, and the row sum is the total rate of loss to other components. Each column shows gains to a particular component, and the column sum is the total rate of gain from all other components. Here again many entries will be zero or negligible. An advantage of this format is that it can handle any degree of complexity within the system. No assumption of discrete trophic levels is needed, and the confusion of a complicated flow diagram can be avoided. The portion of this table that comprises the food web is the "who-eats-whom" matrix.

With this information it is possible to write an equation for the rate of change of each parameter for each of the components. For example, for component 3:

$$dx_3/dt = a_3 - z_3 + (y_{13} + y_{23} + \cdots + y_{n3}) - (y_{31} + y_{32} + \cdots + y_{2n}) .$$

If component 3 were leaf litter, this equation might represent the rate of change with time in the amount of phosphorus due to: litter moved inward across the ecosystem boundary, minus that moved outward, plus litter received from the various plant components (other y_{i3} terms may be zero), minus litter removed by consumers or decomposers, and minus litter transformed into humus (other y_{3j} terms may be zero).

Now assume that we have made a study based on this approach, and have estimated the average amounts of phosphorus in all of the components and also all of the average rates of movement in and out of the system and among components. Then, Tables 1 and 2 can be filled with simple numbers representing amounts and rates. What have we learned? First, we have discovered where activity does and does not occur, and for most ecosystems this alone may be a considerable increase in information. We also can examine the study for completeness, searching for unevaluated components and rates or those subject to doubt. Finally, we can set up this system in the computer, beginning with all the x_i's at their estimated levels, and letting them change through time according to the equations, dx_i/dt.

At this point we will discover that these data are not at all adequate for the goals of the program. Each component will change in the same direction at the same rate for as long as the computer simulation continues. In the real world the ecosystem *responds* to change with changes in the rates, and the nature of these responses is what we are after. The use of an average rate (simple number) rather than a variable rate (mathematical function) destroys the system properties that are one goal of the study.

It turns out in fact that the new problem is an order of magnitude more complicated than the estimation of averages. In particular, the rates must be expressed as *functions* of the system, and not as simple numbers, if we wish to learn anything about the system. Each transfer rate, y_{ij}, and each outflow, z_i, is a *set* of functions, an equation, which relates the rate to all those direct causal factors that govern it. It is only then that we can predict how each rate will change if the system is changed. Furthermore, if all of these are combined in the computer, we can predict how the system will respond to change. This is the only way that we can discover how ecosystems operate.

This complication invokes a vast amount of experimental research, much of it at the level of physiological and behavioral ecology. Again, the system model helps define where the research must be done. We cannot decide casually to study the behavior of this or the physiology of that. We must discover and study *that which is relevant* in the system. For example, the rate at which grasshoppers eat grass may depend upon both the amount and quality of the grass, the number and size of the grasshoppers, the temperature, rainfall, humidity, wind, harassment from predators, and many other factors. Success of the study depends very strongly on locating and analyzing the more important processes relating cause and effect in the system. For this reason it is necessary for investigators to work together in constant communication sharing creative ability and data, and to work as closely with synthesists as with field researchers. Each investigator must concentrate on his particular problem with aims for maximal information input to all related problems. The increased efficiency with which each can pursue his research is the individual benefit from teamwork; the coordination of work permits synthesis and the study of whole ecosystems — in effect the creation of a whole new level of ecological science.

The expressions of rates as sets of functions require the estimation of many parameters other than amounts and rates of phosphorus. In a simplified form these can be arranged as shown in Table 3. For each component, in addition to x_i, a_i, z_i, and y_{ij},

Table 3. *Additional descriptors needed for the specification of functional relations in an ecosystem of n components*

Component	Component attributes (open list)					
1	w'_{11}	w'_{12}	w'_{13}	.	.	.
2	w'_{21}	w'_{22}	.	.	.	
3	w'_{31}	w'_{32}	w'_{33}	w'_{13}	.	.
.						
.						
.						
n	w'_{n1}	w'_{n2}	w'_{n3}	.	.	.

External attributes (viz., climate, weather; open list)
A_1, A_2, A_3, A_4, etc.

There may be a list of descriptors, additional attributes that are needed. These w_{ij}'s may include the average size and number of individuals in a component, age structure, distribution in space, etc. These, like the y_{ij}'s and the z_i's, may be sets of functions. In addition, there will be a set of inputs (A_i) to the system that are not included in Table 1. These are external variables such as those of climate and season that effect the system as a whole. These, together with the inflows, a_i, are the externally controlled variables that influence the system but are not in turn affected by the system. The remaining rates, y_{ij} and z_i, measure system processes that may be functions of any or all of the other amounts and rates that have been presented.

In general terms, the immediate goal of many of the research projects is to find appropriate mathematical functions for the *effect of external variables* and for *relationships among internal variables*. A complete set of these comprise the ecosystem model. The validity of the model can then be estimated by beginning with an initial distribution of amounts, x_i, and a program through time of the input variables (A_i and a_i) and observing how well the component amounts predicted from the computer follow observations in the field. The effect of a treatment such as fertilization or irrigation can also be followed both in the computer and in the field. Once a valid model has been constructed it becomes a powerful tool for prediction and management, as well as for obtaining insight into the governing properties of the ecosystem.

A System Model

So much for the grand scheme. An extremely simple hypothetical ecosystem will show how the methods of systems analysis can contribute to the design of research and the synthesis of results. Imagine a controlled ecosystem of three components: water, an aquatic plant population, and an herbivore population. Let it be internally homogeneous with the external environment held constant. Diagrammatically the system can be presented as a box (Fig. 1) with three compartments. This emphasizes that the ecosystem is a physical whole, a three-dimensional piece of the biosphere whose components must add up to that whole. The only inflow is water (with its phosphorus), while the only outflows are water and herbivores (with both outflows independent of each other). Assume that the herbivores do not take phosphorus from the water, and that the plants do not eat the animals. The ten remaining non-zero quantities are shown on the figure and are listed again in Table 4 in the format used for Tables 1 and 2.

Let us run this system with a constant inflow of 100 mg phosphorus per day, and wait until the outflow equals the inflow — i.e., until steady state has been achieved. This requires monitoring the outflow in water and in herbivores. We might find that, after equilibration, 19 mg/day flow out in the water while 81 mg/day are lost through herbivore emigration. Next, we might ask how the phosphorus is distributed within the system, and find that 9.5 mg are in the water, 1.4 in the plants, and 9.0 in the herbivores, giving a total "standing crop" of 19.9 mg of phosphorus in the ecosystem. Finally, with careful study using tracers we might find that plants are taking up phosphorus from the water at the rate of 133 mg/day, and losing it back to the water through excretion and similar processes at the rate of 7 mg/day. Similarly, we may find that herbivore grazing amounts of 126 mg/day, with 45 mg/day returned to

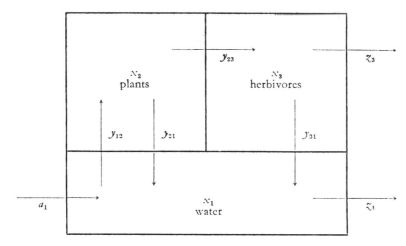

x_1 = amount of P in water
x_2 = amount of P in plants
x_3 = amount of P in herbivores
a_1 = rate of inflow of P in water
z_1 = rate of outflow of P in water
z_3 = rate of outflow of P in herbivores
y_{12} = rate of uptake of P from water by plants
y_{21} = rate of loss of P from plants to water
y_{23} = rate of uptake of P from plants by herbivores
y_{31} = rate of loss of P from herbivores to water

Fig. 1. Flow diagram for a very simple ecosystem composed of three components

Table 4. *The amounts and rates from Fig. 1 arranged in the format of Tables 1 and 2*

| | Component | | |
	Water	Plant	Herbivore
Amount	x_1	x_2	x_3
Inflow	a_1	0	0
Outflow	z_1	0	z_3
Transfers:			
Water	—	y_{12}	0
Plant	y_{21}	—	y_{23}
Herbivore	y_{31}	0	—

the water. These are a complete set of the data, and are summarized in Fig. 2. The reader will appreciate that if this were a real ecosystem with several hundred components an enormous amount of research just has been described.

As stated earlier, while this tells where the action is, it does not tell how it happens. We do not know how the system operates and cannot answer such questions as: What would happen if the inflow rate were doubled? What would happen if an herbivore with a greater tendency to emigrate were substituted? Will an herbivore that more efficiently utilizes plants tend to increase herbivore production and therefore reduce phosphorus outflow in the water? Another set of questions relates to the experimental design. How important are the various quantities? How accurately should each be estimated?

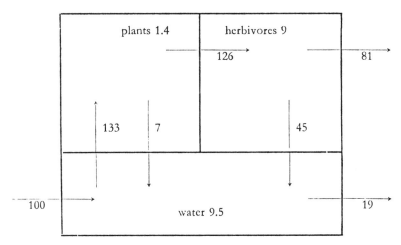

Fig. 2. The distribution (mg) and rates of movements (mg/day) of phosphorus in the three-component ecosystem after equilibration to a constant input rate of 100 mg/day (see Table 5)

To discover more about the system it has to be observed under varying conditions, through controlled experiment or natural variation. One approach is to run the system at different rates of phosphorus inflow, and observe the other nine quantities after equilibration has been achieved. Let us use the inflow rates of 25, 100, and 400 mg/day. (In the real world we should probably use more levels, but this is an imaginary system worked out without sampling error, and three are sufficient.) The results which might be obtained are shown in Table 5. All quantities increase with an increasing rate of flow, but not proportionally. Three quantities (the transfer from water to

Table 5. *Response of a three-component system to three different levels of input (a). Data taken after equilibration has been achieved (see Fig. 2)*

a_1	z_1	z_3	x_1	x_2	x_3	y_{12}	y_{21}	y_{23}	y_{31}
25	9	16	4.5	0.9	4	40.5	4.5	36	20
100	19	81	9.5	1.4	9	133	7	126	45
400	39	361	19.5	2.4	19	468	12	456	95

plants, the transfer from plants to herbivores and the outflow of herbivores) are very sensitive to the input rate, while the other six are much less sensitive.

Now we can proceed further with the analysis, even though this is but one of several kinds of treatments which might have been used. In real ecosystems, both local treatments affecting selected processes and overall treatments that affect the whole system are useful for eliciting changes (responses) in ecosystem processes.

Deriving Ecosystem Hypotheses

The next step is how the study of acquired data leads to the formulation of new concepts or *hypotheses*. Functions that relate the processes of the ecosystem to their most immediate causal factors need to be developed. This step occurs in all scientific work, and depends primarily on intuition or creativity on the part of the investigator. Since this particular problem was worked out backwards, beginning with functions and deriving the data, it is of interest to see how easily the reader can recover the functions given the data. Certainly a powerful influence is the "frame-of-reference", the past training or bias, with which the investigator works.

In this case it is too easy to write too many simple relations. For example, we can write:

$$z_1 = 2x_3 + 1 \quad \text{or} \quad z_1 = 4y_{21} - 9.$$

The best procedure, however, is to begin with proximate factors, and neither herbivore abundance nor the plant excretion rate have an obvious direct effect on the outflow rate of phosphorus in the water. The most obvious direct factor influencing z_1 is x_1, the amount of phosphorus in the water. Here we find that the simple relation $z_1 = 2x_1$ is an adequate and complete account for variations in z_1.

All of the rates have very simple functions that relate them to proximate factors (Table 6). The most simple are the negative exponentials of phosphorus outflow in the water, and in the return of phosphorus to water from plants and from herbivores. The two cases of active uptake or feeding (water to plants and plants to herbivores) are represented by the simple prey-predator notation of Lotka and Volterra. The emigration of herbivores is a simple expression of social interaction, the rate being proportional to the rate of contact between individuals. The functions for these rates are not defended as representative of the real world; they are used merely to produce a very simple system and to challenge the reader to consider other alternatives.

Table 6. *Functional relations of the six process rates in the simple three-component ecosystem (see Table 5)*

Numerical solution	Parameter functions	Parameter estimates
$z_1 = 2x_1$	$z_1 = c_1 x_1$	$c_1 = 2$
$z_3 = x_3^2$	$z_2 = c_2 x_3^2$	$c_2 = 1$
$y_{12} = 10x_1x_2$	$y_{12} = c_3 x_1 x_2$	$c_3 = 10$
$y_{21} = 5x_2$	$y_{21} = c_4 x_2$	$c_4 = 5$
$y_{23} = 10x_2x_3$	$y_{23} = c_5 x_2 x_3$	$c_5 = 10$
$y_{31} = 5x_3$	$y_{31} = c_6 x_3$	$c_6 = 5$

Since the numbers in the functions are only estimates and will vary from species to species, they should be represented by symbols, as shown in the middle column of the table. In this form we see that the system has been defined with six parameters ($c_1 \ldots c_6$) in association with the ten variables. A real-world ecosystem with several hundred components will probably have thousands of parameters, since each process is itself a very complex function, certainly not as simple as those used here.

In Fig. 1 one can observe the arrows in and out of each component, and write the three differential equations for changes in the amount of phosphorus in the three components:

$$dx_1/dt = a_1 + y_{21} + y_{31} - y_{12} - z_1,$$
$$dx_2/dt = y_{12} - y_{21} - y_{23},$$
$$dx_3/dt = y_{23} - y_{31} - z_3.$$

Substituting the functions for these rates, produces a mathematical model for this system:

$$dx_1/dt = a_1 + c_4 x_2 + c_6 x_3 - c_3 x_1 x_2 - c_1 x_1$$
$$dx_2/dt = c_3 x_1 x_2 - c_4 x_2 - c_5 x_2 x_3$$
$$dx_3/dt = c_5 x_2 x_3 - c_6 x_3 - c_2 x_3^2.$$

Sensitivity Analysis

We are now ready to ask how accurately each of the six parameters should be estimated, since this will guide the continuing research on this system (assume that in the real world we are much less certain about the values of these parameters). For this problem we have techniques available in systems analysis collectively called sensitivity analysis. The current version of the model is set up in the computer, and then the value of each parameter in turn is varied upward and downward while the performance of the system is observed. If the system shows little response, it is not sensitive to the precise value of that parameter, which therefore need not be estimated with great accuracy. If the system or one important part of it responds strongly, the value of the parameter is important and must be estimated with precision. In this way future effort can be allocated to improve the accuracy of the sensitive parameters. Even in a system as simple as our example, the sensitivity of response varies considerably from one parameter to another.

Table 7. *Sensitivity analysis. The percentage error in predicted outflow of phosphorus in water (z_1) for an overestimation of one percent in each parameter. The relative sensitivity is obtained from the middle column by setting the largest error equal to 1.0*

Parameter	% error	Relative sensitivity
c_1	+0.900	1.00
c_2	−0.426	0.47
c_3	−0.900	1.00
c_4	+0.047	0.05
c_5	+0.853	0.95
c_6	0.000	0.00

This procedure of observing performance requires first a definition of the criterion for performance. If this were a potential system for removing phosphorus from water, our criterion could be the rate of outflow of phosphorus in water. The amount of change, after equilibration, in the rate of outflow of phosphorus in water (z_1 or $c_1 x_1$) following a given small change in each of the six parameters of our model are shown in Table 7. For a change of one percent in the value of each parameter, the rate of outflow changes from zero to 0.9%. This means that, if any of the parameter estimations were in error by 1%, the predicted rate of phosphorus outflow would be in error by the indicated amount. Sensitivity is zero for the parameter, c_6. This implies that the rate of return of phosphorus to water from the herbivores, y_{31} or c_3x_3, had no effect on the outflow of phosphorus in water, and would not need to be estimated at all! Conversely, the three parameters c_1, c_3, and c_5 are all very important, indicating that the majority of experimental effort should be expended in reducing their sampling errors. Moderate effort should be spent estimating c_2, but all that is needed for c_4 is the order of magnitude.

The relative sensitivities, obtained from dividing the second column by its largest entry, gives a good guideline for the allocation of effort in the estimation of these parameters. This assumes, of course, that we are interested only in the prediction of phosphorus outflow in water. Allocation then becomes an attempt to equalize the contributions from the several parameters to the final error of prediction.

A different criterion may produce different results. In ecosystem analysis, we do not have a single applied goal in mind, but wish to "understand ecosystems". In general, such a model would be exercised to predict the entire array of component amounts in the system. If this approach is applied to our simple system, the results are those of Table 8. Percentage responses of all three components are given for a

Table 8. *Sensitivity analysis. The percentage errors in predicted amounts (x) in the three components for an overestimation of 1% in each parameter. The sum is taken without respect to sign. The relative sensitivities are obtained from the sums by setting the largest equal to unity*

| | Component | | | | Relative |
Parameter	Water	Plant	Herbivore	Sum	sensitivity
c_1	—0.100	—0.079	—0.106	0.285	0.14
c_2	—0.426	+0.354	—0.405	1.185	0.59
c_3	—0.900	+0.068	+0.106	1.074	0.53
c_4	+0.047	—0.004	—0.006	0.057	0.03
c_5	+0.853	—1.064	—0.100	2.017	1.00
c_6	0.000	+0.357	0.000	0.357	0.18

change of one percent in each parameter. One method (perhaps not the best) for summarizing the effects on the several components is to add up the responses without respect to sign. This reflects the summed percentage error of prediction for the system, and is shown on the fifth column. Dividing these by their largest entry yields the relative sensitivities shown in the last column.

Using the sum of the absolute component amounts as our criterion of sensitivity, we find that c_5, the parameter associated with herbivore grazing, is the most important, and should be estimated with the smallest percentage error. Parameters c_2, associated with herbivore outflow, and c_3, associated with plant uptake of phosphorus, are about half as important. The remaining three parameters are relatively unimportant, and need not be accurately estimated.

Of course, this model is not realistic, and its results should not be applied to any related real system. The intent here is to demonstrate that, even in a model this simple, sensitivity to error in parameter estimation varies greatly from one parameter to another. In more complex systems it can be expected that such variations will be at least as great, and that many parameters will not need be known with any great degree of precision. Using this knowledge in the allocation of further research will produce considerable savings in both time and effort. Thus, such a model is useful early in the research program.

Even a crude, tentative model can be useful as a tool for guidance in the allocation of further effort, but this approach has faults. Allocation of effort is based upon the assumption that the model is more or less correct. If it is wrong, the allocation of effort and future research direction also may be wrong. There is the possibility that the research group may become "model-bound", digging a deeper and deeper rut on the wrong road. (It is important, therefore, that critics and skeptics also be included.) Major disasters of this kind may be avoided by developing several studies in different places, each with its own effort toward synthesis and proceeding in a quasi-independent fashion, while at the same time maintaining compatibility between studies and providing for mutual evaluation and self-correction.

Assuming that our model is realistic enough to permit more detailed evaluation of the system, let us now return to a series of questions that was asked earlier concerning: What would happen if the inflow rate were doubled? Using our model, setting the inflow rate at 200 mg/day, and letting it run in the computer until the system stabilized, we would find that:

$$x_1 = 13.6 \qquad z_1 = 27.3 \qquad y_{21} = 9.1$$
$$x_2 = 1.8 \qquad z_3 = 172.7 \qquad y_{23} = 238.4$$
$$x_3 = 13.1 \qquad y_{12} = 247.4 \qquad y_{31} = 65.7$$

What would happen if an herbivore with a greater tendency to emigrate were substituted. Here we look at the sensitivity analyses (Table 7), and consider the effects of an increase in the parameter, c_2. The signs associated with the responses indicate the direction of change for an increase in the size of the parameter. Thus, Table 7 shows that the outflow of phosphorus in the water, z_2, would decrease. Table 8 suggests that the amount of phosphorus in the water inside the system also would decrease, while plant density would increase and herbivore density decrease.

Will an herbivore with a more voracious feeding habit tend to increase herbivore production and, therefore, reduce phosphorus outflow in the water? This involves an increase in the parameter, c_5. Table 8 indicates that phosphorus in plants and herbivores would *decrease*, while the phosphorus in water (and its outflow, Table 7) would *increase*. This effect may not be obvious, and is a good example of the value of using whole systems for drawing such conclusions. Further thought shows that the result is logical. If herbivores inflict greater plant damage, plants are less able to grow and take up phosphorus. This not only implies more phosphorus flowing out in water, but a lower plant density and, after equilibration, also a lower herbivore density. In food chain models it is generally found that, the more efficient in feeding the top trophic level becomes, the less common it becomes!

These questions and answers may help to explain what ecologists mean by "understanding" ecosystems. In part, it is being able to predict what would happen if some changes were induced in the system. It is more, however. After describing several systems, a reasonably large set of properties may be found which are common to many, if not all types of ecosystems, and which are *system* properties having little to do with the particular evolutionary adaptations of the species involved. If this should happen, then there is a discipline called the "principles of ecosystems", rather than a subject area which is merely a collection of studies of particular ecosystems. The following chapters on woodlands will take on significance reaching far beyond the particular cases they treat.

Postscript

Although a simple example was presented and analyzed as though it were set up in the computer, using simulation and various techniques of systems analysis, we were cheating. The model is so simple that it has a mathematical solution, making numerical methods unnecessary. After equilibration, the steady state values for the three components are:

$$c_2 c_3 x_3^2 + c_1 c_5 x_3 + c_1 c_4 = a_1 c_3$$

(solved for x_3 using the quadratic formula),

$$c_3 x_1 = c_4 + c_5 x_3,$$
$$c_5 x_2 = c_6 + c_2 x_3.$$

These can be differentiated for each variable (x) with respect to each parameter (c). For relative errors (percentage changes), an example would be: $(dx_1/x_1)/(dc_1/c_1)$. Simple mathematical solutions will probably not be possible for real ecosystems, and systems analysts will need to depend upon numerical methods using computers.

Suggested Reading

Forrester, J. W.: Industrial dynamics. Cambridge: M. I. T. Press 1961.
Holling, C. S.: The functional response of invertebrate predators to prey density. Mem. Entomol. Soc. Can. 48, 1—85 (1966).
Watt, K. E. F.: Ecology and resource management. New York: McGraw-Hill Book Co. 1968.

Reprinted from *Ecology* **57**:1244–1253 (1976), by permission of the publisher,
Duke University Press, Durham, North Carolina

ECOSYSTEM PERSISTENCE AND HETEROTROPHIC
REGULATION

R. V. O'NEILL

Environmental Sciences Division, Oak Ridge National Laboratory,[3]
Oak Ridge, Tennessee 37830 USA

INTRODUCTION

Trends in recent literature (McMullin 1973) indicate a growing interest in the ecosystem as an object of research. Studies on microcosms (Patten and Witkamp 1967) and on small, unique systems (Wiegert 1973) have made significant contributions, but major stimulus has been provided by large-scale ecosystem projects such as Hubbard Brook (Likens et al. 1970) and the biome studies of the U.S. International Biological Program (Auerbach 1971, Van Dyne 1972). A recent review of the field (Reichle and Auerbach 1972) highlights major trends in ecosystem analysis and points to potential contributions in environmental assessments and land use planning. As data on total ecosystems become available in the literature (e.g., Odum and Pigeon 1970, Satchell 1971, Whitfield 1972, Reichle et al. 1973) theoretical investigations of ecosystem function become more and more feasible. Initial attempts at theory (e.g., Whittaker and Woodwell 1972, Holling 1973, May 1973) have already established the value of studying homeostatic mechanisms at the ecosystem level.

An analysis of ecosystems can be approached from numerous viewpoints. In some instances, emphasis is on animal food webs (Wiegert and Owen 1971, May 1973), while in other cases, plant communities may be the most important consideration

(Whittaker and Woodwell 1972). In the present study, the ecosystem (Fig. 1) is abstracted to three major components (O'Neill et al. 1975): (1) active plant tissue (the photosynthetic or energy capturing base for the system), (2) heterotrophs (hypothesized to play a role in rate regulation), and (3) inactive organic matter (such as soil organic matter, sediments or structural plant tissue, which serve as major buffers against short-term environmental changes and permit retention and recycling of nutrients). These components are suggested as a minimal set of state variables for the ecosystem. My approach explored the conceptualization shown in Fig. 1 with published data on a set of contrasting ecosystems. The purpose will be to investigate general patterns of ecosystem dynamics rather than focusing in detail on a single system.

MATHEMATICAL STATEMENT OF THE CONCEPT

The conceptualization in Fig. 1 can be expressed in a corresponding mathematical representation. The rigor of the mathematics permits greater precision in the statement of the concept and its limitations. Since the concern is with dynamic properties, a set of differential equations is chosen to represent the system.

Three state variables will be emphasized: active plant tissue (P), heterotrophs (R), and inactive organic matter (S). The variables are expressed in units of calories per square metre. For present purposes the choice of energy units can be understood simply as a convenience. Because of the influence of several workers, such as R. L. Lindeman (1942) and H. T. Odum (1957), much of the data available in the literature is expressed in units of energy. The generalizations developed here should

[1] Manuscript received 29 April 1975; accepted 29 April 1976.

[2] Contribution No. 260, Eastern Deciduous Forest Biome, US-IBP. Publication No. 978, Environmental Sciences Division, Oak Ridge National Laboratory.

[3] Operated by Union Carbide Corporation under contract with the Energy Research and Development Administration.

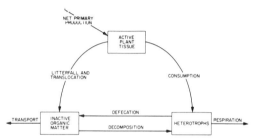

FIG. 1. A simplified concept of the ecosystem. The three boxes represent state variables of the system and arrows represent transfers of energy between the system components.

hold for any single parameter (such as mass or carbon) which can be taken as a measure of the biological (or biologically derived) tissue in the system. The time unit is a year with transfers expressed in calories per square metre times year. This time scale is enforced by data limitations, but also permits a degree of simplification consistent with the present effort. No attempt is made to consider seasonal dynamics.

Basing the present mathematical construct on energy flow has the advantage of permitting analysis of published studies on ecosystems. However, this decision also prevents the inclusion of factors such as species diversity and other sources of heterogeneity in the system which have been clearly shown (e.g., May 1973) to influence system dynamics.

A set of equations for the rate of change in the three variables can be derived on the basis of conservation of energy. Each term in the equations (separated by a + or −) represents a single input or output as represented in Fig. 1.

$$\dot{P} = j - aP - bcPR/(cP + R), \qquad P(t_o) = P_o,$$

$$\dot{R} = (1 - f)bcPR/(cP + R) + gqSR/(qS + R)$$
$$- dR(1 + R), \qquad R(t_o) = R_o, \qquad (1)$$

$$\dot{S} = fbcPR/(cP + R) + aP - gqSR/(qS + R)$$
$$- hS, \qquad S(t_o) = S_o,$$

where

  j is net primary production (gross production minus autotrophic respiration) in units of energy fixed per square metre times year,

  a is the fraction of P transferred to S per year (litterfall + translocation + mortality, etc.),

  b is the maximum consumption of P per unit R times year under conditions of no food limitation,

  c is the fraction of P available for consumption by the heterotrophs,

  f is the fraction of ingested material that is not assimilated,

  g is the maximum decomposition rate per unit R times year under conditions of no food limitation,

  q is the fraction of S available for decomposition,

  d is the fraction of R lost per year in respiration, and

  h is the fraction of S lost from the system per year in leaching or physical transport.

Net primary production (j) has been chosen as the driving force for pragmatic reasons. While a more dynamic representation might consider gross production and include autotrophic respiration as a loss, in fact, very few ecosystem studies are available with sufficient information to proceed in this manner. By utilizing net primary production, it is possible to include more data sets in the analysis.

The expression for respiration of heterotrophs (final term of second equation) is composed of a linear term (dR) and a quadratic term (dR^2). The quadratic term is utilized in the tradition of the Lotka-Volterra (Lotka 1923, Volterra 1938) equations to represent density dependent factors which increase respiration as heterotroph density becomes very large. The use of the quadratic term implies a direct interaction between individual heterotrophs or heterotrophic populations. DeAngelis et al. (1975) have shown that this quadratic term can make a significant contribution to system dynamics. A term of the form:

$$k_1 k_2 X_1 X_2/(k_2 X_1 + X_2) \qquad (2)$$

where k_i represents some parameter and X_i represents the value of a state variable, appears in all equations for the nonlinear feedback relationship implied in consumption and decomposition processes. Since this expression appears in all equations, some brief explanation of its properties seems appropriate. Derivation of models of this form can be found in Smith et al. (1975) and Kerr (1971). A particularly elegant derivation is available (Timin et al. 1973) based on predator search time. DeAngelis et al. (1975) have provided extensive justification and mathematical analysis. Expression (2) has been successfully applied in three ecosystem models (Shugart et al. 1974, Park et al. 1974, MacCormick et al. 1972), as well as a fish population model (Kitchell et al. 1974) and a study of resource competition (Smith et al. 1975).

At values of the X_i for which $k_2 X_1$ is approximately equal to X_2, the rate calculated from Expression (2) depends on the magnitude of both variables. At very large values of X_1 (i.e., superabundant food, $k_2 X_1 \gg X_2$), the rate depends only on the value of X_2 (consumers available to feed on the superabundant food). At very large values of X_2, (i.e., $X_2 \gg k_2 X_1$) there are more consumers

TABLE 1. Parameter values for Eq. (1) derived for six ecosystems. In many instances parameters are shown to three significant figures, but all calculations discussed in the paper utilized the parameters to seven figures to reduce rounding errors. Sources for table headings as follows: Tundra (Whitfield 1972); Tropical forest (Odum and Pigeon 1970); Deciduous forest (Reichle et al. 1973); Salt marsh (Teal 1962); Spring (Tilly 1968); Pond (Emanuel and Mulholland 1975)

	Tundra	Tropical forest	Deciduous forest	Salt marsh	Spring	Pond
P	32,010	6597	5568	6227	1017	20
R	128	61.1	86.86[a]	166.6	77	108.8
S	143,980	164,283	198,509	3606	3696	400
a	0.0183	0.846	1.180	1.154	9.395	162
b	0.0627	9.043	3.566	6.311	0	53.982
c	1.0[b]	1.0[b]	1.0[b]	0.87[b]	0	1.0[b]
d	0.0360	1.616	0.855	0.162	0.900	0.348
f	0.7[c]	0.7[c]	0.7[c]	0.7[c]	0.9[d]	0.9[d]
g	4.759	99.942	76.996	26.551	71.697	47.476
h	0	0	0	1.018	1.109	0
j	594	6132	6873.6	8205	9508	4152
q	0.03[c]	0.016[e]	0.011[c]	1.0	1.0	1.0

[a] B. S. Ausmus (*personal communication*).
[b] All active plant tissue was assumed to be available for consumption, except for Salt Marsh where author considered that the algal portion was not consumed.
[c] Assimilation efficiency assumed to be 30%.
[d] Assimilation efficiency assumed to be 10%.
[e] Only upper layers of soil organic matter were considered to be available for decomposing heterotrophs.

than can be supported by the system and the rate depends only on the value of X_1. Note that for fixed X_1 or X_2, the expression takes on the general form of the Michaelis-Menten equation. The presence of X_2 in the denominator of Expression (2) implies that heterotrophs show a mutual interference or competition in feeding. Thus, Eq. (1) incorporates both direct interference (implied in the quadratic term) and indirect interference through feeding.

A number of experimental studies (Salt 1967, Hassell 1971, Wynn-Edwards 1962) suggest that mutual interference in feeding is an important phenomenon. Of particular interest is a recent study (Salt 1974) in which food supply and consumer populations were varied independently. Expression (2) reduces to a Michaelis-Menten equation with respect to either food supply or consumer when the other variable is held constant. This property is required to fit the data of Salt, and is not found in other models presently available in the literature.

Expression (2) appears to have some generality when applied to the interaction between pairs of species. However, Eq. (1) clearly extrapolates to interactions between entire trophic levels. No strong justification can be made for this extrapolation at present but the approach appears reasonable. Further research will be required before it is possible to derive a more adequate expression.

DERIVATION OF MODEL PARAMETERS

The simplicity of the present concept (three variables and nine parameters), permits the parameters of Eq. (1) to be estimated for six systems by analyzing data from literature sources (Table 1). In many cases it is necessary to accept assumptions developed and justified by the original authors. Calculation of the parameter values in Table 1 was straightforward since the papers present values for the state variables (P, R, and S) and each of the fluxes shown in Fig. 1. Individual terms of Eq. (1), representing each flux, were then set equal to the value given in the paper and solved for the parameter values. The only additional assumptions required are indicated in the footnotes to the table.

All of the state variable values given in Table 1 represent equilibrium values with the exception of the Oak Ridge deciduous forest. The literature cited presented the state variable values shown in the table. However, these values represent a transient condition. The equilibrium values are $P = 5,561$, $R = 89.17$, and $S = 637,000$ (values rounded to four significant figures). Thus, while the autotroph and heterotroph components are near equilibrium, the inactive organic matter, S, can be expected to continue to grow in this system. The equilibrium values have been used in the analyses presented in the remainder of the paper.

Values for state variables and fluxes given by the original authors have been accepted here without judgment on their accuracy or reliability. In many cases the original fluxes were not directly measured but were calculated based on equilibrium or other assumptions. While each set of numbers is internally consistent and relevant to the present study, the effects of data inaccuracies on conclusions in the present study are difficult to assess accurately. Therefore, some care has been taken to present only those results which appear consistent among all the data sets or which show orders of magnitude differences which are unlikely to be negated by measurement errors.

ECOSYSTEM RECOVERY FROM PERTURBATION

The temporal behavior of the six systems (Table 1) was simulated on the computer to compare their relative ability to recover from perturbations. The same perturbation (a 10% decrease in the initial condition for P) was applied to each system and an index of the recovery of each system was taken as the square root of the sum of squares of the deviations between the perturbed transient behavior and the equilibrium. The squared deviations were summed over time (25-yr simulations) and over the three variables, and each deviation was divided by the value of the variable at equilibrium. The sum-

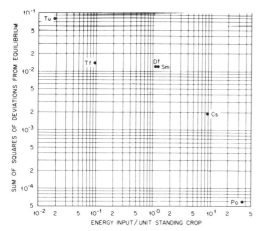

Fig. 2. Comparison of system recovery ability estimated from two independent concepts. The abscissa represents the power parameter ($j/P + R$) suggested by Odum and Pinkerton (1955). The ordinate represents deviations from equilibrium during recovery from perturbation. The values on the ordinate were calculated by numerical simulation of Eq. (1) using the parameter values given in Table 1. Points on the graph correspond to the systems in Table 1: Tu, tundra; Tf, tropical forest; Df, deciduous forest; Sm, salt marsh; Cs, cold spring; Po, pond.

mations were made annually since the basic time resolution of Eq. (1) is 1 yr.

In all of the simulations performed, the model displayed asymptotic stability, i.e., the perturbed system returned asymptotically toward the equilibrium. The systems differed in the rate at which they reapproached equilibrium as well as in the maximum deviation from equilibrium which occurred in each state variable during the simulations. By using the sum of squared deviations as an index, it was possible to express in a single number both aspects of the recovery behavior.

This analysis provided the following ranking of the systems: the pond system displayed the smallest values for the recovery index followed by the spring, salt marsh, deciduous forest, tropical forest, and tundra. From this analysis, it appears that the aquatic systems with rapidly changing phytoplankton populations or annual grasses show greater recovery ability than terrestrial systems. It should be carefully noted that this ranking does not account for the probability of perturbations actually occurring in each environment. The recovery potential of the tropical and deciduous forest systems might be an evolutionary consequence of the mesic environment in which they are found. The greater recovery capability of the aquatic systems may be a result of highly probable perturbations, particularly in nutrient availability.

Odum and Pinkerton (1955) suggested the concept of power as a basic parameter of system function. For ecological systems, they defined power capacity as the quantity of energy processed per unit time per unit living tissue (net primary production per unit autotroph plus heterotroph). They hypothesized that greater power capacity would result in greater capability to counteract change. The power parameter can be estimated for each of the systems in Table 1, permitting an independent ranking of their relative recovery ability.

Figure 2 shows that there is a monotonic relationship between the power parameter and the calculated recovery index. The same relationship holds for a 10% reduction in net primary production and for random fluctuations in temperature with all fluxes (except litterfall) represented as Q_{10} functions of temperature.

Figure 2 clearly shows that the power parameter and the simulations produce the same ordering of the relative recovery ability of the six systems. The correspondence cannot be taken as evidence of the validity of either parameter, but does suggest that the ordering of the systems in Fig. 2 may reflect real system properties worthy of further investigation.

Equilibrium Solutions

Examination of the equilibrium solution of Eq. (1) reveals some of the implications contained in the present ecosystem concept. A solution can be determined relatively easily for systems with $h = 0$ (see Table 1). In this case, the only input to the system is net primary production and the only output is heterotrophic respiration. Since the input and output must be equal at equilibrium, we can write

$$j = dR(1 + R).$$

Solving for R we obtain:

$$R = \frac{1}{2}(-1 + \sqrt{1 + 4j/d}). \tag{3}$$

Having obtained an expression for R as a function of the parameters, one can solve the first equation of Eq. (1) for P at equilibrium:

$$P = 1/(2\alpha)[\beta + (\beta^2 + 4\alpha\gamma)^{\frac{1}{2}}] \tag{4}$$

where

$\alpha = ac$
$\beta = jc - aR - bcR$
$\gamma = jR.$

And having obtained an expression for P, one can utilize the third equation to obtain

$$S = zR/(1 - qz) \tag{5}$$

where

$$z = [(fbcPR/cP) + aP]1/gqR.$$

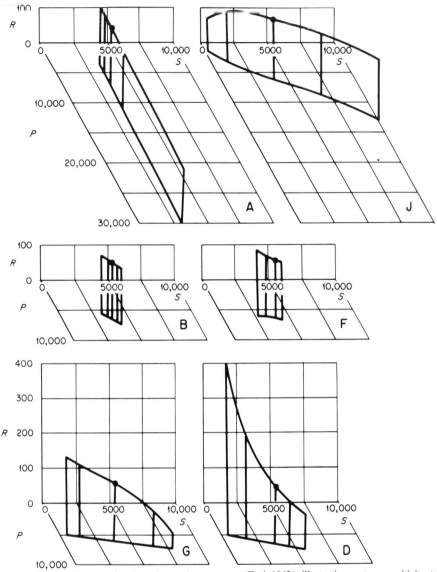

Fig. 3. State space diagrams for the salt marsh ecosystem (Teal 1962) illustrating system sensitivity to parameter changes. The parameters are defined following Eq. (1). The three axes of the graphs correspond to the three state variables of Eq. (1). Points in the state space represent values of the state variables with the parameter set at 20%, 60%, 100% (black dot), 140%, and 180% of their tabled values.

The equations are more complex than they appear at first, since values of R and P appearing in Eqs. (4) and (5) are actually equilibrium values which must first be calculated from the preceding equations.

Equation (3) is particularly interesting since it indicates a direct relationship between net primary production and heterotrophic standing crop. The relationship suggests the mechanism by which hetero-

trophs could exert effective control on the system (Golley 1973, Lee and Inman 1975). Since net primary production is considered as input to the system, heterotrophic respiration represents the energy loss from the system with the potential to compensate for changes in input.

A control mechanism must be able to monitor inputs to a system and adjust outputs to achieve

TABLE 2. Values of the partial derivative of the equilibrium value of heterotroph standing crop with respect to the j (net primary production). The tabled value is the change in heterotroph standing crop required to compensate for a unit change in net primary production

Tundra (Whitfield 1972)	1.1×10^{-1}
Tropical forest (Odum and Pigeon 1970)	5.0×10^{-3}
Deciduous forest (Reichle et al. 1973)	6.5×10^{-3}
Pond (Emanuel and Mulholland 1975)	1.3×10^{-2}

some objective. The objective of ecosystem persistence requires resistance to and recovery from perturbations. If it is assumed that the parameter d is a constant, then the heterotrophs might compensate for changes in system input by changes in energy content and consequent changes in respiration output from the system. In fact, given the existence of density dependent factors, the change in R required to re-establish equilibrium is proportional to the square root of the change in net primary production.

An expression for the change in the equilibrium value of R following a change in j can be derived by taking the partial derivative of Eq. (3) with respect to j.

$$\frac{\partial R}{\partial j} = 1/\{d[1 + (4j/d)]^{\frac{1}{2}}\}. \qquad (6)$$

Table 2 gives values for this partial derivative for the four ecosystems in Table 1 for which $h = 0$. The values range from 1.1×10^{-1} for the tundra system to 5.0×10^{-3} for the tropical forest. The smaller the value of the partial, the smaller the change in heterotroph energy content resulting in an equilibrium following a change in j. For all systems considered here, the change in R is at least an order of magnitude smaller than the change in input. Thus, there is considerable amplification of the signal in the system. Notice that change is not required in the number of heterotrophs, with associated reproductive time delays. Given that respiration rate is directly proportional to the size of individuals, output could be increased by increased body weight of higher organisms or rapid reproduction of microbial heterotrophs. Because of differences in turnover rates and density dependent phenomena, only small changes in heterotroph standing crop result. Because the changes given in Table 2 are quite small, the control mechanism appears very efficient and it appears feasible for the heterotrophs to exert a control function in spite of their relatively small standing crops.

PARAMETER SENSITIVITIES

Parameter sensitivity permits further insight into system behavior as represented by Eq. (1). The

equations were quantified for the salt marsh system (Teal 1962) and solved numerically. By varying parameter values and examining model output after 25 yr of simulated time, the results graphed in Fig. 3 were obtained. Each parameter was set at 20%, 60%, 100%, 140%, and 180% of the values given in Table 1. A black dot indicates the point in state space obtained with the tabled value.

Changes in the litterfall parameter, a, cause the greatest change in P as might be expected. Decreasing the parameter rapidly increases the value of P and very slightly decreases S while having little effect on R. Increasing net primary production, j, increases both autotrophs and detritus but affects heterotrophs only when greatly reduced production begins to limit the food supply. Changing the value of the maximum feeding rate, b, and the fraction of the food which is not assimilated, f, cause much smaller changes in the total system. The changes are quite predictable, with an increase in b or a decrease in f reducing both P and S, but increasing R. The small absolute changes reflect the relatively small fraction of the total energy pool of autotrophs which is directly consumed by heterotrophs. In contrast, changes in decomposition rate, g, cause much larger effects on S and R because of the larger absolute magnitude of decomposition fluxes. Changes in respiration rate, d, cause the most dramatic changes in heterotroph energy pool. Examination of Eq. (1) shows that this results from the quadratic term in the second equation, i.e., the density dependent term. Because of the density dependent factors, the response of the heterotrophs to change in d is curvilinear.

Sensitivity of the active tissue $(P + R)$ at equilibrium to perturbations in the parameters can be estimated by the partial derivatives of the equilibrium equations, Eqs. (3) and (4), with respect to the parameters of interest. For example,

$$\frac{\partial(P + R)}{\partial a} = \frac{1}{2\alpha a}\left[\frac{a(2\gamma c - \beta R)}{\phi} - aR - \beta - \phi\right]$$

$$\frac{\partial(P + R)}{\partial b} = -\frac{R}{2\alpha}\left[1 + \frac{\beta}{\phi}\right]$$

$$\frac{\partial(P + R)}{\partial c} = \frac{1}{2\alpha c} \qquad (7)$$

$$\frac{\partial(P + R)}{\partial d} = \frac{j}{d^2\theta}\left[\frac{1}{2\alpha}\left(a + bc - \frac{2\alpha j - a\beta - \beta bc}{\phi}\right) - 1\right]$$

where

$$\phi = (\beta^2 + 4\alpha\gamma)^{\frac{1}{2}}, \quad \text{and}$$
$$\theta = (1 + 4j/d)^{\frac{1}{2}}.$$

TABLE 3. Equilibrium sensitivities for three ecosystems. The tabled values indicate the percentage change in active living tissue $(P + R)$ to be expected from a 10% change in the parameter. The sensitivities were calculated from Eq. (7)

	Tundra (Whitfield 1972)	Tropical forest (Odum and Pigeon 1970)	Pond (Emanuel and Mulholland 1975)
Litterfall (a)	9.96	9.90	1.25
Maximum feeding rate (b)	0.14	0.97	0.353
Autotroph availability (c)	0.00054	0.0089	0.298
Heterotroph respiration rate (d)	0.048	0.438	4.215

Table 3 gives values for these partials for three contrasting systems. Examination of the table indicates some of the same trends seen in Fig. 3. However, a comparison of the three systems demonstrates some further differences in these ecosystems.

Of the parameters considered, changes in transfer rate from autotrophs to detritus cause the greatest change in standing crop of living tissue. In terrestrial systems, there is an annual turnover of photosynthetic issue $(a \approx 1.0)$ and a 10% change in the parameter a causes approximately a 10% change in standing crop. In the aquatic system, phytoplankton turnover is far more rapid and changes in the transfer rate have an effect which is almost an order of magnitude smaller. Changes in heterotroph respiration, d, or availability of autotrophs for consumption, c, cause order of magnitude differences in the responses of the three systems. These differences reflect the differences in the relative intensity of heterotrophic processes. Differences in sensitivity to changes in feeding rate, b, follow the pattern of the partial derivatives in Table 2.

ROLE OF HETEROTROPHS IN ECOSYSTEMS

These analyses suggest a control function for heterotrophs which may not be directly reflected in the magnitude of their standing crop. Table 3 indicates orders of magnitude differences in response to changes in heterotrophic respiration, d, suggesting important differences in the intensity of heterotroph activity in these systems.

Earlier work (O'Neill et al. 1975) suggested that ecosystems display interesting differences in their energy-capturing bases. Some systems possess a number of species, each with a different spectrum of responses to the environment, and each capable of rapid reproduction (e.g., phytoplankton communities in a lake). Other systems, such as forests, have large structural components to maintain the plants through unfavorable periods. O'Neill et al. (1975) also pointed out that the "phytoplankton" system permits rapid reproduction with potential exhaustion of a critical resource and subsequent depression of plant populations. If system persistence involves constancy in spite of environmental changes, the need for explicit regulation of the auto-

trophic base would be more apparent in a pelagic system than in a forest. Therefore, it is not surprising that a larger biomass of consumer organisms is typically found in the aquatic ecosystem.

The hypothesis offered by O'Neill et al. (1975) and the analyses in previous sections suggest an examination of the systems in Table 1 for patterns determining heterotrophic standing crop. If the ecosystem is conceived as merely the chance coincidence of populations with overlapping environmental requirements, one might expect the heterotroph standing crop to be proportional to net primary production. With a larger supply of available energy, one might expect more heterotrophs to be supported without disrupting the system. Figure 4 clearly indicates that the present data do not support the premise that heterotrophs are simply "parasitic" on the autotrophs.

If the ecosystem is conceived as an interacting system, then heterotrophs might be assigned a control function. If heterotrophs can operate to control the rate of energy production by autotrophs as suggested by O'Neill et al. (1975) and Lee and Inman (1975), this function should be more important in systems with a greater potential for rapid changes, i.e., systems with greater "power" capacity (Odum and Pinkerton 1955). Thus, one would expect

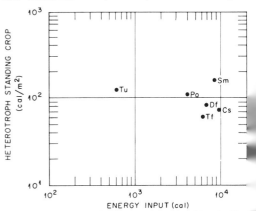

FIG. 4. Relationship between heterotrophic standing crop and energy input (net primary production) for the six systems in Table 1. The abbreviations for the eco systems are given in the legend to Fig. 2.

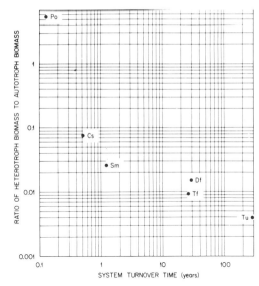

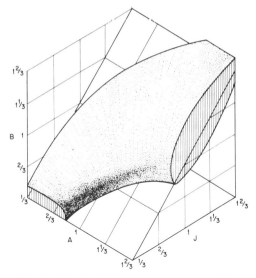

FIG. 5. Ratio of heterotroph to autotroph standing crop graphed against system production time $(P + R + S/j)$ for the six systems in Table 1. The abbreviations for the ecosystems are given in the legend to Fig. 2.

FIG. 6. Graph of the ecosystem response for the tropical forest ecosystem (Odum and Pigeon 1970). The axes represent parameter values of Eq. (1) A = litterfall and translocation, B = maximum grazing rate, and J = net primary production. The volume enclosed by the solid figure contains combinations of parameter values which do not change the state variables by > 20% at equilibrium.

greater heterotroph standing crop (regulators) per unit autotroph standing crop (regulated) as the production turnover time $(P + R + S/j)$ of the system increases, i.e., as the system potential for rapid change increases. Note that turnover time is defined relative to net primary production, rather than the sum of losses from the system. Although the turnover times are nearly identical, use of net primary production seems a more relevant index of the system's capability for rapid increase. Figure 5 indicates that the expected relationship exists. While such a simplistic relationship may not stand the test of more extensive data analysis, present evidence supports the hypothesis that heterotrophs play a rate regulation role in the ecosystem.

Ecosystem Response Volume

A primary motivation for the development of ecosystem theory is the desire to understand the full range of system responses to perturbations. Such understanding would indicate which perturbations were unacceptable because they move the system too far from the current operating state. To illustrate this concept, one can define an arbitrary criterion: equilibrium values for the three state variables in Eq. (1) are not changed by more than 20%. Then one can define the region in parameter space within which this criterion remains valid. If the criterion represents a judgment about what constitutes "serious damage" to the system, this region

in parameter space shows the system manipulations which can be tolerated. We will designate this region the ecosystem response volume.

Figure 6 shows the response volume for the tropical forest of Table 1. The three parameters (a, b, and j) were set at their tabled value and then increased and decreased by $1/3$ and $2/3$. The volume enclosed within the space represents values of the parameters which result in equilibrium values for active tissue (P and/or R) that do not differ more than 20% from the values in Table 1.

The figures show a plane running at 45° from the "j-axis." All parameter combinations resulting in no change in the state variables lie on this plane. This is simply explained by our analysis of the role of heterotrophs. A change in net primary production could be balanced by a change in heterotroph feeding rate causing a zero net effect on the system. The response volume tends to form along the diagonal of this plane with unacceptable regions associated with high transfer from autotrophs to detritus combined with low productivity and low values of the translocation combined with high productivity.

The significance of Fig. 6 is not that the present conceptualization is capable of drawing realistic bounds on the ecosystem response volume. The significance lies in the potential to refine this type of analysis as ecosystem theory becomes better developed. Eventually one should be able to establish the regions around the current operating state of ecosystems beyond which alterations must be regarded as intolerable.

Conclusions

Analysis of the conceptualization represented by Fig. 1 and Eq. (1) has led to several interesting insights into system behavior. Analysis of the model permitted a ranking of the relative recovery ability of six contrasting ecosystems. A similar ordering results from application of the independent theory of Odum and Pinkerton (1955). The results suggest a relationship between the predictability of the habitat and selective pressure for the ability to recover from perturbations that merits further consideration and investigation.

The analyses also considered the potential regulatory role of heterotrophs in the ecosystem. This hypothesis seems consistent with the magnitude of heterotroph standing crop across a variety of systems (Fig. 5). Analysis reveals that only small changes in heterotroph biomass could re-establish system equilibrium and counteract perturbation. Therefore, the regulatory function of the heterotrophs are probably not adequately represented by standing crop or energy flow alone.

The present study emphasizes aspects of system dynamics which can be represented by analyzing energy flow. Obviously, the dynamics of the system are also affected by nutrients, spatial heterogeneity, and diversity both of species and genetic composition of populations. Continued explorations into ecosystem dynamics should attempt to broaden the phenomenological base on which homeostatic mechanisms are founded, rather than relying on only a few of the many possible mechanisms.

Acknowledgments

I thank W. F. Harris for stimulating discussions and J. B. Mankin, R. S. Booth, O. L. Smith, F. B. Golley, L. Maguire, O. L. Loucks, S. A. Levin, and M. Rosenzweig for their comments on earlier drafts of the manuscript. This research was supported by the Eastern Deciduous Forest Biome, US-IBP, funded by the National Science Foundation under Interagency Agreement AG-199, BMS 69-01147 A09 with the Energy Research and Development Administration–Oak Ridge National Laboratory.

Literature Cited

Auerbach, S. I. 1971. The Deciduous Forest Biome Programme in the United States of America, p. 677–684. In P. Duvigneaud [ed.] Productivity of forest ecosystems. UNESCO, Paris.

DeAngelis, D. L., R. A. Goldstein, and R. V. O'Neill. 1975. A model for trophic interaction. Ecology 56:881–892.

Emanuel, W. R., and R. J. Mulholland. 1976. Energy based model for Lago Pond, Georgia. IEEE Trans. Automatic Control.

Golley, F. B. 1973. Impact of small mammals or primary production, p. 142–147. In J. A. Gessaman [ed.] Ecological energetics of homeotherms. Utah State University Press, Logan.

Hassell, M. P. 1971. Mutual interference between searching insect parasites. J. Anim. Ecol. 40:473–486.

Holling, C. S. 1973. Resilience and stability of ecological systems. Ann. Rev. Ecol. Syst. 4:1–23.

Kerr, S. R. 1971. Prediction of fish growth efficiency in nature. J. Fish. Res. Board Can. 28:809–814.

Kitchell, J. F., J. F. Koonce, R. V. O'Neill, H. H. Shugart, J. J. Magnuson, and R. S. Booth. 1974. Model of fish biomass dynamics. Trans. Am. Fish. Soc. 103:786–798.

Lee, J. J., and D. L. Inman. 1975. The ecological role of consumers—an aggregated systems view. Ecology 56:1455–1458.

Likens, G. F., F. H. Bormann, N. M. Johnson, D. W. Fisher, and R. S. Pierce. 1970. Effects of forest cutting and herbicide treatment on nutrient budgets in the Hubbard Brook Watershed Ecosystem. Ecol. Monogr. 40:23–47.

Lindeman, R. L. 1942. The trophic dynamic aspect of ecology. Ecology 23:399–418.

Lotka, A. J. 1923. Contribution to quantitative parasitology. J. Wash. Acad. Sci. 13:152–158.

MacCormick, A. J. A., O. L. Loucks, J. F. Koonce, J. F. Kitchell, and P. R. Weiler. 1972. An ecosystem model for the pelagic zone of Lake Wingra. Eastern Deciduous Forest Biome, EDFB-IBP-74-7. Univ. Wisconsin, Madison. 93 p.

May, R. M. 1973. Stability and complexity in model ecosystems. Princeton Univ. Press, Princeton, New Jersey. 235 p.

McMullin, B. B. 1973. A survey of papers on ecosystems analysis from 1947 to 1971 in the journal Ecology. ORNL-EIS-72-19, Oak Ridge National Laboratory, Oak Ridge, Tennessee. 50 p.

Odum, H. T. 1957. Trophic structure and productivity of Silver Spring, Florida. Ecol. Monogr. 27:55–112.

Odum, H. T., and R. C. Pinkerton. 1955. Times speed regulator, the optimum efficiency for maximum output in physical and biological systems. Am. Sci. 43:331–343.

Odum, H. T., and R. F. Pigeon [ed.]. 1970. A tropical rain forest. Division of Technical Information, U.S. Atomic Energy Commission, Oak Ridge, Tennessee. 1650 p.

O'Neill, R. V., W. F. Harris, B. S. Ausmus, and D. E. Reichle. 1975. Theoretical basis for ecosystem analysis with particular reference to element cycling, p. 28–40. In F. G. Howell, J. B. Gentry, and M. H. Smith [ed.] Mineral cycling in southeastern ecosystems. ERDA (Energy Research and Development Administration) Symposium Series, CONF-740513.

Park, R. A., R. V. O'Neill, J. A. Bloomfield, H. H. Shugart, R. S. Booth, J. F. Koonce, M. Adams, L. S. Clesceri, E. M. Colon, E. H. Dettman, J. Hoopes, D. D. Huff, S. Kate, J. F. Kitchell, R. C. Kohberger, E. J. LaRow, D. C. McNaught, J. Petersen, D. Scavia, R. G. Stross, J. Titus, P. R. Weiler, J. W. Wilkinson, and C. S. Zahovack. 1974. A generalized model for simulating lake ecosystems. Simulation 23:33–50.

Patten, B. C., and M. Witkamp. 1967. Systems analysis of cesium 137 kinetics in terrestrial microcosms. Ecology **48**:813–825.

Reichle, D. E., and S. I. Auerbach. 1972. Analysis of ecosystems, p. 260–280. *In* J. Behnke [ed.] Challenging biological problems. Directions toward their solution. AIBS 25th Anniv. Vol., Oxford Univ. Press, New York.

Reichle, D. E., B. E. Dinger, N. T. Edwards, W. F. Harris, and P. Sollins. 1973. Carbon flow and storage in a forest ecosystem, p. 345–365. *In* G. M. Woodwell and E. V. Pecan [ed.] Carbon and the biosphere. Proc. 24th Brookhaven Symp. Biol., Technical Information Center, U.S. Atomic Energy Commission.

Salt, G. W. 1967. Predation in an experimental protozoan population (*Woodruffia-Paramecium*). Ecol. Monogr. **37**:113–144.

———. 1974. Predator and prey densities as controls of the rate of capture by the predator *Didinium nasutum*. Ecology **55**:434–439.

Satchell, J. E. 1971. Feasibility study of an energy budget for Meathop Wood, p. 619–630. *In* P. Duvigneaud [ed.] Productivity of forest ecosystems. UNESCO, Paris.

Shugart, H. H., R. A. Goldstein, R. V. O'Neill, and J. B. Mankin. 1974. TEEM: A terrestrial ecosystem energy model for forests. Oecol. Plant. **9**:231–264.

Smith, O. L., H. H. Shugart, R. V. O'Neill, and R. S. Booth. 1975. Resource competition and an analytical model of zooplankton feeding on phytoplankton. Am. Nat. **109**:571–591.

Teal, J. M. 1962. Energy flow in the salt marsh ecosystem of Georgia. Ecology **43**:614–624.

Tilly, L. J. 1968. The structure and dynamics of Cone Spring. Ecol. Monogr. **38**:169–197.

Timin, M. E., B. D. Collier, J. Fich, and D. Walters. 1973. A computer simulation of the Arctic tundra ecosystem near Barrow, Alaska. U.S. Tundra Biome Report 73-1, San Diego State Univ., San Diego, California. 82 p.

Van Dyne, G. M. 1972. Organization and management of an integrated ecological research program, p. 111–172. *In* J. M. R. Jeffers [ed.] Mathematical models in ecology. Blackwell Scientific Publications, Oxford.

Volterra, V. 1938. Population growth, equilibria and extinction under specified breeding conditions—a development and extension of the theory of the logistic curve. Human Biol. **10**:1–11.

Whitfield, D. W. A. 1972. Systems analysis, p. 392–409. *In* L. C. Bliss [ed.] Devon Island IBP project, high Arctic ecosystem. Dept. Botany, Univ. Alberta, Edmonton. 413 p.

Whittaker, R. H., and G. M. Woodwell. 1972. Evolution of natural communities, p. 137–159. *In* J. A. Wiens [ed.] Ecosystem structure and function. Oregon State Univ. Press, Corvallis. 176 p.

Wiegert, R. G. 1973. A general ecological model and its use in simulating algal-fly energetics in a thermal spring community, p. 85–102. *In* P. W. Geier, L. R. Clark, D. J. Anderson, and H. A. Dix [ed.] Insects: Studies in population management. Ecol. Soc. Australia, Mem. 1, Canberra.

Wiegert, R. G., and D. F. Owen. 1971. Trophic structure, available resources and population density in terrestrial *versus* aquatic ecosystems. J. Theor. Biol. **30**:69–81.

Wynn-Edwards, V. C. 1962. Animal dispersion in relation to social behavior. Hafner Publishing Company, New York.

Editors' Comments
on Papers 19 and 20

19 PATTEN
A Simulation of the Shortgrass Prairie Ecosystem

20 PARK et al.
A Generalized Model for Simulating Lake Ecosystems

DEVELOPMENT OF MAJOR ECOSYSTEM MODELS

Large ecosystem simulation models have received considerable attention in systems ecology. Stimulated by the early conceptualizations of Van Dyne (1966), ecosystem models tried to include all known processes and interactions within a system. With the processes abstracted into mathematical functions, the resultant model could synthesize this entire body of information into a single simulation tool. These models typically contained twenty to fifty variables and several hundred parameters.

Because of the magnitude of the effort involved in developing these models, it is not surprising that most were associated with the large IBP Biome Programs. Each biome program evolved at least one of these models: *Grassland*—PWNEE (Bledsoe et al. 1970), LINEAR (Patten 1972), ELM (Innis 1978), Tundra (Timin et al. 1973), Desert (Goodall 1973), Coniferous Forest (Overton 1972); and *Deciduous Forest*— TEEM (Shugart et al. 1974), CLEAN (Park et al. 1974), WINGRAII (MacCormick et al. 1975).

But ecosystem modeling was not exclusively the province of the Biome Programs. Patten et al. (1975) developed a model for the aquatic system in Lake Texoma. Several models were developed for aquatic systems (Thomann et al. 1970; Chen and Orlob 1972) in connection with resource management problems. Forest stand development models should also be mentioned in this connection (Botkin et al. 1972; Ek 1974; Shugart and West 1977).

The size of these models has made documentation a major

problem. As a result, it is difficult to find papers of reasonable length that could be reprinted in this volume. Most of these models are available only in extensive internal documents, typically over one hundred pages in length. Documentation must include not only the functional forms used in the model, but derivations, justifications, and details of computer implementations as well. Approaches to making the models available have ranged from abstracting into a single journal article (Shugart et al. 1974) to the production of an entire volume (Innis 1978).

REFERENCES

Bledsoe, L. J., R. C. Francis, G. L. Swartzman, and J. D. Gustafson. 1970. *PWNEE, a Grassland Ecosystem Model*, Tech. Rep. 64, Grassland Biome. Natural Resource Ecology Laboratory, Colorado State Univ., Fort Collins, 179 pp.

Bledsoe, L. J., and D. A. Jameson. 1970. Model structure for a grassland ecosystem, pp. 410–35. In Dix, R. L., and R. G. Beidleman (eds.), *The Grassland Ecosystems—A Preliminary Synthesis*, Range Sci. Ser. No. 2. Range Sci. Dep., Colorado State Univ., Fort Collins, 437 pp.

Botkin, D. B., J. F. Janak, and J. R. Wallis. 1972. Rationale, limitations, and assumptions of a northeastern forest growth simulator. *IBM J. Res. Dev.* **16**:101–16.

Chen, C. W., and G. T. Orlob. 1972. *Ecologic Simulation for Aquatic Environments*, Final Report for the Office of Water Resources Research, Washington, D.C., 156 pp.

Clymer, A. B., and L. J. Bledsoe. 1970. A guide to the mathematical modeling of an ecosystem, pp. 175–99. In Wright, R. G., and G. M. Van Dyne (eds.), *Simulation and Analysis of Dynamics of a Semi-desert Grassland*, Range Sci. Ser. No. 6. Range Sci. Dep., Colorado State Univ., Fort Collins, 297 pp.

Ek, A. R. 1974. Nonlinear models for stand table projection in northern hardwood stands. *Can. J. For. Res.* **4**:23–27.

Goodall, D. W. 1973. Ecosystem simulation in the US/IBP Desert Biome, pp. 777–80. In *Proc. 1973 Summer Comput. Simul Conf.*, Montreal, Can.

Innis, G. S. 1978. *Grassland Simulation Model*. Springer-Verlag, New York, 298 pp.

MacCormick, A. J. A., O. L. Loucks, J. F. Koonce, J. P. Kitchell, and P. R. Weiler. 1975. An ecosystem model for the pelagic zone of a lake, pp. 339–82. In Yannacone, V. J. (ed.), *Environmental Systems Science*, vol. 2., Proc. Nat. Instit. Environ. Litigation. Am. Bar Assoc., Washington, D.C.

Overton, W. S. 1972. Toward a general model structure for a forest ecosystem, pp. 37–47. In Franklin, J. F., L. J. Dempster, and R. H. Waring (eds.), *Research on Coniferous Forest Ecosystems*, Proc. Symp. Conif. For. Ecosys., Bellingham, Wash. Pacific Northwest Forest and Range Exper. Sta., U.S. Forest Service, Portland, Ore., 322 pp.

Patten, B. C. 1971. A primer for ecological modeling and simulation with analog and digital computers, pp. 3–121. In Patten, B.C. (ed.), *Systems Analysis and Simulation in Ecology,* vol. 1. Academic Press, New York, 607 pp.

Patten, B. C., D. A. Egloff, T. H. Richardson, et al. 1975. Total ecosystem model for a cove in Lake Texoma, pp. 206–421. In Patten, B. C. (ed.), *Systems Analysis and Simulation in Ecology,* vol. 3. Academic Press, New York, 601 pp.

Shugart, H. H., Jr., R. V. O'Neill, R. A. Goldstein, and J. B. Mankin. 1974. Terrestrial ecosystem energy model. *Oecologia Plant.* **9**:231–64.

Shugart, H. H., Jr., and D. C. West. 1977. Development of an Appalachian deciduous forest succession model and its application to assessment of the impact of the chestnut blight. *J. Environ. Manage.* **5**(3):161–79.

Thomann, R. V., D. J. O'Connor, and D. M. DiToro. 1970. Modeling of the nitrogen and algal cycles in estuaries. In *Fifth Int. Water Pollut. Res. Conf.,* San Francisco, Calif.

Timin, M. E., B. D. Collier, J. Zich, and D. Walters. 1973. A computer simulation of the arctic tundra ecosystem near Barrow, Alaska, *US Tundra Biome Rep. 73-1.* Also in *Proc. 1972 Tundra Biome Symp.,* Univ. of Washington, Seattle.

Van Dyne, G. M. 1966. *Ecosystems, Systems Ecology and Systems Ecologists,* ORNL/TM-3957. Oak Ridge National Laboratory, Oak Ridge, Tenn.

A simulation of the shortgrass prairie ecosystem *

by

Bernard C. Patten
Department of Zoology and Institute of Ecology
University of Georgia
Athens, Georgia 30601

BERNARD PATTEN is an ecologist interested in applications of systems science to ecosystems. Born in New York City in 1931, he has degrees in botany and zoology from Cornell (AB), Michigan (MA), and Rutgers (MS, PhD). Since receiving his doctorate in 1959, he has held positions at the Virginia Institute of Marine Science, Gloucester Point, and Oak Ridge National Laboratory, with concurrent teaching appointments at William and Mary and the University of Tennessee, respectively.

Currently a professor of zoology at the University of Georgia, he is one of the developers of the rapidly expanding new field, systems ecology. His research has been largely in marine and freshwater environments, with some terrestrial experience. He consults on ecological modeling and lectures widely on systems ecology and the future of human society. He has published extensively in biological journals and is on the editorial board of the *Journal of Theoretical Biology*. He is editor of a series, *Systems Analysis and Simulation in Ecology*, published by Academic Press, and he is also completing a textbook, *Systems Ecology*. Dr. Patten is a member of numerous professional societies and is a Fellow of AAAS.

---

* University of Georgia, *Contributions in systems Ecology*, No. 12

The model described herein was produced as a systems ecology class exercise by W. G. Cale, J. R. Carter, G. I. Child, K. Cromack, C. Gist, C. D. Goodyear, R. H. Mural, T. Richardson, H. H. Shugart, C. H. Wasser, N. E. West, R. C. Wiegert, and J. C. Zieman. The subject of a forthcoming book, the model is here presented in digest form for the interested readership of *SIMULATION*.

SUMMARY

Ecosystem modeling for simulation and systems analysis is complicated by (i) size and complexity of ecosystems, (ii) lack of data and methods for obtaining needed measurements, and (iii) inadequacies of modeling theory for macrosystem applications. This paper presents an S/360 CSMP simulation of a generalized shortgrass prairie scaled large enough to preserve meaningful biological detail, but still small enough for computer implementation. Both nominal and small perturbation behaviors are described, together with a modeling rationale tied to the closed-loop recycling structure of ecosystems. The model, which is piecewise linear and piecewise stationary, demonstrates in the context of system complexity and insufficient data the potential of linear methods for successfully characterizing ecosystem dynamics.

INTRODUCTION

The North American Prairie is one of the major continental biomes, occurring in temperate regions where rainfall is too low to support forests but high enough to preclude desert vegetation. The major grassland types are zoned from east to west along the precipitation gradient: tallgrass, mixedgrass, shortgrass, and bunchgrass. Grassland communities, like most ecological assemblages, are complex systems of plant, animal, and microbial organisms engaged in the life processes of mobilizing energy and matter into organic configurations, degrading the end products, and restructuring the resultant inorganic forms again in an endless periodic cycle.

In modeling these processes, the ecologist confronts complexity on a scale not met in engineering and physical science. Pragmatically, ecosystems must be reduced to their essentials if they are to be modeled for simulation and systems analysis purposes. In this study the biomass dynamics of a generalized shortgrass ecosystem was modeled on a scale large enough to preserve meaningful biological detail but still small enough for computer implementation. The objective was to simulate the nominal and small perturbation behavior of principal functional components of this grassland type conceived as a total ecosystem unit.

The model reported is an outgrowth of one developed by the Grasslands Biome study of the U. S. IBP (International Biological Program), named PWNEE[2] after the principal site which it sought to represent, the Pawnee National Grassland in northeastern Colorado.

STRUCTURING THE SYSTEM

Submodels

Following the PWNEE model, four submodel sections (Figure 1) were formed: abiotic, producer, consumer, and decomposer. The abiotic section formulates the system driving functions. The producer and consumer submodels treat growth of vegetation and its subsequent utilization in animal food chains. The decomposer section represents the complex processes of organic matter breakdown resulting in regeneration of inorganic nutrients that feed back to control plant growth.

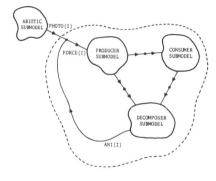

Figure 1 - Structure of the model

State variables

Biomass (total dry weight of organisms in g m^{-2}) was the principal variable of the model. Since nitrogen is the most common limiting nutrient in typical shortgrass ecosystems, total inorganic nitrogen (g N m^{-2}) was used as the variable for several compartments. Table I lists the 40 compartments (state variables) of the model, and Table II denotes their coupling.

Modeling rationale

The nominal dynamics of major ecosystems such as the shortgrass prairie can be construed as near-steady-state behavior. Accordingly, linear models of the form

$$\dot{x}(t) = A(t)\,x(t) + B(t)\,z(t)$$
$$y(t) = C(t)\,x(t) + D(t)\,z(t) \tag{1}$$

are appropriate for their representation. $x \in R^n$ is the state vector, $z \in R^m$ the input vector, and

$y \in R^n$ the output. $A(t)$ and $C(t)$ are $n \times n$ matrices of time-varying coefficients, and $B(t)$ and $D(t)$ are $n \times m$. In the present model $n = 40$, $m = 7$, $C(t)$ is the identity matrix, and $D(t)$ the null matrix. Each compartment is thus described by an equation of the form

$$\dot{x}_i = \sum_{\substack{j=1 \\ j \neq i}}^{40} a_{ij}(t)x_j + z_i(t) - T_i(t)^{-1}x_i, \quad i=1,\ldots,40 \tag{2}$$

expressing mass balance between inflows and outflows of biomass (or nitrogen). The $T_i(t)^{-1}$ are the turnover rates (inverse turnover times, which, when constant, correspond to the time constant) and appear as elements of the principal diagonal [i.e., $a_{ii}(t) = -T_i(t)^{-1}$] in $A(t)$. The units of each term of Equation 2 are g biomass m^{-2}wk^{-1} (or g N m^{-2} wk^{-1}), and the coefficients are in wk^{-1}.

DESCRIPTION OF THE MODEL

Abiotic submodel

Primary input of energy into the ecosystem is through plant photosynthesis. The environmental (i.e., external) variables that most strongly influence photosynthetic production of organic matter in the shortgrass prairie are: sunlight, air temperature, and soil moisture. Simple functions were formulated to represent these driving variables:

SUN = 200.* SIN(.12093*t) + 400. $\qquad$ (3)

TEMP = (9. + 11.* SIN(.12093*t - .3424))*.061 $\quad$ (4)

MOIST = 2. - .06*Y $\qquad$ (5)

The data upon which SUN is based represent the Pawnee site.[3] ω = .12093 is one week in radian measure. SUN provides for a sinusoidal input with 200 ly d^{-1} (gcal cm^{-2} d^{-1}) amplitude around a mean of 400 ly d^{-1} which occurs at the vernal and autumnal equinox. Initial time (t=0) in the model corresponds to March 21.

TEMP was also abstracted from data for the Pawnee site.[4] The function is a sinusoid which lags SUN by about three weeks (.3424 radians), and represents extremes of $\pm11°C$ air temperature around a mean of 9°C. The coefficient .061 g wk^{-1} °C^{-1} is the slope of a linear approximation to a photosynthesis-temperature function.[5]

Soil moisture on the Pawnee grassland is also periodic over a year, being greatest in March and least in October. Only the down side of the curve is involved during the growing season, however, and this is reasonably approximated by the line MOIST representing relative soil moisture storage. Y is a timer counting weeks of the growing season.

Producer submodel

Following the original PWNEE model, vegetation was partitioned into 7 compartments (Table 1). Warm season grasses (VA1), typified by blue grama, compose most of the aboveground biomass of living plants. These grasses begin growth in mid or late spring and continue to grow through most of the hot, dry summer. Cool season grasses (VA2), such as western wheatgrass, start growing in early spring, become semidormant in summer, and resume growth again in the autumn. Forbs (VA3) are the herbaceous plants comprising only a small amount of the total live aboveground biomass of vegetation. Growth of various

Table 1
List of compartments (state variables).
Asterisks denote nitrogen as variable; biomass otherwise.

Compartment	Computer code	Representative organisms
	PRODUCER SUBMODEL	
Warm season grasses	VA1	Blue grama grass (*Bouteloua gracilis*)
Cool season grasses	VA2	Western wheatgrass (*Agropyron smithii*)
Forbs	VA3	Scarlet globemallow (*Sphaeralcea coccinea*)
Cacti	VA4	Prickly pear (*Opuntia polyacantha*)
Belowground plant parts	VB	
Plant standing dead	VS	
Plant litter	VL	
	CONSUMER SUBMODEL	
Wild herbivores	C1	Jackrabbit
Domestic herbivores	C2	Cow
Carnivores	C3	Coyote
Birds	C4	Lark bunting
Insects	C5	Grasshopper
	DECOMPOSER SUBMODEL	
Root fungi	Q1	*Aspergillus, Penicillium, Rhizopus, Chaetomium*
Root bacteria	Q2	*Bacillus, Pseudomonas, Clostridium*
Root nematodes	Q3	
Root mites	Q4	
Root Diptera	Q5	Calliphoridae, Sarcophagidae
Misc. root arthropods	Q6	Coleoptera and cicada larvae
Fossorial mammals	Q7	Ground squirrels, moles, pocket gophers, prairie dogs
Litter bacteria	Q8	*Bacillus, Pseudomonas, Clostridium*
Nitrogen pool*	Q9*	
Litter fungi	Q10	*Aspergillus, Penicillium, Rhizopus, Chaetomium*
Litter mites	Q11	Oribatidae
Litter Collembola	Q12	Entomobryidae, Sminthuridae
Litter diplopods	Q13	Julidae, Polydesmidae
Litter isopods	Q14	*Armadillidium vulgare*
Litter predatory mites	Q15	Bdellidae, Canacidae, Ascidae, Cheyletidae, Rhodocaridae
Macro-predators	Q16	Carabidae, spiders, aphids, robber flies
Carrion bacteria	Q17	
Carrion beetles	Q18	Dermestidae
Carrion Diptera	Q19	Blowflies
Protozoa	Q20	Flagellates, rhizopods, ciliates
Dung fungi	Q21	*Chaetomium, Ascobolus*
Dung bacteria	Q22	
Dung Diptera	Q23	Muscidae
Dung beetles	Q24	Coprinae, Aphrodiinae
Blue-green algae*	Q25*	
Denitrifiers*	Q26*	
Feces	SH	
Carrion	AD	

species occurs throughout the growing season. Cacti (VA4) are slow-growing but persistent components of the plant community. They maintain green biomass throughout even the cold seasons.

The remaining plant compartments are roots and related belowground plant parts (VB), standing dead (VS), and plant litter (VL). Roots of all species were lumped into VB since they cannot be conveniently separated. VS, standing plant material of all kinds remaining after death, is gradually converted through wind action and other forces to litter, VL.

Consumer submodel

Consumer animals were divided into 5 compartments (Table 1) as in the PWNEE model. Wild herbivores (C1) include rabbits, rodents, and other plant-eating mammals of the shortgrass community. Domestic herbivores (C2) represent cattle and sheep, the former being of greater importance on the Pawnee site. Carnivores (C3), exemplified by the coyote, include all flesh-eating animals that obtain food by predation. Migratory birds (C4) arrive in the spring

and depart in the fall. Insects (C5) are present and remain viable throughout the growing season. They are killed by frost and freezing temperatures in the autumn.

Two external inputs are associated with consumers. Grazing management usually involves introducing livestock onto the range in spring and removing them at roundup in the fall. Bird migration similarly involves an inflow of biomass in spring and an outflow in autumn.

Decomposer submodel

Departing from the PWNEE model, an expanded decomposer submodel (28 of the 40 compartments) was developed to underscore the ecological significance of the detritus subsystem and draw attention to the paucity of data and gaps in knowledge concerning decomposition processes. Five sectors were recognized (Table 1): root and soil organisms (Q1,...,Q7, Q20), litter organisms (Q8, Q10,..., Q16), carrion organisms (Q17,..., Q19 and AD), dung organisms (Q21,..., Q24 and SH), and a nitrogen pool subsystem (Q9,Q25,Q26).

219

Table 2 - Connectivity matrix denoting coupling topology of model. Nonzero elements (Boolean ones dots in table) signify biomass (or nitrogen) transfers from compartments listed as columns to compartments listed as rows. Null entries (Boolean zeros) indicate no corresponding coupling in the model.

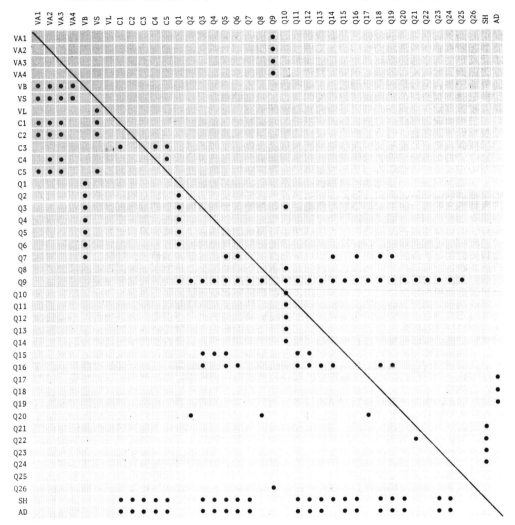

The function of this submodel is to decay organic matter output from producers and consumers and to generate a pool of inorganic soil nitrogen (Q9). The latter was used to close the material recycling loop characteristic of ecosystems through feedback regulation of plant primary production (Figure 1). To achieve the transformation from biomass to nitrogen, it was necessary to estimate the protein content of the compartments and rates of mineralization. Budgeting took into account such factors as precipitation and leaching, and losses due to denitrification and volatilization of ammonia. Also, nitrogen fixation by blue-green algae and symbionts of vascular plants (Q25) was estimated and treated as an external drive.

Connectivity

The Table 2 connectivity matrix summarizes the interactive topology of the model. The system is 9.6% connected (149 nonzero coefficients of 1560 off-diagonal elements). Each nonzero coefficient in the matrix specifies a biomass (or nitrogen) transfer at some time of the model year from compartments listed as columns to compartments listed as rows. Many coefficients were constant, but many were time varying. The approach to time variation of the latter is significant, and is discussed below.

MATHEMATICAL CHARACTERISTICS OF THE MODEL

Piecewise linearity

The model is piecewise linear due to the method employed to regulate plant net production based on the availability of soil nitrogen. Potential net photosynthesis of the four compartments for plants living above ground, VA(I), I = 1,...,4, was formulated from the driving variables of the abiotic submodel:

$$\text{PHOTO(I)} = \text{SUN} * \text{TEMP} * \text{MOIST} * \text{EFF(I)} \quad (6)$$

Specificity to plant group was achieved with constant efficiency coefficients, EFF(I). Availability of soil nitrogen to each VA(I) was taken as proportional to the nitrogen pool Q9:

$$\text{ANI(I)} = \text{FNI(I)} * \text{Q9} \quad (7)$$

where ANI(I) is nitrogen available to VA(I) and FNI(I) are constant coefficients. The plant state transition equations corresponding to Equation 2 are:

$$\dot{\text{VA}}(I) = \text{FORCE(I)} - (\text{HA(I)}+\text{TB(I)}+\text{DA(I)})*\text{VA(I)} \quad (8)$$

FORCE(I) is actual net production of VA(I), and HA(I), TB(I), and DA(I) represent turnover components due to herbivory, translocation to belowground plant parts VB, and natural mortality, respectively.

A nonlinear switch was introduced in the formulation of actual net production:

$$\text{FORCE(I)} = \begin{cases} \text{PHOTO(I)} & \text{if PHOTO(I)} \leq \text{ANI(I)} \\ \text{ANI(I)} & \text{otherwise} \end{cases} \quad (9)$$

This equation sets actual net production of VA(I) equal to potential net production PHOTO(I) if nitrogen ANI(I) is not limiting, or to an amount of photosynthesis equal (in biomass units) to available nitrogen when the latter is limiting. Since ANI(I) are functions of a state variable Q9 (Equation 7), the switching logic causes FORCE(I) and hence the plant state equations (8) to be piecewise linear. This is the only nonlinearity in the entire model.

Piecewise stationarity

Time variation of nonstationary coefficients was discontinuous rather than continuous. The model is thus piecewise stationary, with the time line of the model year partitioned into 16 intervals of time-invariant model operation. At each interval endpoint, one or more coefficients change to a new constant value independently of the state of the system. The model, therefore, consists of a concatenation of 16 time-invariant state descriptions during each annual period. The intervals of stationary operation, in weeks, are: (0,2], (2,4], (4,5], (5,7], (7,8], (8,9], (9,18], (18,19], (19,20], (20,22], (22,23.9], (23.9,24], (24,31], (31,47], (47,50], and (50,52].

VALIDATION

A persistent difficulty in total ecosystem modeling is scarcity of adequate data. This forces particular attention to the philosophy and procedure for validating the model. A major practical advantage of linear modeling is that first degree coefficients with units of inverse time are readily interpretable. In the absence of good measurements and of values reported in the literature, ecologists can often provide reasonable estimates of needed numbers. Such opinions, which result from a mental filtering process keyed to average circumstances in the experts' experience, are consistent with the operating-point rationale appropriate to linear models. The $A(t)$ matrix of Equations 1 provides structural constraints for determining missing values. Turnover rates (the principal diagonal elements) are central. If turnover data are sound along with a few other coefficients in each column, missing data can be reasonably estimated using a combination of biological judgment and knowledge that the off-diagonal coefficients in each column must sum to intrasystem (i.e., nonenvironmental) components of the corresponding turnover rate.

Determination of initial states and constant coefficients constitutes the static phase of model validation. With constant inputs a stationary model can be driven to constant steady states which may be matched to the initial states by parameter adjustments. This begins the process of dynamic validation, in which by a series of simulation trials parameters are time varied and tuned until each submodel by itself converges to known or reasonable behavior. The rule here is to alter coefficients with the poorest basis in actual data first or, ideally, exclusively. When submodel behavior is acceptable, the entire system is coupled together and necessary adjustments made. The linear dynamic model has excellent stability and controllability characteristics which facilitate the entire dynamic validation procedure and final matching of the model's output to whatever empirical time-series data are available.

Best data in the present instance were available for producer and consumer state variables. Model validation was thus focused on these compartments. Many variables of the decomposer submodel have never been measured, even statically much less dynamically, because satisfactory techniques do not exist for microbial and microfaunal ecosystem components. State theory was used as a guide in the validation rationale, as follows:

The generalized state and response equations,

$$x(t_0, t_1] = f(x(t_0), z(t_0, t_1])$$
$$y(t_0, t_1] = g(x(t_0), z(t_0, t_1]) \quad (10)$$

provide state and output time series on the interval $(t_0, t_1]$ given the initial state $x(t_0)$ and an input segment $z(t_0, t_1]$ on the same interval. If

(i) f and g have been reasonably structured from biological information (which comes down to proper representation of turnover time series, $T_i(t)^{-1}$, in Equation 2),

(ii) the initial state $x(t_0)$ of the system has been accurately estimated, and

(iii) producer and consumer outputs (i.e., states, Equations 1) have been validated against good empirical data,

then Equations 10 assert automatic validity of the decomposer submodel if accurate inputs $z(t_0, t_1]$ are provided to the system. That is, validated output from producers and consumers serves as principal input to decomposers (external drives like immigration and nitrogen fixation are comparatively minor decomposer input components), and satisfaction of conditions (i) and (ii) guarantees decomposer behavior.

221

Table 3

Calendar of weekly intervals.   Time zero is taken as midnight, March 20.

Interval (weeks)	Dates	Interval (weeks)	Dates
(0,1]	(Mar 20, Mar 27]	(26,27]	(Sep 18, Sep 25]
(1,2]	(Mar 27, Apr 3]	(27,28]	(Sep 25, Oct 2]
(2,3]	(Apr 3, Apr 10]	(28,29]	(Oct 2, Oct 9]
(3,4]	(Apr 10, Apr 17]	(29,30]	(Oct 9, Oct 16]
(4,5]	(Apr 17, Apr 24]	(30,31]	(Oct 16, Oct 23]
(5,6]	(Apr 24, May 1]	(31,32]	(Oct 23, Oct 30]
(6,7]	(May 1, May 8]	(32,33]	(Oct 30, Nov 6]
(7,8]	(May 8, May 15]	(33,34]	(Nov 6, Nov 13]
(8,9]	(May 15, May 22]	(34,35]	(Nov 13, Nov 20]
(9,10]	(May 22, May 29]	(35,36]	(Nov 20, Nov 27]
(10,11]	(May 29, June 5]	(36,37]	(Nov 27, Dec 4]
(11,12]	(June 5, June 12]	(37,38]	(Dec 4, Dec 11]
(12,13]	(June 12, June 19]	(38,39]	(Dec 11, Dec 18]
(13,14]	(June 19, June 26]	(39,40]	(Dec 18, Dec 25]
(14,15]	(June 26, July 3]	(40,41]	(Dec 25, Jan 2]
(15,16]	(July 3, July 10]	(41,42]	(Jan 2, Jan 9]
(16,17]	(July 10, July 17]	(42,43]	(Jan 9, Jan 16]
(17,18]	(July 17, July 24]	(43,44]	(Jan 16, Jan 23]
(18,19]	(July 24, July 31]	(44,45]	(Jan 23, Jan 30]
(19,20]	(July 31, Aug 7]	(45,46]	(Jan 30, Feb 6]
(20,21]	(Aug 7, Aug 14]	(46,47]	(Feb 6, Feb 13]
(21,22]	(Aug 14, Aug 21]	(47,48]	(Feb 13, Feb 20]
(22,23]	(Aug 21, Aug 28]	(48,49]	(Feb 20, Feb 27]
(23,24]	(Aug 28, Sep 4]	(48,50]	(Feb 27, Mar 6]
(24,25]	(Sep 4, Sep 11]	(-2,-1]= (50,51]	(Mar 6, Mar 13]
(25,26]	(Sep 11, Sep 18]	(-1,0] = (51,52]	(Mar 13, Mar 20]
		(0,1] = (52,53]	(Mar 20, Mar 27]

A check on conditions (i) and (ii) through (iii) is further provided by decomposer control of producer input. The major function of the decomposer submodel is to generate the inorganic nitrogen pool Q9 which, as ANI(I), feeds back to limit plant primary production, FORCE(I) (Equation 9). The entire model structure is thus, in effect, nested within and dependent upon the nitrogen recycling loop (Figure 1), which is generated by the submodel that is least validatable by direct comparison to data. Producers (and subsequently, consumers) receive decomposer-controlled input and generate as output the inputs to decomposers. The loop structure in concert with Equations 10 thus provides a strong validation criterion since it imposes mutual interdependence of every state variable in the model. Therefore, Condition (iii) ↔ Conditions (i) and (ii) ↔ decomposer validation.

A bonus of this validation rationale which the ecologist may be expected to exploit in the future concerns data supplementation through modeling. It will probably always be true that certain ecosystem components, such as many of the decomposer compartments in the present case, will be difficult or impossible to measure under natural conditions. Improved models and stronger validation principles of the future offer the distinct possibility of bypassing such measurements in favor of letting the models generate the desired information.

In its present stage of development, and with the simple major inputs described by Equations 3, 4, and 5, the ecosystem model provides a qualitatively faithful rendering of shortgrass prairie dynamics as understood in the winter of 1971 when the work was completed. Timing is very good to excellent for all but a few compartments, and magnitudes are generally accurate to within 10% or better in cases where time-series data were available for comparison. Small-signal perturbation behavior is reasonable, as discussed later. Since the model was completed, the IBP Grasslands Biome program has generated much new data both for inputs and state variables. Future development of the model will be devoted to bringing its performance into closer conformity with latest information.

COMPUTER PROGRAM

Software for the model was prepared in S/360 CSMP. The program is available from the author. It is beyond the scope of this paper to provide adequate documentation. The program in its present form is complicated by the fact that different teams of people originated and programmed the different submodels. A uniform system of notation has not yet been developed, and difficulties in understanding the program will be experienced due to double notations, implicit parameters, etc. For example, output coefficients from one submodel may be named and/or classified differently when they appear as input parameters to another submodel. These difficulties notwithstanding, the interested user should be able to run the program as listed (nominal simulation) and alter some of the more obvious inputs and coefficients for perturbation trials. Complete documentation and revision will be provided in a later publication.[1] The program is self explanatory in regard to integration method, computation interval, etc.

NOMINAL SIMULATION

In this section the nominal behavior of selected input and state variables will be described. Print-plotted outputs for the first year (see Table 3 for calendar) are provided for illustration. The model, with parameter values as listed, is stable for an indefinite period. The maximum run duration was 20 years, during which time only very small changes from the first year simulation occurred due to gradual increase in soil nitrogen Q9 and corresponding relaxation of nitrogen limitation of producer net production. The first-year dynamics illustrated can be taken as indicative of the generalized shortgrass ecosystem operating point. Units in the figures are g m^{-2} for compartments (biomass or nitrogen, as appropriate; see Table 1), and g biomass m^{-2} wk^{-1} for input variables PHOTO and FORCE.

Plant net production

Figure 2 illustrates potential net production, PHOTO1 (Equation 6), of the most important plant compartment,

warm season grasses VA1. Figure 3 shows the annual cycle of soil nitrogen Q9. When nitrogen becomes limiting, actual net production, FORCE1 (Equation 9), is reduced from the potential value. Figure 4 shows this nitrogen limitation of VA1 production during weeks (7, 19]. Results for the other plant compartments, VA2, VA3 and VA4, were qualitatively similar.

Producer submodel

VA1 and VA4 dynamics are illustrated in Figures 5 and 6, respectively. Note that cacti oscillate only slightly from summer to winter biomass, and do not disappear in winter as do the other living plant compartments. Belowground biomass VB (Fig.7) is high at the beginning of the growing season, declines as root reserves are used in spring growth of plant tops, and is restored during the growing season as photosynthetic products are translocated belowground for storage and root growth. After the growing season a zero-input response is observed until the following spring when the cycle is repeated. Plant stand-

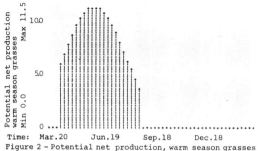

Figure 2 – Potential net production, warm season grasses

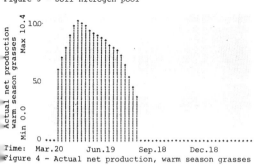

Figure 3 – Soil nitrogen pool

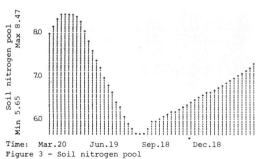

Figure 4 – Actual net production, warm season grasses

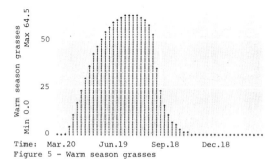

Figure 5 – Warm season grasses

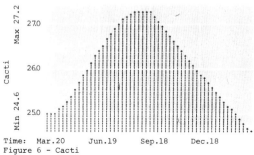

Figure 6 – Cacti

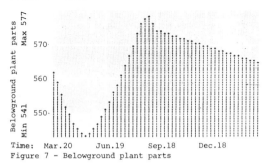

Figure 7 – Belowground plant parts

223

ing dead (Figure 8) is generally low during the growing season, but increases as spring and early summer floras die back. Major increase occurs with cool weather and frost in autumn, and zero-input decay is observed thereafter as standing dead is converted to litter VL (Figure 9) during the nongrowing season. Note the rapid decline of litter in early spring when warming and adequate soil moisture are favorable for microbial and other decomposer activity. This occurs again in the fall when rainfall typically breaks the summer dry period.

Consumer submodel

Dynamics of wild herbivores C1 (Figure 10) essentially reflects behavior of the vegetation and corresponding changes in feeding habits. A switch in dietary preference from green plant material to standing dead VS at the end of the growing season accounts for the biomass increase observed during weeks (26, 31). Gestation times and other reproductive lags cause about a month's delay in spring before C1 biomass begins to build. In the simulation, cattle C2 are

introduced to the range after week 4 and rounded up after week 26. Their weight gain during the growing season is illustrated in Figure 11. Similarly, birds C4 migrate into the system during week (4,5), and out during week (19,20). Bird biomass typically declines at first when food supply is limited, and then increases to maximum values just before egress (Figure 12), reflecting metabolite storage which precedes migration. Growth of insects C5 does not begin until the growing season is well underway, and terminates with autumn frost kills (Figure 13). Unlike birds, insect biomass declines in the latter half of the growing season reflecting the diminishing supply of plant food material.

Decomposer submodel

Fungi of roots and soil Q1 (Figure 14) increase rapidly with spring warming and thawing, subsequently decrease with increased heating and desiccation of the soil, and then increase again with improved temperature and moisture conditions during autumn. Their behavior is zero-input during the nongrowing season. Root-soil bacteria Q2 (Figure 15) have

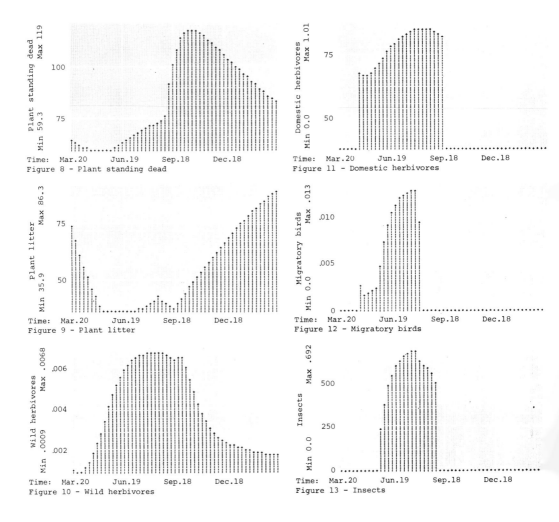

Figure 8 - Plant standing dead

Figure 9 - Plant litter

Figure 10 - Wild herbivores

Figure 11 - Domestic herbivores

Figure 12 - Migratory birds

Figure 13 - Insects

similar dynamics to the fungi and other fast-turn-over components of the soil-rhizosphere sector. A slow-turnover compartment of this sector, fossorial mammals Q7, is illustrated in Figure 16.

Litter bacteria Q8 are shown in Figure 17 for comparison with root-soil bacteria Q2 (Figure 15). Timing is similar, but magnitudes differ somewhat. A slower-turnover compartment of the litter sector, predatory mites Q15, is illustrated in Figure 18. These organisms have diverse diets and are quite stable against fluctuations of other litter populations.

The dynamics of soil nitrogen Q9 (Figure 3) has already been described. Another compartment of the nitrogen pool sector, nitrogen fixers Q25, is illustrated in Figure 19. These organisms, primarily construed as blue-green algae which form mats on the soil surface, show greatest activity at the beginning and end of the growing season. During summer they, like other microbial components, are capable of rapid blooms of activity in response to discrete rainfall events. Since such inputs were not provided (Equa-

tion 5), this behavior capability is not manifested in the present simulation.

PERTURBATION TRIALS
Reduced soil moisture

Moisture stress is an important ecological factor in a xeric (dry) prairie. This was simulated by translating the MOIST function (Equation 5) downward 20% (intercept = 1.6 instead of 2.0), which is still within the normal operating experience of the shortgrass ecosystem. The change is mediated through potential net production PHOTO(I) (Equation 6) to influence plant growth (Equations 8 and 9) and subsequently all other compartments. Some first-year results are summarized in Table 4, which compares peak values (the perturbations influenced magnitudes more than timing) in perturbed vs. nominal behavior.

While aboveground biomass of vegetation declined in proportion to the loss of soil water, roots VB were conserved (only 3% reduction). This is consistent with the known ecology of root systems, which tend to be highly resistent to perturbing influences.

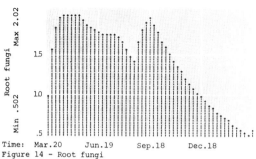

Figure 14 - Root fungi

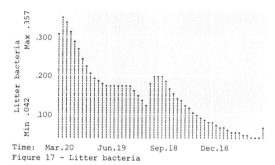

Figure 17 - Litter bacteria

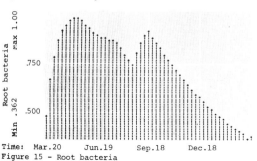

Figure 15 - Root bacteria

Figure 18 - Litter predatory mites

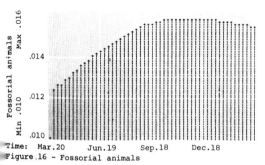

Figure 16 - Fossorial animals

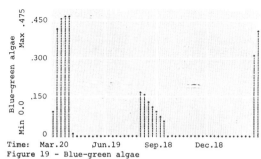

Figure 19 - Blue-green algae

225

Soil nitrogen Q9 increased slightly; nitrogen never became limiting in this experiment due to the reduced values of PHOTO(I) (Equation 9). Carrion AD increased slightly, but fecal material SH more than doubled. Unusual accumulations of feces are symptomatic of range managers that something is out of balance in the ecosystem. The simulation thus appears to have captured this characteristic quite well.

Nitrogen fertilization

Simulation of a nitrogen enrichment experiment was performed by increasing the initial soil nitrogen Q9 by 18.75%. The effects of this treatment were mediated through increased plant production (Equations 7, 8, and 9). Maximum effects were exhibited during the first year for fast-turnover compartments, and during the second year for slower components. Table 4 shows a uniform increase in state variables throughout the system, and predicts an almost 3-fold increase in feces accumulations.

In actual nitrogen fertilization experiments at the Pawnee site (Van Dyne, personal communication), plants grow up earlier in the year as they make more efficient use of soil moisture with abundant nitrogen. The increased herbage yield (e.g., VA1 and VA3 in Table 4) is accompanied by improved chemical composition and hence greater palatability of the plants. Digestibility may be only slightly improved, however, and consequently fecal production increases. Decomposition processes are not greatly influenced under a nitrogen treatment, and hence feces tend to accumulate. The model thus appears to mimic a known effect of nitrogen fertilization.

Overgrazing

Overgrazing is perhaps the most universal human abuse of prairies. This was simulated by doubling the initial density of cows C2 and corresponding ingestion and production rates. A three-year run indicated a long-term trend in degradation of the ecosystem, the full effects of which would not be expressed until after a decade or more. This parallels what really happens to overgrazed grasslands as negative effects accumulate over a long period of time. Some first-year results are noted in Table 4.

Cow biomass increased only 72% although initial weight was doubled; this figure declines in subsequent years. Effects on plants were most noticeable in the cool season grass group VA2, and also standing dead VS and litter VL. The decrease in standing dead and corresponding increases in litter is consistent with expected mechanical effects of large grazing animals on vegetation. Roots VB were only slightly affected, which again is ecologically realistic. Also realistic is the differential effect on the two grass compartments, VA1 and VA2. The growing tissues (meristems) of warm season grasses VA1 are close to the ground and difficult to graze. Growing points of VA2 grasses are more susceptible to grazing damage (Van Dyne, personal communication). Hence the VA2 compartment should be more susceptible to grazing pressure, which the model simulates.

Table 4

Percent change in maximum first-year value of selected state variables

Compartment	Reduced soil moisture (-20%)	Nitrogen fertilization (+18.75%)	Overgrazing (2X)
VA1	-22	+12	-14
VA2	--	--	-27
VA3	-18	+12	-14
VB	-3	+19	-2
VS	-23	+13	-27
VL	-16	+17	+22
C1	-21	+10	-15
C2	-18	+12	+72
C5	-21	+11	-14
Q9	+2	+16	-16
SH	+104	+187	+19
AD	+3	+25	+2

Wild herbivores C1 and insects C5 were reduced in proportion to loss of plant biomass. Soil nitrogen Q9 was also decreased, suggesting a lessened ability of the ecosystem to maintain its nitrogen pool under overgrazing. Carrion AD was unaffected, and animal feces SH increased a much smaller extent than in the two preceding perturbation exercises. The latter is consistent with improved physiological assimilation which occurs in digestion when food or essential nutrients are in short supply.

CONCLUSION

The qualitative and quantitative fidelity of both nominal and perturbation dynamics of the model, even with low resolution inputs, indicates that operating-point methods provide a sound basis for characterizing ecosystem dynamics. Linear or linearized models behave more like ecosystems than do nonlinear ones, as growing experience in ecological simulation is demonstrating.

Linear models may be expanded indefinitely to represent macro-scale systems in fine detail (the limits are data and computer size, not theory), and they offer the additional advantage of making linear system theory and methods available for the large class of developing problems in ecological and environmental systems analysis. If they sacrifice too much biological realism (the criticism of many ecologists), then perhaps at worst they can provide a stable matrix for the introduction of such (nonlinear) realism where it is absolutely needed to understand, illuminate, or solve particular problems.

Just as ecosystems exist in nominal states to be perturbed by man or nature, this study suggests an analogous role for their mathematical homomorphs. It points to initial construction and validation of linear simulations followed by local nonlinear modifications to represent large disturbances as an appropriate working rationale.

REFERENCES

1 PATTEN B C (ed)
   Systems analysis and simulation in ecology
   vol 3 *A state space model for grassland*
   Academic Press   New York   In preparation

2 BLEDSOE L J   FRANCIS R C   SWARTZMAN G L
   GUSTAFSON J D
   PWNEE: A grassland ecosystem model
   US IBP Grasslands Biome Tech Rep 64 1971

3 SELLERS W D
   Physical climatology
   Univ Chicago Press Chicago 1965 pp 27-27

4 HYDER D N
   *The impact of domestic animals on the func-
   tion and structure of grassland ecosystems*
   In DIX R L and BEIDLEMAN R G (eds), *The grassland ecosystem: A preliminary synthesis*
   Colorado State University Range Sci Ser no 2
   1969 pp 243-260

5 GATES D M
   Toward understanding ecosystems
   In CRAGG J B (ed), *Advances in ecological
   research*, vol 5
   Academic Press New York 1968 pp 1-35

6 VAN DYNE G M
   *Organization and management of an integrat
   ecological research program- ...*
   In JEFFERS J N R (ed), *Mathematical models
   ecology (Blackwell, Oxford, 1972, pp 111-1*

20

Reprinted from *Simulation* 21:33–50 (1974)

A generalized model for simulating lake ecosystems*

AUTHORS OF THIS PAPER
The coauthors of this paper are associated with the Eastern Deciduous Forest Biome, U.S. International Biological Program. They represent the two aquatic sites in the project, Lake George, New York, and Lake Wingra, Wisconsin, and the central modeling staff from Biome headquarters at Oak Ridge National Laboratory. Richard Park (see photo left) is a geologist and systems ecologist who has responsibility for coordinating the aquatic modeling effort in the Biome. Robert O'Neill is a systems ecologist and is modeling and science coordinator for the Biome, with responsibility for three terrestrial sites as well as the aquatic sites. He is assisted in the central modeling staff by Henry Shugart and Robert Goldstein, who are also systems ecologists, and by Raymond Booth and J. B. Mankin, who are systems analysts. Jay Bloomfield is an environmental engineer and micro-biologist, Joseph Koonce is a systems ecologist, and Don Scavia is an environmental engineer. The remaining coauthors have contributed by developing specific parts of the lake model and by evaluating the model. They include ecologists, microbiologists, hydrologists, systems analysts, physical limnologists, environmental engineers, statisticians, and mathematicians.

SUMMARY

CLEAN, a generalized lake-ecosystem model with strong ecological realism, has been developed in response to one aspect of the growing need for models suitable for helping man to manage his environment. The model currently consists of twenty-eight ordinary differential equations which represent approximately sixteen compartments, including attached aquatic plants, phytoplankton, zooplankton, bottom-dwelling aquatic insects, fish, suspended organic matter, decomposers, sediments, and nutrients. These equations can be linked in any meaningful combination to simulate a given point in a lake (a separate model for lake circulation is available to represent spatial variations and to couple simulations of different regions of the lake). Subprogram functions exist for each principal physiological and ecological process, and a submodel for lake water balance is presently being implemented. The program is written in FORTRAN for UNIVAC and IBM time-sharing systems.

The model has provided intuitively realistic simulations and has given us insight into the effects of nutrient enrichment on the functioning of the lake ecosystem as a whole. Sensitivity analysis has indicated priorities for further studies to obtain more precise estimates of parameters. Also, evaluation of the logic and organization of the model by experimenting with it are providing information to use in planning new experimental approaches. CLEAN is presently being tested using data from Lake George, New York, and Lake Wingra. Wisconsin.

* Contribution No. 152 from the Eastern Deciduous Forest Biome, U.S. International Biological Program.

AUTHORS AND THEIR AFFILIATIONS

Richard A. Park
Fresh Water Institute and
    Department of Geology
Rensselaer Polytechnic Institute
110-8th Street, G2
Troy, New York 12181

Robert V. O'Neill
Environmental Sciences Division
Oak Ridge National Laboratory
P.O. Box X, Bldg. 2001
Oak Ridge, Tennessee 37830

Jay A. Bloomfield
Department of Biology
Rensselaer Polytechnic Institute
110-8th Street
Troy, New York 12181

H. Henry Shugart, Jr.
Environmental Sciences Division
Oak Ridge National Laboratory
P.O. Box X, Bldg. 3017
Oak Ridge, Tennessee 37830

Raymond S. Booth
Instrumentation and Controls Div.
Oak Ridge National Laboratory
P.O. Box X, Bldg. 3500
Oak Ridge, Tennessee 37830

Robert A. Goldstein
Environmental Sciences Division
Oak Ridge National Laboratory
P.O. Box X, Bldg. 2001
Oak Ridge, Tennessee 37830

J. B. Mankin
Oak Ridge National Laboratory
Oak Ridge, Tennessee 37830

Joseph F. Koonce
Department of Biology
Case Western Reserve University
Cleveland, Ohio 44106

Don Scavia
314 Management Building
Rensselaer Polytechnic Institute
Troy, New York 12181

Michael S. Adams
Department of Botany
University of Wisconsin
Madison, Wisconsin 53706

Lenore S. Clesceri
Department of Biology
Rensselaer Polytechnic Institute
Troy, New York 12181

Emilio M. Colon
Rensselaer Polytechnic Institute
Present address:
Calle G Maranon 339
Urb. El Senorial
Rio Piedras, Puerto Rico 00926

Edward H. Dettmann
532 WARF Building
610 North Walnut Street
University of Wisconsin
Madison, Wisconsin 53706

John A. Hoopes
Department of Civil Engineering
University of Wisconsin
Madison, Wisconsin 53706

Dale D. Huff
Department of Civil Engineering
University of Wisconsin
Madison, Wisconsin 53706

Samuel Katz
Department of Geology
Rensselaer Polytechnic Institute
Troy, New York 12181

James F. Kitchell
Laboratory of Limnology
University of Wisconsin
Madison, Wisconsin 53706

Robert C. Kohberger
Rensselaer Polytechnic Institute
Present address:
Albany Medical School
Albany, New York 12208

Edward J. LaRow
Department of Biology
Siena College
Loudonville, New York 12210

Donald C. McNaught
Department of Biological Sciences
State University of New York
Albany, New York 12203

James L. Peterson
University of Wisconsin
Present address:
National Commission on Water Quality
1111 - 18th Street, N.W.
Washington, D.C. 20036

John E. Titus
Department of Botany
University of Wisconsin
Madison, Wisconsin 53706

Peter R. Weiler
Institute of Environmental Studies
Room 511, WARF Building
610 North Walnut Street
Madison, Wisconsin 53706

John W. Wilkinson
School of Management
Rensselaer Polytechnic Institute
Troy, New York 12181

Claudia S. Zahorcak
Rensselaer Polytechnic Institute
Present address:
Hydroscience Inc.
363 Old Hook Road
Westwood, New Jersey 07675

INTRODUCTION

Many lake ecosystems are endangered by man's activities. Siltation from improper land use, waste heat
from power stations, and pollution by oil, pesticides,
and herbicides are all causing serious alterations
of the natural environment. Probably nutrient enrichment from improper use of fertilizer and from the
discharge of sewage is the most critical for lake
ecosystems. Eutrophication resulting from nutrient
enrichment may improve fishing, but it can also increase the population of undesirable species of fish,
and of foul-tasting and odor-producing algae, and
produce unsightly algal scum. In severe cases, eutrophication can deplete dissolved oxygen to the point of
killing large numbers of fish. Unfortunately, total
removal of nutrients from sewage is costly; further,
as recreational developments spread along lakeshores
there is more sewage to treat.

The pressing need for enlightened management of our
lake ecosystems is made more critical by the difficulty of predicting the consequences of man-induced
perturbations. The lake is a complex system with
large numbers of biological, chemical, and physical
components. Interactions among these components are
characteristically nonlinear and involve intricate
feedback loops. Desirable changes in one component
of the system may trigger a complex chain of effects
which results in deterioration of the total system.

The complexity of the lake ecosystem and the growing
need for proper management of this natural resource
has placed a high priority on large-scale integrated
approaches to the problem. This paper outlines and
presents a set of equations for a generalized lake
ecosystem model. The model is the result of the
joint efforts of a team of aquatic specialists and
systems modelers. It is designed as a diagnostic
tool to study the effects of nutrient enrichment and
other perturbations on the ecosystem. It has been
structured to permit studying a wide range of lake
systems by appropriately changing parameter values
and driving variables.

The present model is not without historical precedent.
Most present-day efforts, including this model, have
been influenced by the work of Riley and his
colleagues.[1-5] Steele[6-8] has also had an important
influence on the development of aquatic models.
Following from these early studies, models have become more complex, paralleling advances in computer
technology and ecological theory. Parker[9] modeled
the relationships among phytoplankton, zooplankton,
and fish. Chen[10] presented a model for simulating an
ecosystem with phytoplankton, zooplankton, fish,
detritus, biological oxygen demand (BOD), nutrients,
and oxygen. Williams[11] demonstrated the applicability
of computer simulation using data from Lindeman's
classic study.[12] Our own modeling effort has been
influenced in particular by the extensive literature
review and formulations of DiToro, O'Connor, and
Thomann.[13]

The model described here is known as CLEAN (Comprehensive Lake Ecosystem ANalyzer). It represents an
advance over previous models in that individual ecological processes are represented in greater detail.
Furthermore, the model's equations encompass a
broader spectrum of the processes of lake-ecosystem
dynamics. Incorporation in the CLEAN model of
detailed information about these processes has been
made possible by the close collaboration among
diverse specialists and generalists within the

Eastern Deciduous Forest Biome, International Biological Program. Many of the formulations used to
describe the biological interactions in CLEAN are
derived from formulations previously developed for
TEEM, a terrestrial ecosystem energy model developed
by Shugart, Goldstein, O'Neill and Mankin.[14]

The model is being developed as a collection of submodels, each focusing on a specific component of the
ecosystem. Specialists in ecology and limnology have
contributed their insights into the processes most
familiar to them and have participated in collecting
data needed to implement the model. Although the
theoretical structure of the model is still greatly
simplified, it conforms to what is currently known in
the applicable sciences. Several of the submodels
are simplifications of more extensive process models
developed in the Biome.

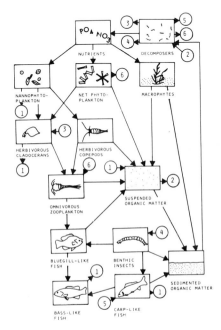

Figure 1 - Ecosystem components forming submodels
in CLEAN (numbers identify common points)

STRUCTURE

The present model is formulated as 28 coupled ordinary
differential equations representing the most important
compartments of the lake ecosystem (Figure 1).
Detailed interactions are indicated in Figure 2.
Allowances were made in programming the model to
accommodate additional components as required; these
will include submodels for lake water balance
(described below) and for additional types of fish
and organisms that live on the lake bottom. Currently
the driving variables include incident solar radiation,
water temperature, nutrient loadings, wind or change
in barometric pressure, and influx of dissolved and
particulate organic material from the terrestrial
system surrounding the lake. In addition, a separate
circulation model is available to be run in conjunction with CLEAN.

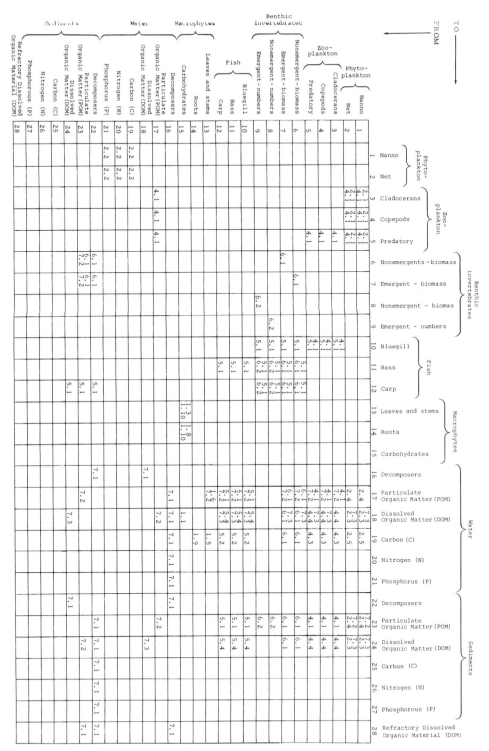

Figure 2 — Transfer matrix for CLEAN (numbers
refer to equations and tables)

229

CLEAN employs modular programming so that a user may execute a specific submodel or link any meaningful combination of submodels as required for a specific simulation. Furthermore, a separate subprogram function is created for each process (such as respiration), permitting the process formulations to be altered quickly and easily. The result is a greatly compressed program as most of the biological processes occur repeatedly in the model with only changes in parameter values.

The modular structure greatly expedites submodel development and testing. As submodels are initially implemented, they represent a wide range of biological sophistication. Therefore, it is important that the submodels be accessible both individually and in varied combinations to permit evaluations of specific terms in them and to make it easier to calibrate them and to use them.

CLEAN has been implemented in interactive mode in FORTRAN for both UNIVAC and IBM time-sharing systems. Parameters and initial conditions can be edited on-line, and driving variables can be changed readily by using an internal CHART function. Thus, the program makes it convenient to run experiments on the model and to update the model as new knowledge develops in this rapidly advancing field.

DESCRIPTION OF COMPONENTS IN *CLEAN*

To simplify the task of presenting a detailed description of the submodels of CLEAN, the equations that make up the submodels are presented as a series of tables. The text presents a simplified description of the important factors in each of the submodels, and cites references for the process models which served as the basis for developing CLEAN.

$$P_{net,m} = (1-X)\left[\frac{P_{max,m}}{\kappa_1} \ln\left(\frac{A + I_0 \exp[-z(\epsilon+\kappa_2 B_p)]}{A + I_0 \exp[-z(\epsilon+\kappa_2 B_p) - \kappa_1 F]}\right) \cdot f(T)g(V)\right] - \rho_f F \qquad 1.1$$

$P_{net,m}$ = Net primary production (quantity of material fixed by photosynthesis and available for plant growth)

X = Fraction of photosynthetic product excreted from plant and lost

$P_{max,m}$ = Maximum photosynthetic rate (under conditions of optimal light, temperature, and nutrients)

κ_1 = Coefficient for light extinction due to plant leaves

A = Light saturation coefficient (at light intensity A, photosynthesis rate = $1/2\ P_{max,m}$)

I_0 = Light intensity at water surface (driving or extrinsic variable which varies during the year)

z = Water depth

ϵ = Coefficient for light extinction due to water

κ_2 = Coefficient for light extinction due to phytoplankton biomass

B_p = Phytoplankton biomass

F = Leaf area index (total surface of leaves suspended over a unit area of lake bottom)

V = Water current velocity

$g(V)$ = Function of water current velocity necessary to renew nutrients in boundary zone of plant, ≤1.0

T = Temperature, °C

$f(T)$ = Function of temperature (see Table 3); ≤1.0

ρ_f = Respiration rate of plant leaves; a function of temperature

$$dL/dt = L*\delta(t_p) + G_L - (R_L + M_L) \qquad 1.2$$

$$\delta = \begin{cases} 1, & t = t_p \\ 0, & \text{otherwise} \end{cases} \qquad 1.21$$

L = Biomass of plant leaf tissue

t = Time

t_p = Time for initial plant growth in spring, based on cumulative effect of temperature during early spring

$L*$ = Initial biomass pulse of arbitrarily small value

G_L = Growth of new leaf tissue

R_L = Leaf respiration

M_L = Leaf mortality (sloughing of entire leaves)

Table 1
Macrophytes

Macrophytes[15] (Table 1)

Macrophytes are aquatic plants which are generally rooted in the lake sediment and which grow toward the lake surface. Growth of these plants depends on photosynthesis (Equation 1.1) which enables macrophytes to utilize solar energy, carbon dioxide, and water to form complex carbohydrate molecules. The model considers a maximum photosynthetic rate P_{max}, which is reduced by suboptimal levels of light, temperature, and available carbon. The shading effects of algae and macrophytes and absorption of light by the water and by material suspended in it attenuate light as depth increases. The model incorporates this effect and the depth distribution of photosynthetic macrophyte tissue by integrating the photosynthetic rate over the leaf-area index, yielding Equation 1.1. A portion of the photosynthetic

material is lost through respiration and extracellular secretion ("excretion"). The remaining photosynthate forms a pool of carbohydrates S, which is then drawn upon for plant growth and metabolism.

Incremental changes (Equation 1.2) in leaf biomass L result from growth of new tissue G_L, metabolic losses R_L, and leaf sloughing M_L. Growth of new tissue (Equation 1.3) is a function of leaf biomass and proceeds asymptotically to an optimal leaf-area index as long as the available carbohydrates exceed a minimum level S_0. The optimal leaf-area index (Equation 1.4) is that ratio of leaf-surface area to lake-surface area which results in maximum net photosynthesis.[16] A leaf-area index greater than the optimum results in reduced net gain due to self-shading and the metabolic cost of maintaining additional biomass. Loss by cellular respiration of leaf

$$G_L = \lambda L \left(1 - \frac{L}{nF_{opt}}\right)\left(\frac{S - S_0}{S_1 + S - S_0}\right) \nu(nF_{opt} - L)\nu(S - S_0) \qquad 1.3$$

$$\nu(a) = \begin{cases} 0, & a \le 0 \\ 1, & a > 0 \end{cases} \qquad 1.31$$

λ = Fractional increase in leaf biomass per unit time

n = Units of leaf biomass per unit leaf area

F_{opt} = Optimal leaf area index

S = Stored pool of carbohydrates resulting from photosynthesis

S_0 = Minimal level of carbohydrates reserved for overwintering

S_1 = Carbohydrate level at which growth is 1/2 maximum

$$F_{opt} = \frac{1}{\kappa_1} D\nu(D) \qquad 1.4$$

$$D = \ln\left(\frac{I_0 [P_{max}(1-\chi)g(V)f(T) - \rho_f]\exp[-z(\epsilon + \kappa_2 B_p)]}{A\rho_f}\right)$$

$$R_L = \rho_L L\nu[\nu(S_0 - S) + \nu(F - F_{opt})] \qquad 1.5$$

ρ_L = Fraction of leaf biomass lost by respiration per unit time, a function of temperature

$$M_L = \gamma_1 L[\nu(t - t_1)\nu(t_2 - t) + \nu(t - t_3)\nu(t_4 - t)] - \gamma_2 L\nu(F - F_{opt}) \qquad 1.6$$

γ_1 = Fraction of leaf biomass sloughed per unit time during sloughing periods

γ_2 = Fraction of leaf biomass sloughed per unit time when F_{opt} is exceeded by F

t_1 to t_2 = Initial period of sloughing

t_3 to t_4 = Subsequent period of sloughing

$$dR/dt = G_R - R_R\nu \qquad 1.7$$

R = Biomass of roots

G_R = Growth of root tissue

R_R = Root respiration

$$G_R = \mu R \left(1 - \frac{R}{R_{max}}\right)\left(\frac{S - S_0}{S_1 + S - S_0}\right) \nu(S - S_0) \qquad 1.8$$

μ = Fractional growth of root biomass per unit time

$$R_R = \rho_R R\nu(S_0 - S) \qquad 1.9$$

$$dS/dt = P_{net,m} - \rho_R R\nu(S - S_0) - \rho_L L\nu(S - S_0)\nu(F_{opt} - F) - \alpha(G_L + G_R) \qquad 1.10$$

α = Inverse efficiency of conversion of labile carbohydrates to leaf and root biomass

Table 1 (continued)
   Macrophytes

structural material (Equation 1.5) is proportional to leaf biomass; it occurs only if the carbohydrate pool is below the minimal level S_0, or if active growth of leaves is occurring. Leaf sloughing for macrophytes in Lake Wingra (Equation 1.6) occurs primarily during two periods of the year, although the mechanisms causing the loss are not well understood; in addition, if leaf area exceeds the optimal leaf-area index, sloughing occurs.

Changes in root biomass (Equation 1.7) also involve growth of new tissue and the respiratory cost of maintaining the existing biomass. Growth of new tissue (Equation 1.8) proceeds at a rate proportional to the root biomass up to a maximum value as long as sufficient carbohydrates are available. Cellular respiration of root structural material (Equation 1.9) is proportional to root biomass, but this form of loss ceases when available carbohydrates exceed the minimal level S_0 (Equation 1.9).

Phytoplankton[17,18] (Table 2)

In addition to the larger attached macrophytes discussed above, there are also many species of floating algae, which are known collectively as phytoplankton. Again, net photosynthesis is considered as the difference between a maximum rate P_{max}, modified by suboptimal conditions and a respiration rate (2.2). The combined limitations of light and nutrients are represented as a normalized factor that is mathematically analogous to the inverse of the total effect of electrical resistors in parallel. Grazing is a function of zooplankton biomass, temperature, and capturability as discussed below (Equation 3.1). Excretion is considered as proportional to net photosynthesis (Equation 2.3), and nonpredatory mortality is taken to be the sinking rate as a function of temperature (Equation 2.4). In the current implementation of CLEAN, two phytoplankton groups are distinguished: nannophytoplankton (extremely small) and net (larger) phytoplankton (Figure 1), corresponding to two major subdivisions of the algal community.

Feeding term[14,16] (Table 3)

Most interactions between living components of the lake ecosystem take the form of feeding relationships, i.e., one population is utilizing another as a food supply. This relationship forms a nonlinear linkage which is extremely important to the overall dynamics of the system. Therefore, it must be carefully formulated.

Equation 3.1 shows the dependence of the feeding rate on the biomass of consumers B_j and the biomass of the food supply B_i. The relationship is modified by a food preference term $w_{i,j}$ to differentiate between feeding rates on different types of food. As food supply becomes very abundant, the $w_{i,j}$ term becomes dependent only on the mass of consumers of each type in relation to their available food supplies. Alternatively, as consumer populations become very large in relation to the food supply, feeding is dependent only upon the available food supply, with a limit on the population of consumers being imposed by a minimum food supply, $B_{min,j}$. At intermediate biomass densities, the feeding rate is dependent on the biomass of both interacting populations. The ecological theory and mathematical implications of the nonlinear linkage have been explored in other papers.[19-21]

Table 2
Phytoplankton

$$dB_p/dt = (P_{net,p} - G_{p,z} - U_p - M_p)B_p \qquad 2.1$$

$G_{p,z}$ = Rate of grazing of zooplankton on phytoplankton

U_p = Excretion rate of phytoplankton

M_p = Nongrazing mortality rate of phytoplankton

$$P_{net,p} = (P_{max,p} n/[\Sigma_i 1/\mu_i] - R_p)f(T) \qquad 2.2$$

$$\mu_1 = \left[\frac{2.71828\Phi_p}{\varepsilon + \kappa_2 B_p}\right]$$

$$\cdot \left[\exp\left(\frac{-I_0}{I_s \Phi_p}\exp[-z(\varepsilon + \frac{\kappa_2 B_p}{z})]\right)\right.$$

$$\left. - \exp\left(\frac{-I_0}{I_s \Phi_p}\right)\right] \qquad 2.21a$$

$$\mu_i = \frac{C_i}{N_i + C_i}; \qquad i = 2, 3, \ldots, n \qquad 2.21b$$

R_p = Respiration

μ_i = ith limiting factor; μ_1 = light; μ_i, $i \neq 1$, = nutrients

C_i = Concentration of nutrient i

N_i = Concentration of nutrient i at which photosynthesis rate = $\frac{1}{2}P_{max,p}$

I_s = Saturated light intensity (light intensity above which photosynthesis is maximum)

Φ_p = Photoperiod (fraction of light in a 24 hour period)

$$U_p = \begin{cases} a\, P_{net,p}; & P_{net,p} \geq 0 \\ 0 & ; P_{net,p} < 0 \end{cases} \qquad 2.3$$

a = Fraction of net photosynthate lost per unit-time

$$M_p = bT \qquad 2.4$$

b = Fraction of algal biomass dying or sinking to lake bottom per unit of time and temperature

$$R_p = R_{max} V^X e^{X(1-V)} \qquad 2.5$$

$$R_{max} = KP_{max,p}$$

See 3.2 for V and X

Table 3
Trophic interaction or feeding term

$$C_{i,j} = C_{max_j} H_{j,C}$$

$$\cdot \left(\frac{w_{i,j}(B_i - B_{min_i})}{Q_j z + r_j B_j + \Sigma w_{i,j}(B_i - B_{min_i})} \right) \quad 3.1$$

$$H_{j,n} = a_{j,n}\left(b_{j,n}\frac{(K_j - B_j)}{K_j} + 1\right)c_{j,n} \quad 3.11$$

$C_{i,j}$ = Consumption of ith prey by jth organism

Q_j, r_j = Environmental and population interaction coefficients

w_{ij} = Coefficient relating preference, availability, capturability, etc., of i as a food for j

K_j = Carrying capacity of ecosystem for predator j

$a_{j,n}$ = Correction factor for behavioral effects on the nth process (e.g., $n=C$ for consumption, R for respiration, M for mortality, F for fishing, G for gamete production) for the jth organism

$b_{j,n}$ = Correction factor for effects of age structure of population on nth process

$c_{j,n}$ = Correction factor for physiological effects on nth process

For example:

$$c_{z,C} = f(T) = V^x \exp[X(1-V)] \quad 3.2$$

where z denotes zooplankton and C consumption

$$V = \frac{T_{max} - T}{T_{max} - T_{opt}} \quad 3.21$$

$$X = \frac{W^2[1 + \sqrt{(1 + 40/W)}]^2}{400} \quad 3.22$$

$$W = (\ell n S_Q)(T_{max} - T_{opt}) \quad 3.23$$

T_{max} = Upper lethal temperature

T_{opt} = Optimum temperature (temperature at which rate is maximum)

S_Q = Q_{10} value (factor in log range by which rate is increased for a $10°C$ increase in temperature)

($C_{j,k}$=Consumption of jth organism by kth predator)

The adult feeding rate $C_{i,j}$ is considered to be a complex function of temperature. The rate increases exponentially up to a potential maximum C_{max} at an optimal temperature T_{opt}. At higher temperatures, the rate rapidly decreases until a lethal temperature is reached and the organism dies. Behavioral characteristics a_j also modify the feeding rate. For example, some animals may go into hibernation below some critical temperature, causing feeding to fall to zero.

The feeding rate $C_{i,j}$ is defined as adult feeding rate and must be modified to account for the presence of immature organisms in the population. The age-structure correction factor is b_j, which appropriately increases the feeding rate to account for juveniles. As the population reaches its maximum possible density K_j, reproduction is greatly reduced, resulting in a population composed primarily of adults. Therefore, the magnitude of the correction factor is decreased to zero as the "carrying capacity" of the ecosystem is approached. This construct has proven to be satisfactory for predators in the upper levels of the predator-prey food chain, but we are presently replacing it with a better formulation for populations subject to heavy predation.

Zooplankton[22] (Table 4)

The next step in the food chain consists of small invertebrate animals known collectively as zooplankton (Figure 1). Three groups are distinguished in the model: calanoid copepods (which feed on phytoplankton), cladocerans (which feed on both phytoplankton and suspended particulate organic matter), and omnivorous zooplankton which can feed on other zooplankton as well as on phytoplankton and particulate organic matter. Multiple food supplies are incorporated in the model through a summation over individual feeding terms as shown in Equation 4.1. In a similar manner, predation on the zooplankton is represented by another summation over all potential consumer populations. Defecation and metabolic excretion are represented by the product of the feeding term and appropriate constants. Respiratory and mortality losses are presented as proportional to the biomass, with appropriate modifications for behavior, age structure, and physiological responses to the environment.

Fish[23] (Table 5)

Fish populations in the lake ecosystem are divided into three categories: a bluegill-like generalized predator feeding on zooplankton and benthic (bottom-dwelling) insect larvae; a carp-like scavenger; and a bass-like predator that consumes other fish. For each of these categories, model equations closely resemble the zooplankton model (4.1). Additional loss terms are included to account for the metabolic cost s_2 of food digestion and utilization, incorporation of material into eggs and sperm, and losses due to fishing pressure by man. Equation 5.2 also introduces the concept of density-dependent stress due to overcrowding.

Table 4
Zooplankton

$$\frac{dB_z}{dt} = \sum_{i=1}^{n} C_{i,z} - \sum_{k=1}^{m} C_{z,k}$$

$$- (R_z + U_z + F_z + M_z') \qquad 4.1$$

B_z = Biomass of zooplankton

R_z = Respiration rate

U_z = Rate of metabolic excretion

F_z = Rate of egestion (elimination of unassimilated materials)

M_z = Rate of nonpredatory mortality

(with $a_{z,C}$, $b_{z,C}$, $c_{z,C}$ in $C_{i,z}$)

$$F_z = \sum_{i=1}^{n} f_{i,z} \, C_{i,z} \qquad 4.2$$

$f_{i,z}$ = Fraction of food supply i consumed but not assimilated by zooplankton

$$R_z = R_{max} H_{z,R} B_z \qquad 4.3$$

(with $a_{z,R}$, $b_{z,R}$, $c_{z,R}$ in $H_{z,R}$)

$$U_z = u_z \sum_i C_{i,z} \qquad 4.4$$

u_z = Fraction of food assimilated and subsequently excreted

$$M_z = Z_M H_{z,M} B_z \left(\frac{d_{z,M} B_z}{K_j} + 1 \right) \qquad 4.5$$

$d_{z,M}$ = Density-dependent term for increased mortality due to overcrowding

Z_M = Natural rate of adult mortality

Table 5
Fish

$$dB_f/dt = \sum_{i=1}^{n} C_{i,f} - \sum_{k=1}^{m} C_{f,k}$$

$$- (F_f + R_f + U_f + G_f + M_f + C_{f,man}) \qquad 5.1$$

B_f = Biomass of fish

R_f = Respiration rate

U_f = Metabolic excretion rate

G_f = Rate of gamete production

M_f = Rate of nonpredatory mortality

$C_{f,man}$ = Mortality rate due to fishing

F_f = Egestion rate

$$R_f = s \sum_i C_{i,f} + H_{f,R} B_f \left(1 + \frac{d_{f,R}}{K_f} B_f \right) \qquad 5.2$$

(with $a_{f,R}$, $b_{f,R}$, $c_{f,R}$ in $H_{f,R}$)

s = Metabolic cost of food digestion and utilization

$d_{f,R}$ = Density-dependent term for increased respiration due to overcrowding

$$F_f = \sum_i f_{i,f} C_{i,f} \qquad 5.3$$

$f_{i,f}$ = Fraction of ith ingested food that is not assimilated

$$U_f = u_f \sum_i C_{i,f} \qquad 5.4$$

u_f = Fraction of assimilated food that is excreted

$$G_f = H_{f,G} \, Z_G \, \kappa_G \, B_f \qquad 5.5$$

(with $a_{f,G}$, $b_{f,G}$, $c_{f,G}$ in $H_{f,G}$)

Z_G = Instantaneous rate of gamete mortality

κ_G = Fraction of adult biomass in gametes at spawning

$c_{f,G}$ = Temperature switch permitting spawning between a maximum and minimum temperature

$$M_f = Z_M H_{f,M} B_f \left(\frac{d_{f,M} B_f}{K_f} + 1 \right) \qquad 5.6$$

(with $a_{f,M}$, $b_{f,M}$, $c_{f,M}$ in $H_{f,M}$)

Z_M = Rate of adult natural mortality

$$C_{f,man} = Z_F H_{f,F} B_f \qquad 5.7$$

(with $a_{f,F}$, $b_{f,F}$, $c_{f,F}$ in $H_{f,F}$)

Z_F = Rate of adult mortality due to fishing

Table 6
Bottom-dwelling organisms (benthos)

$$dB_{b_j}/dt = \sum_i C_{i,b_j} + I_{b_j} - \left(R_{b_j} + F_{b_j} + U_{b_j} + X_{b_j} + P_{b_j} + M_{b_j} + \sum_k LC'_{b_j,k}\right) \quad 6.1$$

R_{b_j} = Respiration

$$= R_{maxb_j} H_{b_j,R} \left(1 + \frac{d_{b_j,R} B_{b_j}}{K_{b_j}}\right) B_{b_j} \quad 6.11$$

$$B_{b_j,n} \begin{cases} a_{b_j,n} & = \text{Physiological effects due to dissolved oxygen for } n = M,C,R \\[4pt] b_{b_j,n} & = \text{Correction term for effect of mean weight on process;} \\ & \quad \text{replaces } \left(b_{j,n} \frac{(K_j - B_j)}{K_j} + 1\right) \\[4pt] & = a\overline{W}_{b_j}^{\beta-1}; \ a,\beta \text{ fitted coefficients} \\[4pt] c_{b_j,n} & = \text{Physiological effects due to } T \text{ for } n = M,C,R \\[4pt] B_{b_j} & = \text{Biomass of benthos in } j\text{th size class; } j = 1 \text{ for instars 1-3; } j = 2 \text{ for 4th instar} \\[4pt] d_{b_j,n} & = \text{Effects due to crowding on } n\text{th process} \end{cases}$$

F_{b_j} = Egestion

$$= \sum_i f_{i,b_j} C_{i,b_j} \quad 6.12$$

f_{i,b_j} = Fraction of ith food not assimilated (i.e., egested)

U_{b_j} = Excretion

$$= u_{b_j} \sum_i C_{i,b_j} \quad 6.13$$

u_{b_j} = Fraction of assimilated food excreted

X_{b_j} = Exuviation

$$= x_{b_j} B_{b_j} \quad 6.14$$

x_{b_j} = Fraction of biomass exuviated

M_{b_j} = Nonpredatory mortality

$$= z_{b_j} H_{b_j,M} \left(1 + \frac{d_{b_j,M} B_{b_j}}{K_{b_j}}\right) B_{b_j} \quad 6.15$$

z_{b_j} = Instantaneous rate of nonpredatory mortality

$d_{b_j,M}$ = Increase in mortality due to density-dependent factors (overcrowding)

K_{b_j} = Carrying capacity of ecosystem

I_{b_j} = Influx into class. For $j = 1$, = egglaying = λW_e; for $j = 2$, = maturation of 3rd instars into 4th = $P_{b_1} W_3$. $\quad 6.16$

λ = Number of eggs that hatch = $f(B_{b_1}, N_{b_2})$

W_e = Average weight of an egg

P_{b_1} = Number of molting 3rd instars (maturing into 4th)

W_3 = The critical weight of a 3rd instar at which molting is induced

P_{b_j} = Promotion out of class. For $j = 1$, maturation of 3rd instar into 4th (= I_{b_2}); for $j = 2$, emergence of 4th instars to adults, = $a_b W_\sigma$. $\quad 6.17$

a_b = Number of emerging insects

$$= \begin{cases} 0 & ; \ \overline{W}_{b_2} < \overline{W}_e \\[4pt] N_{b_2} \kappa_1 (\exp\{\kappa_2 (\overline{W}_{b_2} - W_e)\} - 1)\text{)}; \ \overline{W}_{b_2} \geq \overline{W}_e \end{cases} \quad 6.18$$

$\overline{W}_{b_j}$ = Mean weight of jth size class

$\overline{W}_e$ = Critical population mean weight to begin emergence

W_σ = Actual organism's weight to induce emergence

κ_1, κ_2 = Fitted coefficients

$$dN_{b_j}/dt = \frac{1}{\overline{W}_{b_j}} \left(-M_{b_j} - \sum_k C_{b_j,k}\right) + \frac{I_{b_j}}{W_k} - \frac{P_{b_j}}{Y_k} \quad 6.2$$

N_{b_j} = Numbers of jth class of benthos

$\overline{W}_{b_j}$ = Mean weight of jth class

$$W_k = \begin{cases} W_e; & k=1 \text{ (1st size class)} \\ W_3; & k=2 \end{cases}$$

$$Y_k = \begin{cases} W_3; & k=1 \\ W_\sigma; & k=2 \end{cases}$$

Benthos[24,25] (Table 6)

The changes in biomass (Equation 6.1) and numbers (Equation 6.2) for bottom-dwelling (benthic) insect larvae are considered separately for two size classes. The first three larval stages (instars) are included in one size class and the final larval stage in a second class; adults and eggs constitute two additional size classes.

The formulation for biomass changes resembles the approach used for zooplankton (Table 4) and fish (Table 5), with feeding rate balanced against respiration, excretion, mortality, and other loss terms. An additional term is added to the benthos equation to account for material lost as the larvae shed the outer covering (exoskeleton) during molt. Terms are also added to account for input and output from one size class to another; these factors include egglaying, maturation from one larval size class to the other, and metamorphosis into adults and subsequent emergence from the lake.

The addition of a numbers equation permits the calculation of mean weight so that the processes of respiration, feeding, predation, and promotion (maturation and emergence of adults) can be made functions of mean weight. The numbers equation includes those terms from the biomass equation that represent discrete changes in numbers of individuals; each term is divided by mean weight in order to convert from weight to numbers.

Maturation and emergence are considered a function of a particular size class. As weight increases above a critical level, the probability of promotion increases exponentially. Nonpredatory mortality considers the effect of overcrowding, which causes increased mortality as the population approaches carrying capacity (determined as the maximum number of individuals that can occupy a unit area without touching). Respiration, feeding, and nonpredatory mortality are also affected by the dissolved oxygen level, which is often near critically low levels for maintenance of the bottom-dwelling organisms.

Organic matter and decomposition[26] (Table 7)

Decomposition (conversion of dead plant and animal material to inorganic forms) occurs in the water column and in bottom sediments. The process can be considered as occurring in two stages. First, particles of organic matter B_p are made soluble by hydrolysis. The resultant dissolved organic matter B_{DOM} serves as substrate for decomposers (fungi, bacteria, and protozoans) B_d. As byproducts of their metabolism, these decomposers produce inorganic carbon, nitrogen, phosphorus, and dissolved organic matter that is not readily subject to further bio-degradation. The decomposer-particulate-matter aggregate also serves as a food source for zooplankton, benthic insect larvae, and bottom-feeding fish.

The equations for particulate (Equation 7.2) and dissolved (Equation 7.3) organic matter consist of fluxes associated with animal and plant losses of organic matter (particularly by death, elimination of waste materials, and leaf sloughing), terms for inputs from outside the lake ecosystem, and flux terms due to decomposer activity. The decomposer equation (Equation 7.1) contains uptake, respiration, excretion, and mortality terms. The set of three equations is considered to represent processes both within the water column and within the sediments.

Decomposer uptake of dissolved organic matter follows simple saturation kinetics (Equation 7.4), except that a certain amount DOM_{min} of the dissolved material is deemed unusable. Uptake is

Table 7
Decomposition

$$dB_{d_i}/dt = V_{d_i} - \left(R_{d_i} + U_{d_i} + S_{d_i} + M_{d_i} + \textstyle\sum_j C_{d_i,B_j} \right) \qquad 7.1$$

B_{d_i} = Biomass of the ith decomposer group

V_{d_i} = Rate of uptake of organic materials

R_{d_i} = Respiration rate

U_{d_i} = Rate of excretion of inorganic and refractory organic materials

S_{d_i} = Rate of sedimentation (or resuspension)

M_{d_i} = Rate of mortality due to lysis (disintegration), inactivation, and micropredators

$$dB_{p_i}/dt = I_{p_i} + \textstyle\sum_j F_j + \textstyle\sum_j K_{p,M_j}M_j + S_{p_i} - \left(H_{p_i} + \textstyle\sum_j C_{p_i,j} \right) \qquad 7.2$$

B_{p_i} = Mass of the ith Particulate Organic Matter (POM) compartment

I_{p_i} = External loading of POM

F_j = Defecation rate of the jth consumer group

$K_{p,M_j}M_j$ = Input rate of POM due to nonpredatory mortality

S_{p_i} = Rate of sedimentation (or resuspension)

H_{p_i} = Hydrolysis rate

$C_{p_i,j}$ = Rate of grazing by jth consumer group

$$dB_{DOM_i}/dt = I_{DOM_i} + \textstyle\sum_{j=1}^{n} U_j + \textstyle\sum_j K_{p,M_j}M_j + H_{p_i} + D_{DOM_i} - V_{d_i} \qquad 7.3$$

B_{DOM_i} = Mass of the ith Dissolved Organic Matter (DOM) compartment

I_{DOM_i} = External loading of ith DOM

U_j = Excretion rate of organics by the jth biotic group

$K_{p,M_j}M_j$ = Input rate of DOM due to nonpredatory mortality

D_{DOM_i} = Diffusion rate of ith organic between sediment and water

$$V_{d_i} = V_{max,d_i} \; \alpha_{d_i,V} \; \sigma_{d_i,V} \left(\frac{B_{DOM_i} - DOM_{min_j}}{K_{DOM_i} + B_{DOM_i} - DOM_{min_j}} \right) B_{d_i}$$

$$\nu \left(B_{DOM_j} - DOM_{min_j} \right) \qquad 7.4$$

where i = ith decomposer group, j = jth DOM group

V_{max,d_i} = Maximum uptake rate

$\alpha_{d_i,V}$ = Effect of temperature on uptake (see Equation 3.2)

$\sigma_{d_i,V}$ = Effect of dissolved oxygen

$$= 1 + K_{V,O_2}\nu(O_{2,min} - O_2) \qquad 7.41$$

DOM_{min_j} = DOM level below which uptake is negligible

K_{DOM} = Half-saturation constant for uptake

$O_{2,min}$ = Oxygen level for anaerobic uptake

K_{V,O_2} = Change in uptake due to anaerobic conditions

$$R_{d_i} = R_{max,d_i} \; \sigma_{d_i,R} \; \alpha_{d_i,R} \; B_{d_i} + K_{resp,d_i} V_{d_i} \qquad 7.5$$

R_{max,d_i} = Maximum endogenous respiration rate

K_{resp,d_i} = Constant for effect of uptake on respiration

$$U_{d_i} = \textstyle\sum_j u_j \; \nu(O_{2,min} - O_2)(V_{d_i} - R_{d_i})\nu(V_{d_i} - R_{d_i}) \qquad 7.6$$

u_j = Percent of net assimilation which is excreted for jth compound

$$H_{p_j} = H_{max,p_j} \; \sigma_{p_j,H} \; \alpha_{p_j,H} \; h_{p_j,H} \left(\frac{B_{d_i}}{K_{sv_j}B_{p_j} + B_{d_i} + K_{POM_j}} \right) B_{p_j} \qquad 7.7$$

H_{max,p_j} = Maximum hydrolysis rate

$h_{p_j,H}$ = Effect of pH on hydrolysis

$$= \exp \left[- \left(\frac{pH - pH_{opt}}{\sigma} \right)^2 \right] \qquad 7.71$$

K_{sv_j} = Available surface area parameter

K_{POM_j} = Saturation constant for hydrolysis

pH = pH of water or interstitial water in sediment

pH_{opt} = Optimum pH for hydrolysis

σ = pH "bandwidth" constant for hydrolysis

POM = Particulate Organic Matter

$$S_{d_i} = \nu_{d_i} \; \exp[K_{sed_{d_i}} (\Delta T/\Delta z)] \; \frac{\Delta P_{bar}}{\Delta t} \; B_{d_i} \qquad 7.8$$

$\Delta T/\Delta z$ = Maximum vertical thermal gradient

$\Delta P_{bar}/\Delta t$ = Change in barometric pressure over past day

$K_{sed_{d_i}}$ = Thermal gradient constant

where B_{d_i} is either water or sediment compartment depending on the sign of $\frac{\Delta P_{bar}}{\Delta t}$

ν_{d_i} = Sedimentation rate constant

$$D_{DOM_i} = K_{DIF} \left(B_{DOM_i} - B_{DOM_{i+1}} \right) \qquad 7.9$$

K_{DIF} = Boundary layer resistance coefficient

DOM = Dissolved Organic Matter

$$M_{d_i} = K_{d_i,M} \; \alpha_{d_i,M} \; B_{d_i} \; C_{d_i,M} \qquad 7.10$$

$K_{d_i,M}$ = Nonmacropredatory mortality

changed by a fraction K_{V,O_2} under anaerobic conditions. Respiration losses from the decomposers are considered proportional to decomposer biomass, modified by temperature and the amount of dissolved oxygen. Grazing rates on the decomposer-particulate substrate have already been considered in the models for the individual grazers (see Tables 3–6).

The rate of hydrolysis of particulate organic matter is dependent on water temperature, acidity/alkalinity (pH), dissolved oxygen, and the amount of organic matter and decomposers (Equation 7.7). The parameter K_{gv} is a function of the mean available surface area of particulate matter. The complex hydrolysis term produces a flux proportional to particulate concentration when decomposers are abundant, proportional to decomposers when organic matter is abundant, and proportional to both at intermediate concentrations.

Through metabolic excretion the decomposers produce organic byproducts and remineralize phosphorus and nitrogen. The rate of this process is considered a function of uptake and respiration (Equation 7.6), and the excretion of organics increases significantly with low concentrations of oxygen. The refractory dissolved organic matter might be considered as a carbon source for methane-producing bacteria, but this process is ignored in the present version of the model.

The rate of sedimentation of particulate organic matter is a function of the potential for vertical mixing, as indicated by the thermal gradient, and wave agitation (Equation 7.8). Wave agitation is a result of wind stress; however, in order to make the model more general, we have considered using the change in barometric pressure as a driving variable for wave agitation instead of wind.[27] This also permits simulation of the resuspension of sediment when wave base intersects the lake bottom at a particular time. Further study is being conducted to determine if this construct is adequate, or if it is an oversimplification.

Lake-water balance[28],[29] (Table 8)

The hydrology of the lake basin is of interest in that it affects transport of nutrients into the lake, the nutrient concentrations in the lake, and the wash-out from the lake of both nutrients and plankton. A separate hydrological transport model has been developed to predict nutrient loadings[29]; output from this model can be used to drive CLEAN. Experience with WNGRA2, the site-specific version of CLEAN for Lake Wingra, has been quite encouraging.[31] At present, a water-balance submodel for lakes is being incorporated into CLEAN.

Positive terms include precipitation, surface inflow, and groundwater inflow. Negative terms include surface outflow, evaporation, and groundwater outflow. Inflow and outflow rates may be expressed as functions of lake stage so that the effects of water-level fluctuations, particularly in recent impoundments, can be simulated. Several formulations for calculating evaporation have been used, including one that corrects for the horizontal transport of energy by water currents. Groundwater flow may be determined empirically using data from observational wells. In lake basins having little contribution from groundwater (e.g., rock basins) groundwater flow can be

Table 8
Lake water balance

$$\frac{dV}{dt} = (p-e)A(E) + Q_s f(E)$$
$$+ \sum_{i=1}^{n} \left[K_i \left(\frac{w_i - E}{D_i} \right) L_i [m_{0i} + \alpha (w_i - w_{0i})] \right] - K(t)$$

8.1

V = Volume of lake water

$A(E)$ = Lake surface area as a function of lake stage, E

p = Precipitation rate

e = Evaporation rate

Q_s = Surface inflow rate

$f(E)$ = Elevation-dependent correction factor for inflow

n = Number of observation wells

K_i = Saturated hydraulic conductivity at observation well i

w_i = Elevation of groundwater table at well i

E = Lake level

D_i = Distance of well i from lake

L_i = Effective length of aquifer along the shoreline associated with well i

m_{0i} = Base value for effective saturated thickness

α = Parameter which characterizes dependence of m on w

w_{0i} = Index elevation

$K_3(t)$ = Known time-series of outflow

simplified or neglected. The outflow term shown in Equation 8.1 refers to conditions of known outflow. For other situations, such as flow over an ungated spillway or natural stream drainage, an appropriate expression can be substituted.

Lake-circulation and substance-transport

Wind-driven currents, lake oscillations, and heat transfer across the air-water boundary are of importance to the aquatic ecosystem in that they transport and diffuse nutrients and particulate materials (including influents, organic matter, plankton, and resuspended sediments) from one region of the lake to another. Therefore, the modeling program would not be complete without some means of predicting lake motions. A separate circulation model has been developed as a part of the research effort at Lake Wingra[32] (see Table 9). Because CLEAN was formulated as a point model to minimize

Table 9

Lake circulation
and substance transport

Mass conservation

$$\frac{\partial}{\partial x}\left((UH)_{k-\frac{1}{2}}\right) + \frac{\partial}{\partial y}\left((VH)_{k-\frac{1}{2}}\right) + \omega_{k-1} - \omega_k + \frac{\partial Z_{k-1}}{\partial t} - \frac{\partial Z_k}{\partial t} = 0 \qquad 9.1$$

Momentum conservation

a) x-component

$$\frac{\partial}{\partial t}(UH)_{k-\frac{1}{2}} + \frac{\partial}{\partial x}\left(\alpha_1(U^2H)_{k-\frac{1}{2}}\right) + \frac{\partial}{\partial y}\left(\alpha_2(UVH)_{k-\frac{1}{2}}\right) - f(VH)_{k-\frac{1}{2}} + (\omega\bar{U})_{k-1} - (\omega\bar{U})_k$$

$$= - H_{k-\frac{1}{2}}\left(\frac{\partial}{\partial x}\left[gn + \frac{P_s}{\rho_o}\right] + P_{x_{k-\frac{1}{2}}}\right) + \frac{\partial}{\partial x}\left(\left[A_{m_h}\right]_{k-\frac{1}{2}}\frac{\partial}{\partial x}(UH)_{k-\frac{1}{2}}\right) + \frac{\partial}{\partial y}\left(\left[A_{m_h}\right]_{k-\frac{1}{2}}\frac{\partial}{\partial y}(UH)_{k-\frac{1}{2}}\right)$$

$$+ \left(A_{m_v}\frac{\partial \bar{U}}{\partial z}\right)_{k-1} - \left(A_{m_v}\frac{\partial \bar{U}}{\partial z}\right)_k \qquad 9.2$$

b) y-component

$$\frac{\partial}{\partial t}\left((VH)_{k-\frac{1}{2}}\right) + \frac{\partial}{\partial x}\left(\alpha_2(UVH)_{k-\frac{1}{2}}\right) + \frac{\partial}{\partial y}\left(\alpha_3(V^2H)_{k-\frac{1}{2}}\right) + f(UH)_{k-\frac{1}{2}} + (\omega\bar{V})_{k-1} - (\omega\bar{V})_k$$

$$= H_{k-\frac{1}{2}}\left(\frac{\partial}{\partial y}\left[gn + \frac{P_s}{\rho_o}\right] + P_{y_{k-\frac{1}{2}}}\right) + \frac{\partial}{\partial x}\left(\left[A_{m_h}\right]_{k-\frac{1}{2}}\frac{\partial}{\partial x}(VH)_{k-\frac{1}{2}}\right) + \frac{\partial}{\partial y}\left(\left[A_{m_h}\right]_{k-\frac{1}{2}}\frac{\partial}{\partial y}(VH)_{k-\frac{1}{2}}\right)$$

$$+ \left(A_{m_v}\frac{\partial \bar{V}}{\partial z}\right)_{k-1} - \left(A_{m_v}\frac{\partial \bar{V}}{\partial z}\right)_k \qquad 9.3$$

Heat conservation

$$\frac{\partial}{\partial t}\left((TH)_{k-\frac{1}{2}}\right) + \frac{\partial}{\partial x}\left(\beta_1(UTH)_{k-\frac{1}{2}}\right) + \frac{\partial}{\partial y}\left(\beta_2(VTH)_{k-\frac{1}{2}}\right) + (\omega\bar{T})_{k-1} - (\omega\bar{T})_k$$

$$= \frac{\partial}{\partial x}\left(\left[A_{h_h}\right]_{k-\frac{1}{2}}\frac{\partial}{\partial x}(TH)_{k-\frac{1}{2}}\right) + \frac{\partial}{\partial y}\left(\left[A_{h_h}\right]_{k-\frac{1}{2}}\frac{\partial}{\partial y}(TH)_{k-\frac{1}{2}}\right) + \left(A_{h_v}\frac{\partial \bar{T}}{\partial z}\right)_{k-1} - \left(A_{h_v}\frac{\partial \bar{T}}{\partial z}\right)_k + \phi_{k-\frac{1}{2}} \qquad 9.4$$

Substance (dissolved or particulate) conservation

$$\frac{\partial}{\partial t}\left((CH)_{k-\frac{1}{2}}\right) + \frac{\partial}{\partial x}\left(\delta_1(UCH)_{k-\frac{1}{2}}\right) + \frac{\partial}{\partial y}\left(\delta_2(VCH)_{k-\frac{1}{2}}\right) + (\omega\bar{C})_{k-1} - (\omega\bar{C})_k + (\omega_{ss}\bar{C})_{k-1} + (\omega_{ss}\bar{C})_k$$

$$= \frac{\partial}{\partial x}\left(\left[A_{s_h}\right]_{k-\frac{1}{2}}\frac{\partial}{\partial x}(CH)_{k-\frac{1}{2}}\right) + \frac{\partial}{\partial y}\left(\left[A_{s_h}\right]_{k-\frac{1}{2}}\frac{\partial}{\partial y}(CH)_{k-\frac{1}{2}}\right) + \left(A_{s_v}\frac{\partial \bar{C}}{\partial z}\right)_{k-1} - \left(A_{s_v}\frac{\partial \bar{C}}{\partial z}\right)_k + S_{k-\frac{1}{2}} \qquad 9.5$$

where

$Z_o = \eta(x,y,t)$ = free water surface

$Z_k = -h_k$ = location of rigid intermediate levels (layer boundaries)

$Z_n = -h(x,y)$ = lake bottom

$H_{k-\frac{1}{2}} = Z_{k-1} - Z_k$ = layer thickness

$k = 1,2,3,\ldots,N$ = index denoting particular level (layer boundary) below water surface ($k = 0$)

x,y = horizontal coordinates in plane of equilibrium lake water surface

z = vertical coordinate perpendicular to equilibrium water surface, positive upwards

t = time

$U_{k-\frac{1}{2}} = \frac{1}{H_{k-\frac{1}{2}}}\int_{Z_k}^{Z_{k-1}} u dz$ = layer-averaged x-component of velocity

$V_{k-\frac{1}{2}} = \frac{1}{H_{k-\frac{1}{2}}}\int_{Z_k}^{Z_{k-1}} v dz$ = layer-averaged y-component of velocity

u,v, and ω = horizontal and vertical velocity components in x,y, and z directions at (x,y,z)

$\alpha_1,\alpha_2,\alpha_3$ = coefficients resulting from variation of velocity over layer thickness

$\omega_k = \omega_k - \left[u_k\frac{\partial Z_k}{\partial x} + v_k\frac{\partial Z_k}{\partial y}\right] - \frac{\partial Z_k}{\partial t}$ = velocity (vertical) relative to level k

note that $\omega_o = 0$, $\omega_n = 0$

f = Coriolis parameter = $2\Omega \sin\phi$ (where Ω = angular rotation rate of earth, ϕ = latitude)

$\bar{U}_k = \frac{1}{2}\left(U_{k-\frac{1}{2}} + U_{k+\frac{1}{2}}\right)$ = average velocity of x-component between layers

$\bar{V}_k = \frac{1}{2}\left(V_{k-\frac{1}{2}} + V_{k+\frac{1}{2}}\right)$ = average velocity of y-component between layers

g = gravitational constant

P_s = atmospheric pressure on water surface

ρ_o = reference water density

$P_{x_{k+\frac{1}{2}}} - P_{x_{k-\frac{1}{2}}} = g\left(\dfrac{H_{k+\frac{1}{2}} + H_{k-\frac{1}{2}}}{2}\right) \dfrac{\partial}{\partial x}\left[\dfrac{\sigma_k}{\rho_o}\right]$ = x-component of horizontal pressure gradient due to temperature variations

where

$P_{x_{\frac{1}{2}}} = gH_{\frac{1}{2}} \dfrac{\partial}{\partial x}\left[\dfrac{\sigma_{\frac{1}{2}}}{\rho_o}\right]$; note that the same expression holds for the y-component when x is replaced by y.

$\sigma_k = \psi(\bar{T}_k)$ = equation of state (relating density changes to temperature changes)

A_{m_h} = horizontal turbulent-diffusion coefficient for momentum

A_{m_v} = vertical turbulent-diffusion coefficient for momentum

$\left[A_{m_v}\dfrac{\partial \bar{U}}{\partial z}\right]_k = 2A_{m_v}\left[\dfrac{U_{k-\frac{1}{2}} - U_{k+\frac{1}{2}}}{H_{k-\frac{1}{2}} + H_{k+\frac{1}{2}}}\right]$ = x-component of vertical momentum transfer due to turbulent shear;

note that $\left[A_{m_v}\dfrac{\partial \bar{U}}{\partial z}\right]_0 = \dfrac{\tau_{s_x}}{\rho_o}$ and $\left[A_{m_v}\dfrac{\partial \bar{U}}{\partial z}\right]_N = \dfrac{\tau_{b_x}}{\rho_o}$.

τ_{s_x}, τ_{s_y} = x,y components of wind shear stress on water surface

τ_{b_x}, τ_{b_y} = x,y components of bottom shear stress

$\left[A_{m_v}\dfrac{\partial \bar{V}}{\partial z}\right]_k = 2A_{m_v}\left[\dfrac{V_{k-\frac{1}{2}} - V_{k+\frac{1}{2}}}{H_{k-\frac{1}{2}} + H_{k+\frac{1}{2}}}\right]$ = y-component of vertical momentum transfer due to turbulent shear;

note that $\left[A_{m_v}\dfrac{\partial \bar{V}}{\partial z}\right]_0 = \dfrac{\tau_{s_y}}{\rho_o}$ and $\left[A_{m_v}\dfrac{\partial \bar{V}}{\partial z}\right]_N = \dfrac{\tau_{b_y}}{\rho_o}$.

$T_{k-\frac{1}{2}} = \dfrac{1}{H_{k-\frac{1}{2}}}\displaystyle\int_{Z_k}^{Z_{k-1}} T\,dz$ = layer-averaged temperature

T = water temperature at (x,y,z)

$\bar{T}_k = \frac{1}{2}\left(T_{k-\frac{1}{2}} + T_{k-\frac{1}{2}}\right)$ = average temperature between layers

$\beta_1, \beta_2, \beta_3$ = coefficients resulting from variation of temperature and velocity over layer thickness

A_{h_h} = horizontal turbulent-diffusion coefficient for heat

A_{h_v} = vertical turbulent-diffusion coefficient for heat

$\left[A_{h_v}\dfrac{\partial \bar{T}}{\partial z}\right]_k = 2A_{h_v}\dfrac{T_{k-\frac{1}{2}} - T_{k+\frac{1}{2}}}{H_{k-\frac{1}{2}} H_{k+\frac{1}{2}}}$ = vertical heat transfer due to turbulent diffusion;

note that $\left[A_{h_v}\dfrac{\partial \bar{T}}{\partial z}\right]_0 = Q_o$ and $\left[A_{h_v}\dfrac{\partial \bar{T}}{\partial z}\right]_N = 0$.

Q_o = net rate of heat absorption at water surface

$\Phi_{k-\frac{1}{2}} = \dfrac{1}{H_{k-\frac{1}{2}}}\displaystyle\int_{Z_k}^{Z_{k-\frac{1}{2}}} \phi\,dz$ = net rate of heat absorption within a layer

ϕ = rate of internal radiation absorption at any depth

$C_{k-\frac{1}{2}} = \dfrac{1}{H_{k-\frac{1}{2}}}\displaystyle\int_{Z_k}^{Z_{k-1}} C\,dz$ = layer-averaged concentration of substance

C = concentration of dissolved or particulate material

$\delta_1, \delta_2, \delta_3$ = coefficients resulting from variation of concentration and velocity over layer thickness

$\bar{C}_k = \frac{1}{2}\left(C_{k-\frac{1}{2}} + C_{k+\frac{1}{2}}\right)$ = average concentration between layers

A_{s_h} = horizontal turbulent-diffusion coefficient for substance

A_{s_v} = vertical turbulent-diffusion coefficient for substance

$\left[A_{s_v}\dfrac{\partial \bar{C}}{\partial z}\right]_k = 2A_{s_v}\left[\dfrac{C_{k-\frac{1}{2}} - C_{k+\frac{1}{2}}}{H_{k-\frac{1}{2}} + H_{k+\frac{1}{2}}}\right]$ = vertical transfer of substance due to turbulent diffusion;

note that $\left[A_{s_v}\dfrac{\partial \bar{C}}{\partial z}\right]_0 = -\left(\omega_{ss}\bar{C}\right)_0$, no flux through water surface, and that

$\left[A_{s_v}\dfrac{\partial \bar{C}}{\partial z}\right]_N + \left(\omega_{ss}\bar{C}\right)_N = E_N$

E_N = net flux of material into suspension at lake bottom

ω_{ss} = settling velocity of substance

$S_{k-\frac{1}{2}}$ = net rate of substance production in layer computational time, it does not seem advisable to couple it directly to a two- or three-dimensional circulation model such as this, except where the spatial distributions of ecosystem components are of critical interest.

At present the lake circulation model is being used to predict transient two-dimensional motion in Lake Wingra for different wind patterns, assuming homogeneous (isothermal) conditions. The predicted transport rates can then be plugged into CLEAN or the site-specific version, WNGRA2, in order to couple simulations for different regions of the lake. However, the circulation model is formulated in such a way that CLEAN can be run as a part of it and used to generate net production rates and concentrations of substances. Thus, if needed, there could be a hierarchy of models, with CLEAN operating as a part of the lake circulation model.

To model lake motions, the conservation laws for mass (Equation 9.1), momentum (Equations 9.2 and 9.3), heat (Equation 9.4), and substance mass (Equation 9.5) are written for a lake which is comprised of an arbitrary number N of horizontal layers of different thicknesses. These equations include horizontal and vertical transport of energy and turbulent diffusive transport of momentum, heat, and substance throughout the lake and describe the temporal and spatial variations in velocity, temperature, and concentration averaged over each layer at a particular horizontal location. The lake motions represented in the model are driven by wind shear, atmospheric pressure gradients, and heat absorption and are subject to shoreline and bottom boundary conditions. The model equations must be solved simultaneously because of the coupling between the conservation laws.

EXPERIMENTATION AND ANALYSIS

The mathematical equations are instructive in that they represent a formal, unambiguous statement of our understanding of ecosystem dynamics. Experimentation with the model and sensitivity analysis then assist in drawing out the implications of these equations. In particular, running simulations and checking the results for intuitive reasonableness have been invaluable in the continuing process of evaluating the model and improving it.

Individual submodels, in some cases more detailed than those versions incorporated in CLEAN, have been subjected to intensive testing for internal consistency and for accuracy of prediction. The results of these detailed studies will be reported in a series of technical publications.[20,33,34] Numerous combinations of submodels have also been investigated. The most extensive experimentation has involved components representing the open-water portion of the ecosystem. These include two phytoplankton, two herbivorous zooplankton, omnivorous zooplankton, bluegill-like fish, bass-like fish, particulate and dissolved organic matter, decomposers, and phosphate. This open-water version of CLEAN demonstrates the utility of the more important nonlinear feedback terms in the model; it also exhibits important functional dissimilarities between two groups of phytoplankton and among three groups of zooplankton — groups that are lumped together in many simulation models.

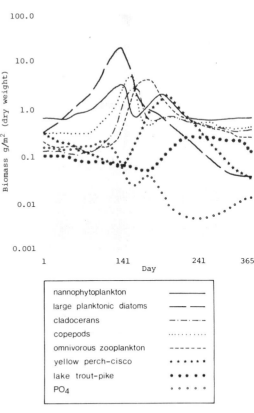

Figure 3 - Simulated seasonal changes in the open-water ecosystem. This and the following figures are for the south end of Lake George, New York.

CLEAN has been run using data from a number of lakes; however, most experimentation has been with data from the intensive research sites, Lake George and Lake Wingra. Biomass values predicted for a typical year at the south end of Lake George are presented in Figure 3. Parameters for the state variables are set to correspond to the principal groups endemic to Lake George; thus, the "net phytoplankton" are large planktonic diatoms, "bluegill-like fish" are principally yellow perch and cisco, and "bass-like fish" are principally lake trout and pike.

The predicted levels are of the same magnitude as the observed levels, but at the present stage of calibration the model seems not to be predicting seasonal patterns with sufficient accuracy to be satisfactory (e.g., Figures 4a and 4b). Unfortunately, given the inherent sampling and transformation errors in the available data, it is not possible to test the adequacy of the model rigorously. However, inspection of the output and consideration of the ecologic basis for the model suggests some general conclusions about the way in which the model behaves.

In spring, as light ceases to be a limiting factor in the model, the phytoplankton can utilize the nutrient that have accumulated during the winter with the

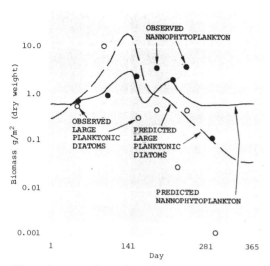

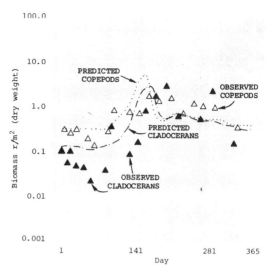

Figure 4a – Comparison of predicted and observed nannophytoplankton and large planktonic diatoms. Open circles represent observed large diatom biomass and closed circles represent observed nannophytoplankton biomass; legend for predicted patterns same as in Figure 3. Data courtesy of H. H. Howard.

Figure 4b – Comparison of predicted and observed herbivorous zooplankton. Open triangles represent observed herbivorous copepod biomass and closed triangles represent observed cladoceran biomass; legend for predicted patterns same as in Figure 3.

large diatoms increasing most rapidly because the optimum water temperature for them is lower than for smaller species and they require less light. The large diatom and nannophytoplankton populations both reach a peak in the latter part of May and then "crash" as available phosphate is depleted and grazing by zooplankton is intensified.

During the summer the biomass moves up the food chain, with the model exhibiting fairly realistic time lags, depending on the position of each group in the food chain. Herbivorous cladocerans and copepods differ somewhat in their seasonal abundances because of differences in their maximum growth rates and their ability to feed on large diatoms. Omnivorous zooplankton feed on the herbivorous zooplankton and are eaten in turn by yellow perch and cisco. Lake trout and pike increase in biomass during the latter part of summer, feeding on yellow perch and cisco, which decline precipitously in the simulation in response to the increased predation. Nannophytoplankton increase following the decline of the herbivorous zooplankton, but the large diatoms continue to decline because the available phosphate level is suboptimal for their requirements. Thus, the model output represents the combined effects of a number of ecological relationships. This behavior of the model is reasonable and is useful for environmental management, although continued refinement is clearly desirable.

Whereas on-line experimentation with the model can give a "feel" for the dynamics of the model, sensitivity analysis is more useful in indicating specific parameters which must be known with greater accuracy to ensure fully satisfactory simulations. A range of values, rather than precise estimates, is available for most parameters because of the difficulty of

measuring biological processes. Within the estimated range, sensitivity can be determined readily by response-surface methodology.[35]

A normalized parameter for rate of change,

$$K(t) = \frac{\dfrac{dB_i}{dt}}{B_i(t)}$$

is determined from field data by a spline function technique[36] and compared to the value of $K(t)$ predicted by the model. Sums of squares of differences between measured and predicted values of $K(t)$ are calculated utilizing a fractional factorial combination of the largest and smallest values in the range of each parameter. This produces a set of sums of squares and corresponding parameter values. Stepwise linear regression then is used to fit the sum of squares to a full quadratic model of individual parameters. The relative sensitivities of the simulation model to the parameters are indicated by the F-values of the parameters in the regression equation.

Such an analysis has revealed the open-water model to be particularly sensitive to maximum photosynthesis rates, respiration constants, phytoplankton sinking coefficients, and phosphate limitations. Grazing components are most sensitive to food preference, maximum feeding rates, and optimal temperature. On the other hand, the model appears relatively insensitive to the level of light saturation, metabolic excretion rates, and correction factors for the age structures of the several populations. This information is now being used in setting priorities for further studies to obtain more precise estimates of the most sensitive parameters.

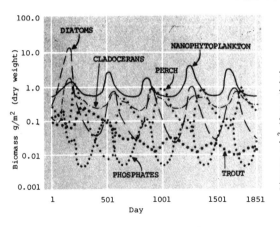

Figure 5 - Simulated 5-year patterns of selected
components of the open-water ecosystem.
Legend same as in Figure 3.

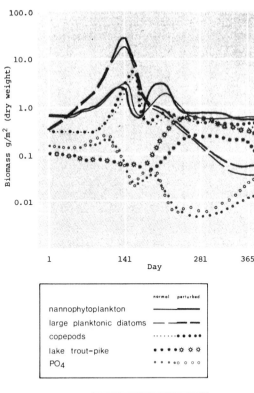

Figure 6 - Comparison of simulated seasonal patterns
of selected ecosystem components for normal and doubled phosphate loadings.
Inset shows 5-year patterns for lake
trout – pike.

The stability of the model can be established
readily through examination of a five-year simulation
based on the same parameters and driving variables as
used for the one-year simulation described above
(Figure 5). However, there is transient response in
the first year. Large diatoms exhibit a large
spring peak only in the first year, and lake trout-
pike eventually reach a lower level in equilibrium
with the yellow perch-cisco population.

The diatom response is a manifestation of an unrealistic time lag in the regeneration of available
phosphate: without suitably high phosphate levels
early in the spring, the simulated diatoms cannot
reach the observed levels using reasonable values
for the parameters. Presumably this difficulty with
the model will be resolved when more precise parameter estimates are available for the decomposer
component, causing phosphate to be simulated more
realistically. It is likely that the lake trout-
pike simulation seeks a lower equilibrium because
the model is over-simplified. The open-water version
of CLEAN does not provide for seasonal differences in
predation brought about by the growth of macrophytes
that serve as cover, nor does it presently account
for the massive restocking of lake trout from
hatcheries. With these additions the two groups of
fish will not be as intensely coupled and will
probably maintain more realistic long-term dynamics.

Experimentation with CLEAN in forecasting lake
eutrophication is instructive. For example, by
symbolically doubling the input of phosphate into
the south end of Lake George, we get some rather
interesting results compared to the "normal" simulation (Figure 6). In the perturbed simulation large
diatoms become half again as abundant, whereas the
nannophytoplankton (which compete with them) are
slightly less abundant than "normal" in the spring.
However, the nannophytoplankton become significantly
more abundant than "normal" in the late summer peak.
The predicted spring increase in large diatoms seems
to be realistic and is of particular interest in
that these diatoms include taste- and odor-producing
forms that seriously degrade the quality of the
water as perceived by tourists and cottage owners.

On the plus side, lake trout and pike become almost
three times as abundant (or big) as before, representing the transfer of biomass up to the higher
levels of the food chain. This, too, is realistic
for a lake in an intermediate stage of eutrophication.

Over the course of a five-year simulation (see inset
in Figure 6) the discrepancy between levels of fish
in the normal and perturbed systems becomes even more
pronounced, although this probably reflects the inadequacies of the present model, as discussed above.

Interestingly enough, available phosphate levels are
almost the same in the normal and perturbed simulations, reflecting the rapid uptake of phosphate by
phytoplankton and underscoring the fact that available phosphate levels are not a suitable index to
eutrophication. Thus, even in its present stage of
development, CLEAN exhibits good potential as a tool
for environmental management.

REFINEMENTS AND EVALUATION

Individual submodels have been tested and found capable of satisfactorily predicting the dynamics of their respective subsystems. However, model constructs are imperfect representations of the real world, and a continuing effort is therefore required to improve these models. For example, various formulations for the limitations of light and nutrients on photosynthesis are being examined. In addition to the mean resistance construct, used in Equation 2.2, a minimum function

$$\min\ (\mu_{light},\ \mu_{PO_4},\ \mu_N,\ \mu_C)$$

has produced excellent results. We have also experimented with a multiplicative function

$$\mu_{light} \cdot \mu_{PO_4} \cdot \mu_N \cdot \mu_C$$

which seems to limit photosynthesis more severely than is actually observed in nature. Ultimately, the choice of a single formulation may depend on the development of new laboratory procedures to determine the nature of the limitation process more accurately.

A principal advantage of large-scale interdisciplinary research is feedback from model analysis to provide redirection for research. This is particularly important in working with complex ecosystems where interaction among components can often lead to counterintuitive results. Initial research has emphasized the dynamics of individual components. Much has been learned from the subsystem models, and this knowledge is rapidly being incorporated into the model. In turn, on-line simulation with the model is yielding valuable insight into whole-system functionalities and is setting the stage for new laboratory approaches and field programs which emphasize the interfaces between components.

From the inception of the present study there has been a well-defined strategy for evaluating the ecosystem model. Lake Wingra is a shallow, nutrient-rich lake and is partially surrounded by a city. However, Lake George is a large, relatively nutrient-poor lake with extensive undeveloped shorelines—although the south end is undergoing eutrophication as the area is increasingly developed. By changing site constants and substituting the appropriate values of the driving variables, various hierarchical versions of CLEAN can be used for either lake (as well as for other lakes for which there are suitable data). Therefore, a model calibrated for Lake Wingra can be evaluated for the totally different conditions at Lake George, as well as for intermediate conditions. From such testing we are presently identifying and generalizing those formulations and parameter values which appear to be unique to a single lake or stage or eutrophication. Ideally, the model should be able to describe the dynamics of both lake ecosystems merely by using the appropriate site constants and driving variables.

Data have been collected from both lakes for four years; so the model can be evaluated on a yearly basis, yielding a model responsive to minor variations in driving variables. We have already found that ecosystem dynamics can be affected by seemingly unimportant annual events that trigger replacement of one organism by another, with important effects on the ecosystem. Recognition of these time-varying effects as well as site-specific effects permits us to develop versions of the model with high resolution but limited generality and versions that are less accurate but more widely applicable. For example, WNGRA2[36] has proved to be useful in examining the impact that changes in the drainage basin have on Lake Wingra. At the other end of the scale, we are presently generalizing the parameters so that CLEAN can be applied to arctic and alpine lakes as well. Such flexibility, which is inherent in the modular structure of CLEAN, is necessary if simulation is to be used as an effective tool in answering the diverse problems facing environmental management.

ACKNOWLEDGEMENT

The research has been supported by the Eastern Deciduous Forest Biome, United States International Biological Program, funded by the National Science Foundation under Interagency Agreement AG 199, 40-193-69 with the Atomic Energy Commission, Oak Ridge National Laboratory. Encouragement and support provided by Stanley I. Auerbach, Biome Director and Robert L. Burgess, Deputy Director, is gratefully acknowledged.

REFERENCES

1 RILEY, G.A.
Factors Controlling Phytoplankton Populations on Georges Bank
Journal Marine Research vol. 6 1946 pp. 54-73

RILEY, G.A.
Seasonal Fluctuations of the Phytoplankton Populations in New England Waters
Journal Marine Research vol. 6 1947 pp. 114-125

RILEY, G.A. STOMMEL, H. BUMPUS, D.F.
Quantitative Ecology of the Plankton of the Western North Atlantic
Bulletin Bingham Oceanographic Collection vol. 12 1949 pp. 1-169

RILEY, G.A.
Theory of Food-Chain Relations in the Ocean
In Hill, M.N., ed., *The Sea*, Interscience, New York, 1963

RILEY, G.A.
Mathematical Model of Regional Variations in Plankton
Limnology and Oceanography vol. 10 supplement 1965 pp. R202-R215

STEELE, J.H.
Plant Production on Fladen Ground
Journal Marine Biological Association vol. 35 1956 pp. 1-33

7 STEELE, J.H.
A Study of Production in the Gulf of Mexico
Journal Marine Research vol. 22 1964 pp. 211-222

8 STEELE, J.H.
Notes on Some Theoretical Problems in Production Ecology
In Goldman, C.R., ed., *Primary Production in Aquatic Environments*
Memorie, Istituto di Idrobiologia vol. 18 supplement University of California Press Berkeley 1965 pp. 383-397

9 PARKER, R.A.
Simulation of an Aquatic Ecosystem
Biometrics vol. 24 1968 pp. 803-822

10 CHEN, C.W.
Concepts and Utilities of Ecologic Model
Journal Sanitary Engineering Division American Society Chemical Engineering vol. 96 1970 pp. 1085-1097

11 WILLIAMS, F.M.
Dynamics of Microbial Populations
In Patten, B.C., ed., *Systems Analysis and Simulation in Ecology*, vol. 1, Academic Press, New York, 1971, pp. 198-267

12 LINDEMAN, R.L.
The Trophic-Dynamic Aspect of Ecology
Ecology vol. 23 1942 pp. 399-418

13 DiTORO, D.M. O'CONNOR, D.J. THOMANN, R.V.
A Dynamic Model of the Phytoplankton Population in the Sacramento-San Joaquin Delta
In Hem, J.D., ed., *Nonequilibrium Systems in Natural Water Chemistry*, American Chemical Society, *Advances in Chemistry* series, vol. 106, 1971, pp. 131-180

14 SHUGART, H.H. GOLDSTEIN, R.A. O'NEILL, R.V. MANKIN, J.B.
TEEM, A Terrestrial Ecosystem Model for Forests
Oecologia Plantarum in press

15 TITUS, J.E. ADAMS, M.S. WEILER, P.R. O'NEILL, R.V. SHUGART, H.H. Jr. BOOTH, R.S. GOLDSTEIN, R.A.
Production Model for Myriophyllum Spicatum L.
US IBP EDFB memo report 72-108 1972 17 pp.

16 O'NEILL, R.V. GOLDSTEIN, R.A. SHUGART, H.H.,Jr. MANKIN, J.B.
Terrestrial Ecosystem Energy Model
US IBP EDFB memo report 72-19 1972 39 pp.

17 PARK, R.A. WILKINSON, J.W.
Lake George Modeling Philosophy
US IBP EDFB memo report 71-19 1971 62 pp.

REFERENCES (continued)

18 KOONCE, J.F.
    *Seasonal Succession of Phytoplankton and a Model
    of the Dynamics of Phytoplankton Growth and
    Nutrient Uptake*
    University of Wisconsin   PhD thesis   1972   192 pp

19 BLOOMFIELD, J.A.   PARK, R.A.   SCAVIA, D.
    ZAHORCAK, C.S.
    *Aquatic Modeling in the Eastern Deciduous Forest
    Biome, U.S. International Biological Program*
    In Middlebrooks, E.J., D.H. Falkenborg, T.E.
    Maloney, eds., *Modeling the Eutrophication Process*,
    Utah State University, 1973, pp. 139-158

20 KITCHELL, J.F.   KOONCE, J.F.   O'NEILL, R.V.
    SHUGART, H.H. Jr.   MAGNUSON, J.J.   BOOTH, R.S.
    Model of Fish Biomass Dynamics
    Transactions American Fish Society   in press

21 de'ANGELIS, D.   GOLDSTEIN, R.A.   O'NEILL, R.V.
    A Model for Trophic Interaction
    Ecology   in review

22 PARK, R.A.   WILKINSON, J.W.   BLOOMFIELD, J.A.
    KOHBERGER, R.C.   STERLING, C.
    *Aquatic Modeling, Data Analysis, and Data
    Management at Lake George*
    US IBP EDFB memo report 72-70   1972   32 pp.

23 KITCHELL, J.R.   KOONCE, J.F.   O'NEILL, R.V.
    SHUGART, H.H.   MAGNUSON, J.J.   BOOTH, R.S.
    *Implementation of a Predator-Prey Biomass Model
    for Fishes*
    US IBP EDFB memo report 72-118   1972   57 pp.

24 PETERSON, J.L.   HILSENHOF, W.L.
    *Preliminary Modeling Parameters for Aquatic
    Insects and Standing Crop Comparisons with Other
    Lake Wingra Biota*
    US IBP EDFB memo report 72-116   1972   7 pp.

25 ZAHORCAK, C.S.
    *Formulation of a Numbers-Biomass Model for
    Simulating the Dynamics of Aquatic Insect
    Populations*
    Rensselaer Polytechnic Institute   MS thesis
    1974   68 pp.

26 CLESCERI, L.S.   BLOOMFIELD, J.A.   O'NEILL, R.V.
    SHUGART, H.H. Jr.   BOOTH, R.S.
    A Model for Aquatic Microbial Decomposition
    US IBP EDFB memo report 72-40   1972   13 pp.

27 FOX, W.T.   DAVIS, R.A. Jr.
    *Computer Simulation Model of Coastal Processes
    in Eastern Lake Michigan*
    Williams College technical report 5 of ONR task
    no. 388-092/10-18-68 (414)   1971   114 pp.

28 DETTMANN, E.H.   HUFF, D.D.
    A Lake Water Balance Model
    US IBP EDFB memo report 72-126   1972   22 pp.

29 COLON, E.M. Jr.
    Hydrologic Study of Lake George, New York
    Rensselaer Polytechnic Institute   Doctor of
    Engineering thesis   1972   190 pp.

30 HUFF, D.D.
    *HTM Program Elements, Control Cards, Input
    Data Cards*
    US IBP EDFB memo report 72-13   1972   46 pp.

31 HUFF, D.D.   KOONCE, J.F.   IVARSON, W.R.
    WEILER, P.R.   DETTMANN, E.H.   HARRIS, R.F.
    *Simulation of Urban Runoff, Nutrient Loading,
    and Biotic Response of a Shallow Eutrophic Lake*
    In Middlebrooks, E.J., D.H. Falkenborg, T.E.
    Maloney, eds., *Modeling the Eutrophication
    Process*, Utah State University, 1973, pp. 33-55

32 HOOPES, J.   PATTERSON, D.   WOLOSHUK, M.
    MONKMEYER, P.   GREEN, T.
    *Investigations of Circulation, Temperature, and
    Material Transport and Exchange in Lake Wingra*
    US IBP EDFB memo report 72-117   1972   8 pp.

33 McNAUGHT, D.C.   BLOOMFIELD, J.A.
    *A Resource Allocation Model for Herbivorous
    Zooplankton Predicts Community Changes during
    Eutrophication*
    Presentation at American Association for the
    Advancement of Science annual meeting   December
    26-31, 1972

34 SMITH, O.L.   SHUGART, H.H. Jr.   O'NEILL, R.V.
    BOOTH, R.S.   McNAUGHT, D.C.
    *Resource Competition and an Analytical Model
    of Zooplankton Feeding on Phytoplankton*
    In preparation

35 MYERS, R.H.
    Response Surface Methodology
    Allyn and Bacon   Boston, Massachusetts   1971

36 GREVILLE, T.N.E.
    Mathematical Methods for Digital Computers
    vol. 2
    Wiley   New York   1967

37 MacCORMICK, A.J.A.   LOUCKS, O.L.   KOONCE, J.F.
    KITCHELL, J.F.   WEILER, P.R.
    *An Ecosystem Model for the Pelagic Zone of
    Lake Wingra*
    US IBP EDFB memo report 72-177   1972   103 pp.

Part III

PRESENT DIRECTIONS IN
SYSTEMS ECOLOGY

At the present time, systems ecology is a widely dispersed and very active field of research. In Part III we attempt to summarize the present status and offer some prognostications about the future. The discussion is divided into three areas: applications where modeling appears to offer unique advantages; continued work on technical problems; and developments in new directions.

UNIQUE ADVANTAGES OF MODELS

Mathematical models offer unique advantages for the exploration of some ecological problems. Because of the considerable time, effort, and ecological expertise applied to their development, the present generation of ecosystem models presumably contains many insights into the dynamics of ecosystems. Under the constraints of previous programs, little exploration of the dynamics of these total models has been feasible. Therefore, one current area of development utilizes available models and explores their behavior as applied to a variety of problems. An excellent example is the effort by Cooper et al. (Paper 21), in which major biome models were applied to assess the potential impact of operating a supersonic transport fleet over the United States.

245

Models synthesize current understanding of system processes and can analyze this understanding to yield new insight and theory. We identify four areas where the unique properties of mathematical models could be fruitfully exploited: (1) model parameters and structure represent a concise summarization of the rate processes and interconnections of the system; (2) models explore long-time scales and large spatial scales that are not conducive to direct measurement; (3) models draw out the logical implications of accepted ecological concepts (e.g., niche theory); and (4) models can explore new methodological approaches (e.g., spectral analysis).

The Model as a Summary

A model is a complex conceptualization of the dynamics and interdependencies of an ecosystem. As a result, the set of model parameters represents a concise summarization of rate processes. Thus, models permit an extensive data base to be summarized onto a single printed page (McGinnis et al. 1968; Odum 1970; Reichle et al. 1973) and allow the ecologist to search for relationships that might otherwise be lost in the large volume of information.

Since the model represents the dynamics of a class of systems (e.g., forests), the parameters for a specific system describe certain unique dynamic characteristics of that system. It is because of this summarization capability that one can, for example, compare populations on the basis of their intrinsic rates of increase (r) or carrying capacity (K). A good example of this use of models to compare ecosystems is given by O'Neill (Paper 18).

Models for Large Spatio-Temporal Scales

Problems involving very large spatio-temporal scales can be attacked most easily by modeling. First, there are phenomena that due to their long time scale are difficult to observe directly. In this case the model can synthesize one's understanding of the processes involved and extrapolate trends. A second area involves problems on a spatial scale so large that data collection and direct observation are encumbered. A good example is the complex succession or forest dynamics model presented below (Paper 26).

If actual field experimentation is difficult or infeasible, modeling may provide an alternative approach to certain ecological problems. For example, O'Neill and Styron (1969) investigated radiation effects on a field population of soil invertebrates. The problem concerned effects of high levels of ionizing radiation on an old-field ecosystem at numerous periods during the year. Such drastic experiments could not be performed. In the absence of direct field experimentation, the problem was addressed by assembling information on components of the old-field ecosystem and developing a model to extrapolate this information to the high radiation level case.

Models as Tools for Drawing Logical Implication

Ecologists are in the process of developing a codification of several decades of observations in the form of ecological generalities. These codifications (e.g., trophic food-chains rarely have more than five links) lack the rigor of physical laws and often conflict one with another. Ecological models can test the implications of accepted generalities for self-consistency and realism. The papers presented earlier on the diversity/stability problem are good examples of this use of models.

Models for Testing New Methodology

The applicability of techniques developed in other fields can be tested by using an appropriate detailed model. The explorations of frequency response analysis presented by Shugart et al. (1976) is a good example of this approach.

TECHNICAL PROBLEMS IN SYSTEMS ECOLOGY

A number of technical questions remain concerning the validity and utility of large-scale models (Oak Ridge Systems Ecology Group 1974). Four problem areas appear particularly important: (1) the propagation of errors and the uncertainty of model predictions; (2) the validation of models; (3) the implications of alternative model structures; and (4) the effects of spatial patterning on system dynamics.

Error Propagation

If models are to be useful in management applications, it is necessary to understand the uncertainty associated with model output (i.e., confidence limits about a specific prediction). Previous work (O'Neill 1973) has shown that errors may be associated both with the measurement of parameter values and with the lumping process by which populations are aggregated into model variables. This problem area is being investigated by a number of workers (O'Neill and Rust, in press; Gardner et al. 1976; Harrison, in press; Cale and Odell 1978).

Validation

A major impediment to the application of ecosystem models has been the question of validation—that is, how closely does model output simulate the dynamics of the real system. The general problem has been clarified (Mankin et al. 1977) by changing the question from a Boolean good–bad decision to a measure of the accuracy and reliability of the model and an analysis of how this information can be utilized. Because of the importance of this problem area, we have chosen to lay considerable emphasis on validation and have included three relevant papers (Papers 23, 24, and 25) in this part.

Model Structures

Implicit in any ecosystem model is an underlying structure (e.g., Forrester 1961) of interactions between system components. Yet we have seldom considered the strong assumptions involved in this underlying structure (Goodall 1973, 1975). In most cases, alternative structures could be proposed with unknown effects on model dynamics. We are aware of no major studies investigating this question and it remains an important problem area for the future.

Spatial Patterning

The present generation of ecosystem models depends heavily on an assumption of spatial homogeneity. The models assume that processes occurring with the borders of a system are sufficiently uniform that one can adequately describe dynamics by considering only total or mean properties of the system. While this assumption has been useful, considerale theoretical work (Dixon-Jones 1975; Levin, in press, Whittaker and Levin, in press; Pielou 1969) demonstrates the inadequacy of this assumption applied to ecological work. These theoretical studies have emphasized population or community levels of organization. While studies have been conducted on population dynamics, to our knowledge, no work has been published on the effect of spatial heterogeneity on total ecosystem function. This remains as another area for fruitful exploration in the future.

As noted, the papers in this part are organized into three subdivisions. The first illustrates applications of system models to the solution of environmental problems. The second subdivision emphasizes the validation of ecosystem models. The third section outlines some new approaches to ecological problems that utilize the advantages of mathematical models.

REFERENCES

Cale, W. G., and P. L. Odell. 1978. Concerning aggregation in ecosystem modeling. In Halfon, E. (ed.), *Theoretical Systems Ecology.* Academic Press, New York.

Dixon-Jones, D. C. 1975. *The Application of Catastrophe Theory to Ecological Systems,* Int. Institute for Applied Systems Analysis Report RR-75-15, Vienna.

Forrester, J. W. 1961. *Industrial Dynamics.* The M. I. T. Press, Cambridge.

Gardner, R. H., J. B. Mankin, and H. H. Shugart. 1976. *The COMEX Computer Code: Documentation and Description,* Oak Ridge National Laboratory Report, EDFB/IBP-76/4. Oak Ridge, Tenn.

Goodall, D. W. 1973. Building and testing ecosystem models, pp. 173–194. In Jeffers, J. N. R. (ed.), *Mathematical Models in Ecology.* Blackwell Scientific Publications, Oxford, 398 pp.

Goodall, D. W. 1975. Ecosystem modeling in the Desert Biome, pp. 73–94. In Patten, B. C. (ed.), *Systems Analysis and Simulation in Ecology,* vol. III. Academic Press, New York, 601 pp.

Harrison, G. W. Compartmental lumping in mineral cycling models. *Proc. Environ. Chem. Cycl. Processes Symp.* Savannah River Ecology Laboratory (in press).

Levin, S. A. Population dynamic models in heterogeneous environments. *Ann. Rev. Ecol. Syst.* **7** (in press).

McGinnis, J. T., F. B. Golley, R. G. Clements, G. I. Childs, and M. J. Dueve. 1968. Elemental and hydrologic budgets of the Panamanian tropical moist forest. *BioScience* **19**:697–700.

Mankin, J. B., R. V. O'Neill, H. H. Shugart, and B. W. Rust. 1976. The importance of validation in ecosystem analysis. In Innis, G. (ed.), *The Future of Systems Ecology.* Simulation Council of America, La Jolla, Calif. (in press).

Oak Ridge Systems Ecology Group. 1974. Dynamic ecosystems models: Progress and challenges, pp. 280–93. In Levin, S. A. (ed.), *Ecosystem Analysis and Prediction.* SIAM Press, Philadelphia.

Odum, H. T. 1970. Summary: An emerging view of the ecological system at El Verde, pp. I-191–I-298. In Odum, H. T., and R. F. Pigeon (eds.), A Tropical Rain Forest. USAEC, Division of Technical Information, Oak Ridge, Tenn.

O'Neill, R. V. 1973. Error analysis of ecological models, pp. 898 908. In Nelson, D. J. (ed.), *Radionuclides in Ecosystems*, USAEC-CONF-710501. Oak Ridge National Laboratory, Oak Ridge, Tenn., 1268 pp.

O'Neill, R. V., and B. W. Rust. Aggregation error in ecological models. *Ecological Modelling* (in press).

O'Neill, R. V., and C. E. Styron. 1969. Applications of compartment modeling techniques to Collembola population studies. *Amer. Midl. Nat.* **83**:489–95.

Pielou, E. C. 1969. *An Introduction to Mathematical Ecology*. Wiley-Interscience, New York, 286 pp.

Reichle, D. E., R. V. O'Neill, and J. S. Olson. 1973. *Modeling Forest Ecosystems*, EDFB/IBP–73–7. Oak Ridge National Laboratory, Oak Ridge, Tenn. 339 pp.

Shugart, H. H., D. E. Reichle, N. T. Edwards, and J. R. Kercher. 1976. A model of calcium cycling in an east Tennessee *Liriodendron* forest: Model structure, parameters, and frequency response analysis. *Ecology* **57**: 99–109.

Whittaker, R. H., and S. A. Levin. The role of mosaic phenomena in natural communities. *Proc. Symp. on Patch Phenomena in Ecol.*, Santa Catalina Island, Calif. (in press).

Editors' Comments
on Papers 21 and 22

21   COOPER et al.
Simulation Models of the Effects of Climatic Change on Natural Ecosystems

22   VAN WINKLE
The Application of Computers in an Assessment of the Environmental Impact of Power Plants on an Aquatic Ecosystem

APPLICATIONS TO REAL-WORLD PROBLEMS

In an earlier section we pointed out that systems ecology has always been viewed in a problem solving context (e.g., Paper 7 by Van Dyne and Paper 8 by Reichle and Auerbach). This emphasis has encouraged the use of mathematical models as a vehicle for the application of ecological information to the solution of practical environmental problems.

In some areas, such applications have become routine. Population models of the stock-recruitment type (e.g., Ricker 1954) have been used in fisheries management. Optimization of stand growth models is used in the management of forest harvesting (Hool 1966; Leak 1964; Myers 1968). Environmental foodchain models, linked with dose assessment models (Booth et al. 1971; Booth and Kaye 1971), have been used for the environmental and health assessment of nuclear power facilities.

Many theoretical studies stress model analysis as a tool for developing management strategies. This has included optimization of both linear (Emanuel and Mulholland 1976) and nonlinear models (Van Dyne et al. 1969). The studies by Shugart et al. (Paper 10) and Martin et al. (Paper 9) presented in an earlier section are both examples of the application of control theory to ecological models.

The two papers reprinted here represent different points along a broad spectrum of model applications. Paper 21 by Cooper et al. applies major ecosystem models to assess the potential impact of operating a supersonic transport fleet over the continental United States.

Paper 22 by Van Winkle concerns the environmental impact assessment for the Indian Point Power Station, begun in 1971, on the Hudson River. Initially, an effort was made to assess the impact of entrainment and impingement on the striped bass population that spawns in the river. The assessment involved a fish population model.

As a result of hearings on licensing the power station, the original striped bass model has been extensively expanded. Both the power company and the Nuclear Regulatory Commission have produced separate versions of the model that yield conflicting results. Through testimony at the various hearings, the model has been subjected to detailed examination at a level unprecedented in the history of ecology. Litigation continues in 1978 and is expected to go on at least for another five years. Already, perhaps 30,000 pages of testimony and 50–100,000 pages of supporting material have been produced. The model continues to be extremely influential and has already had an impact measureable in millions of dollars.

The case history of the striped bass model can be expected to be repeated in the future. The striped bass model can be regarded as a precedent in which the utility of ecological models for real-world decision making is exemplified. It is perhaps the best example of the potential contribution of the systems approach in solving environmental problems.

REFERENCES

Booth, R. S., O. W. Burke, and S. V. Kaye. 1971. Dynamics of the forage-cow-milk pathway for transfer of radioactive iodine, strontium and cesium to man. *Proc. Nuclear Methods Environ. Res.,* Univ. of Missouri, Columbia.

Booth, R. S., and S. V. Kaye. 1971. *A Preliminary Systems Analysis Model of Radioactivity Transfer to Man from Deposition in a Terrestrial Environment,* ORNL/TM-3135. Oak Ridge National Laboratory, Oak Ridge, Tenn.

Emanuel, W. R., and R. J. Mulholland. 1976. Linear periodic control with applications to environmental systems. *Int. J. Control* **24**:807–20.

Hool, J. N. 1966. A dynamic programming Markov chain approach to forest production control. *For. Sci. Monogr.* **12**.

Leak, W. B. 1964. *Estimating Maximum Allowable Timber Yields by Linear Programming,* USDA Forest Servic Res. Pap. NE-17. Upper Darby, Penn., 11 pp.

Myers, C. A. 1968. *Simulating the Management of Even-Aged Timber Stands,* USDA Forest Service Res. Pap. RM-42. Upper Darby, Penn., 32 pp.

Ricker, W. E. 1954. Stock and recruitment. *J. Fish. Res. Board Can.* **11**:559–623.

Van Dyne, G. M., W. E. Frayer, and L. J. Bledsoe. 1969. *Some Optimization Techniques and Problems in the Natural Resource Sciences*, Preprint 1. Natural Resource Ecology Laboratory, Colorado State Univ., Fort Collins, 51 pp.

21

Reprinted from pages 550–562 of *Third Conf. Climatic Impact Assess. Program Proc.*, U.S. Dept. Trans., Cambridge, Mass., 1974

SIMULATION MODELS OF THE EFFECTS OF CLIMATIC CHANGE ON NATURAL ECOSYSTEMS

CHARLES F. COOPER
San Diego State University, San Diego, California

T.J. BLASING AND H.C. FRITTS
University of Arizona, Tucson, Arizona

OAK RIDGE SYSTEMS ECOLOGY GROUP
Oak Ridge National Laboratory, Oak Ridge, Tennessee

FREEMAN M. SMITH AND WILLIAM J. PARTON
Colorado State University, Fort Collins, Colorado

GERARD F. SCHREUDER AND PHILLIP SOLLINS
University of Washington, Seattle, Washington

JON ZICH AND WAYNE STONER
San Diego State University, San Diego, California

Direct experimentation to determine the long-term effects of a sustained climatic change on natural biological communities is clearly impractical. Computer simulation has proved to be an effective surrogate in many instances where time, expense, or uniqueness of the system being investigated preclude direct experimentation. These instances range from testing the effects of novel fiscal and monetary policies on the national economy to designing moon rockets. Ecosystem responses to climatic change can also be simulated by computers, even though eco-system models have not yet attained the precision characteristic of econometric or aerospace models.

Several simulation models of complex ecosystems have been developed recently, mostly under the aegis of the International Biological Program. Five of these models, each designed for a different purpose, were operated in parallel under perturbations representing the climatic changes assumed by CIAP. Simultaneously, the predictions of a statistical model of climate and tree growth were tested for congruence with

254

those of the ecosystem models.

This experiment was not intended to yield quantitative predictions of ecological change in natural ecosystems due to climatic change. Rather, it was designed to test whether common qualitative patterns could be discerned in the behavior of the simulated ecosystems as a result of climatic stress. Several consistent trends were in fact observed, not all of which would necessarily have been predicted by the more traditional methods of ecological analysis.

Several major conclusions emerged from the computer simulations:

- Primary plant production is expected to decrease linearly with temperature decrease throughout the range of temperatures tested, except in truly arid regions. This decrease will be greatest in those ecosystems more or less well supplied with water, and will be less under moisture stress conditions characteristic of shortgrass prairies. Plant production in deserts and semi-arid coniferous forests is indicated to be accelerated slightly by temperature decrease.

- Climatic perturbation may have a nonlinear effect on consumer organisms that is more severe and less predictable than the associated effect on primary producers.

- Rates, pathways, and the seasonal course of microbial decomposition are likely to be altered in relatively complex ways by climatic change; this could significantly influence long-term successional responses.

- Total ecosystem standing crop may be either increased or decreased as a result of climatic change, depending in part on the initial climate and the successional status of the community being investigated.

- Water stress in vegetation is likely to be slightly reduced by a decrease in temperature even if that temperature decrease is accompanied by a reduction in precipitation.

- Interactions among climatic variables may generate responses whose signs differ from those which would be pre-

dicted from superposition of the independent effects of the several climatic forcing functions.

THE MODELS

Development of ecosystem-simulation models has been a major feature of U.S. participation in the International Biological Program. Since about 1969, many man-years of research effort have gone into construction and testing of ecosystem models. Each of the existing models has been extensively operated and validated, but this is apparently the first time that several comprehensive ecosystem models have been operated in concert to address a single set of questions. The results are therefore of some significance in the development of the science of ecology as well as in the specific task of predicting the ecological consequences of increased stratospheric flight.

The ecosystem models described here have all been developed as a part of the Biome Studies of the U.S. IBP, so called because many separate investigations of ecosystem processes have been grouped to cover the broad groups of ecosystems which are similar in vegetational physiognomy, fauna, and climate, known as biomes. The participating groups include the Deciduous Forest Biome, with headquarters at Oak Ridge National Laboratory, Tennessee; the Coniferous Forest Biome, at the University of Washington, Seattle, and Oregon State University, Corvallis; the Grassland Biome, at Colorado State University, Fort Collins; the Desert Biome, at Utah State University, Logan; and the Tundra Biome, with modeling efforts (but not overall direction of the research program) centered at San Diego State University. One non-IBP group, the Laboratory of Tree-Ring Research at the University of Arizona, Tucson, also participated. The research reported here was supported by a contract from the U.S. Department of Transportation, but the great bulk of the earlier research at the participating institutions which made this study possible was sponsored by the National Science Foundation.

The ecosystem models discussed here are largely comprised of linked sets of non-linear differential or difference equations, often with discontinuities, and subject to constraints imposed by biological limitations. The initial state of the ecosystem is described by a large number of variables which express the numbers, weights, age and sex distribution, and composition of each of the major groups of organisms, and similar values which specify the organisms' non-living environment. The model incorporates mathematical specifications of the rates of change of each of the state variables in terms of the various factors influencing them. These factors may be other state variables, or may be variables whose values are not derived from within the model. Values for these exogenous variables, or forcing functions, which normally include local meteorological conditions, must be supplied for the period being simulated. Forcing functions may be entered as constants, as tables with rules for interpolation, as time-dependent functions, or as stochastic functions, but in any case are independent of events within the system. Climatic or other forcing functions may be modified in a specified manner to test the effect of a new assumed condition, such as climatic change. The equations are then solved on a computer. Even though the full complexity of an entire ecosystem may still be beyond the capacity of the current generation of computers, the way has been opened in recent years for a mathematical representation of ecosystem dynamics which can be cast into a form suitable for digital solution. An ecosystem can thus be simulated in the computer (Goodall, 1972).

It is not intended to document here the detailed structure of each of the ecosystem models used in this study, but only to indicate their general nature and probable range of validity. Although all of the models follow the general pattern described above, each has been developed independently and has numerous unique features. All but the University of Arizona model are basically deterministic process models, built up from observations and assumptions about organism responses and interactions in the field. The Arizona Tree-Ring Model is a multivariate statistical model of observed relationships among climatic variables and growth responses of tree rings; it has been described in detail by Fritts et al. (1971).

The tundra simulations were based on two separate models. The first is a model of physical processes affecting primary production in the Arctic (Miller and Tieszen, 1972). It was developed from the energy-budget equation for single leaves. It includes parameters defining the relationships between photosynthesis and light, photosynthesis and temperature, respiration and temperature, and leaf resistances, all calculated from field data collected in 1969 and 1970. The data on single leaves were translated into eco-system responses by means of a weighted summation based on the relative surface area of leaves of each species in the community, their inclination to the sun, shading and light extinction, and other pertinent structural features. The model has been validated with observational data for 1965 and 1970 from Point Barrow, Alaska.

The other tundra model is a preliminary simulation of the tundra ecosystem developed by Timin et al. (1973). The state variables are a subset of those in the real system. Those of interest for this study include above-ground plant biomass density, and population densities of brown lemmings, least weasels, and pomarine jaegers (Figure 1). The environmental inputs or driving variables are solar-energy flux (global short-wave), mean daily temperature, day length, and fraction of the ground that is snow-free. Photosynthesis and growth rates for the plant community are functions of solar radiation, temperature, and day length. The comprehensive tundra photosynthesis model described above has not yet been coupled into the total ecosystem model. A significant defect of this aspect of the model is that snowmelt, and implicitly the beginning of the growing season, are functions of time and not of temperature. Spring melt occurs at the same time each year regardless of temperature, whereas under a real climatic change it would presumably be delayed by reduced temperature.

Most aspects of lemming population dynamics (birth rates, recruitment rates, death rates, food-consumption rates, etc.) are simulated as functions of food supply. Lemmings are assumed to eat primarily live green plants in the summer and frozen edible plant matter in the winter. The model is programmed to allow for the transfer of an appropriate fraction of material from the live-green to the frozen-edible compartments at the end of the growing season. Least weasels prey on lemmings, and weasel population densities in the model are based primarily on lemming numbers. In-migration and out-migration of weasels are triggered by critical

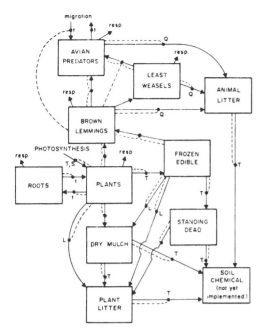

Figure 1. Block diagram of state variables now included in the Barrow simulation model. Solid lines denote flow of matter or energy and dashed lines denote an influence on rate of flow. Denoted by letters are: t, time of year; T, temperature; S, solar radiation; L, lemmings; and Q, various consumption rates.

lemming population densities. Pomarine jaegers immigrate each spring and emigrate each fall. Their summer population densities are modeled as functions of lemming abundance, as is the survival of their offspring. When lemming densities are low enough, jaegers fail to establish territories and mate. The large size of their territories sets an upper limit on their numbers in the simulations regardless of the size of the lemming population.

The tundra ecosystem model includes about 150 separate equations, and a single year can be run in about 0.28 seconds on the 360/91 computer at UCLA. Out of necessity, it contains many simplifying assumptions. An inherent deficiency is the lack of adequate justification for many of the parameter values, simply because of the unavailability of pertinent data. Field research at the Barrow, Alaska site since the model was developed has in fact shown that several of the assumptions about biological interactions are partially incorrect, at least with

respect to that location. Nevertheless, the tundra ecosystem model is considered to have significant heuristic value in providing clues about the direction and nature of the probable response of the real system to climatic change, even if it does not yield quantitatively accurate results.

The Coniferous Forest Biome model simulates transfers and accumulation of carbon among biological compartments in a 10.3-hectare watershed dominated by old-growth Douglas fir on the H.J. Andrews Experimental Forest of the U.S. Forest Service in the Cascade Mountains, Oregon. The primary-production model is an outgrowth of one developed earlier by Sollins et al. (1973) for a yellow-poplar stand in the southern Appalachian Mountains of Tennessee as a part of the Deciduous Forest Biome research.

The current model consists of eight non-linear first-order difference equations operating on a weekly level of resolution. Input variables are mean weekly solar radiation (Ly min^{-1}), mean weekly air temperature ($^\circ$C), mean weekly daylength, and mean weekly soil water. Flows and compartments are shown in Figure 2. The units of storage are metric tons of carbon per hectare. The entire 10.3 ha is assumed in the present model to be homogeneous spatially and to contain but a single species of vegetation.

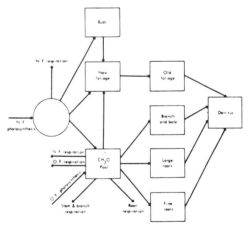

Figure 2. Flows and storages of carbon in Coniferous Forest Biome primary-production model. N.F. – new foliage; O.F. – old foliage.

The other Biome models, which are not described in detail here, are of the same general nature and order of complexity as those previously described. They all have been extensively

validated with field data from the specific ecosystems which they purport to describe.

The models are not designed to predict
long-term successional trends or changes in plant
and animal community composition; ecological
modeling has not yet reached the stage at which
meaningful predictions can be made about the
potential replacement of one species group by
another. The simulation runs are therefore
limited to a 5-year period and emphasize changes
in primary production and in associated aspects
of community structure. It is questionable how
far beyond this 5-year period the results can
validly be extrapolated.

By definition, all of the conclusions and
predictions of the simulations are built into the
models, some more explicitly than others. For
example, all of the simulated systems are more
sensitive to temperature change than to an
equivalent change in solar radiation when each is
tested separately. This primarily reflects the
interplay between the hyperbolic relation of net
assimilation to light intensity (at high light
intensity the curve is flat) and the exponential
effect of temperature on assimilation. Explanations for some other patterns are not so
apparent, however. One of the principal values of
ecological simulation may be the uncovering of
assumed but unrecognized relationships implicit
in model structure and their exposure to explicit
study and analysis.

Complete comparability was not achieved
among all the models. Neither as input driving
variables nor as output variables could the same
components be included in all of them. Thus, the
direct, independent effect of precipitation could
not be treated in the tundra model, nor could
that of solar radiation alone in the tree-ring
model. And, of course, each was designed to deal
with a distinct climatic region having plant and
animal species quite different from those of the
others. Nevertheless, the degree of convergence is
encouraging.

In carrying out this experiment, each model
was run separately in the respective laboratory;
results were synthesized at San Diego State by
C.F. Cooper in consultation with the other
investigators. The Desert Biome participated only
to the extent of providing a simula-
tion of primary production in a desert shrub
community.

PRIMARY PRODUCTION

Net primary production — total photosynthesis minus plant respiration, but not
including consumption by herbivores or death of
plant parts — was simulated in all of the models.
This was the only output variable reported by
every one of the participating groups. All of the
ecosystem simulations except the Desert Biome's
predicted a reduction in net primary production
as temperature and associated climatic parameters were decreased. This reduction varied from
about 18% for each 1°C in the deciduous-forest
model to 6%/°C in the grassland model (Figure
3). The Desert Biome model suggested that
alleviation of water stress through lowering of
temperature might actually increase primary
production in extreme deserts by as much as
3%/°C. The University of Arizona statistical
model of tree growth demonstrated the same
effect in conifers on arid sites.

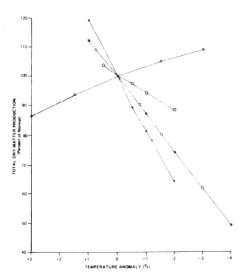

Figure 3.   Simulated response of plant productivity
in five regional ecosystems to hypothetical
climatic change. Associated with the temperature anomalies are the accompanying
changes in solar radiation, wind field, and
precipitation specified by CIAP. △, desert;
○, grassland; *, tundra; □, coniferous
forest; •, deciduous forest.

The response of primary production to
simulated climatic change was essentially linear
throughout the range of temperatures investigated. Because of the great differences in productivity among the several ecosystems, all results

are reported as percentage deviations rather than as absolute values.

Several reservations must be taken into account in interpreting the results of the simulations. Temperature and radiation effects on plant production were explicitly built into each of the models. (There was more or less regular interchange among the several modeling groups during the development period; similarity of equations and of some of the results might therefore have been expected.)

Furthermore, many long-term features of the response of plant production to climatic change cannot be predicted from existing models. For instance, some plants may become acclimated to new conditions if the new climate remains stable for some length of time. A change in temperature or water relations may alter physiological processes other than photosynthesis and respiration, and these, by a feedback mechanism, will indirectly affect plant production. Certain threshold values, such as those for temperature induction of enzymes, may be exceeded or not reached — an effect not covered in the models. Finally, long-continued climatic change will influence the outcome of competition among species, leading perhaps to a new community structure with changed productivity patterns. This problem likewise cannot be solved with a model of photosynthesis or primary productivity. Nevertheless, the similarities and differences in the responses of the simulated ecosystems to climatic change may yield significant clues about the possible reactions of real systems to alterations in climate.

The differing slopes of the temperature-response curves can presumably be explained by the climatic conditions of the several biomes, particularly by their differences in water regime. Plant production in the tundra region, for instance, is almost wholly temperature- and radiation-dependent. Apparent soil-water deficiencies of a magnitude likely to inhibit growth are not normally observed on the experimental sites in the wet tundra at Barrow, Alaska. Accordingly, precipitation and soil-water variation were not programmed into the tundra simulation model. These factors were accounted for in the coniferous forest model, but decreases in soil water do not markedly inhibit tree growth in the cool-temperate climate of the high-elevation Douglas-fir forest simulated in this

experiment. The slopes of the curves produced by the tundra and the coniferous-forest simulations were almost identical, predicting a relative decrease of 13% in net primary production for each 1°C reduction in mean temperature.

The deciduous-forest model predicted an even sharper decrease in primary production with temperature than the two preceding simulations: 18%/°C. The deciduous-forest region, although well supplied with moisture, does suffer occasional periods of deficiency during the growing season. The TEEM model (O'Neill et al., 1972) used in the deciduous-forest simulation is responsive to variation in temperature and radiation, but includes no explicit dependence on the water cycle or wind velocity. These effects are incorporated in a separate model of atmosphere-soil-plant relations known as PROSPER (Goldstein and Mankin, 1972), but the two have not yet been coupled. Therefore, PROSPER was run separately and subjective observations were made of the biotic response to water stress. The results, as will be discussed in more detail in the later section on evapotranspiration and water stress, indicated that decreased temperature would decrease water stress on the vegetation. The reduced growth due to lower temperature might be partially compensated for by decreased water stress. The subjective judgment of the Deciduous Forest Biome group as a result of running the PROSPER model is that the system changes predicted by the TEEM growth model may be somewhat exaggerated, and would be partially compensated for by changes in water stress which are not incorporated in the TEEM model. If this should prove to be the case, the proportional decrease in primary production with temperature would be lessened, bringing the deciduous-forest line in Figure 3 into closer accord with those for the tundra and the coniferous forest. This points up a significant area for future investigation and for inter-biome comparison.

Plant production in the grassland region, in coniferous forests on relatively arid sites, and especially in the desert, is more dependent upon seasonal and annual patterns of rainfall and soil water than it is in regions of higher precipitation. Grassland plant production is highly sensitive to spring and early summer soil moisture, which in turn is affected by the interplay between air temperature, solar radiation, and precipitation. Reduced temperatures in this moisture-deficient region may reduce soil-water losses more than they depress photosynthesis. Consequently, the simulated net loss of primary production in the Grassland Biome model is less for a given temperature decrease than it is in those biomes more abundantly supplied with water.

Water deficiency is even more critical in the desert. The Desert Biome model was run for a simulated 100 days during the 180-day main growing season of a single species, *Hammada scoparius*, a native of the Negev Desert of Israel. The growing season of this species extends from about April 1 to September 1. There is normally no rain in the Negev from mid-April to the end of October. Field data for the simulation were collected by E.D. Schultze and associates of the University of Würzburg, Germany, and used by him in development of a photosynthesis sub-model as part of the U.S. Desert Biome research.

This series of comparative simulation experiments with ecosystems in different climatic regions suggests that the response of net primary production to changes in temperature and solar radiation, and associated precipitation and wind variables, is likely to be nearly linear within the range of temperature and radiation changes expected from increased stratospheric flight (Figure 3). In regions relatively well supplied with water during the growing season, a $1°C$ decrease in temperature is likely to reduce net primary production in the neighborhood of 12-15%. The similarity in response of the tundra, coniferous-forest, and deciduous-forest simulations is striking. (A 12% decrease in plant production is of course much larger in a forested region of high production than in the low-production tundra.)

Where water for plant growth is more limiting, the reduction in primary production with temperature will be less, chiefly because lower temperature and reduced solar radiation will result in lowered leaf-water stress. In true desert climates, this effect may be so strong that plant production will actually increase as a consequence of reduced temperature.

The validity of the numerical values suggested here can be established only after additional field research and observation. The qualitative pattern suggested by the comparison is probably valid; logical and reasonable explanations for this pattern can be identified. It is, however, doubtful that this specific prediction of the ecological effects of climatic change could have been made without some sort of comparative modeling effort.

CONSUMER ORGANISMS

When consumer organisms were added to the simulated systems, interactions among trophic levels made the system response appreciably more complex and less predictable than in the relatively simple case of primary production alone. In the Arctic tundra simulation, primary production is reduced so much by a $2°C$ decrease in mean temperature that the assumed cyclical pattern of lemming population numbers breaks down. Lemming numbers fail to build up enough to support appreciable numbers of predators. The typical crashes in lemming numbers due to predation do not occur. The lemmings thereafter persist in low but essentially constant numbers, determined primarily by their food supply. This food supply is in turn determined by the interplay between reduced primary production and overgrazing. Overgrazing is more or less continuous in the absence of significant predation.

The response of lemming populations to temperature decrease is distinctly nonlinear, in contrast to the nearly linear effect on primary production. At the point where the self-regulating system effectively breaks down, the percentage decrease in lemming numbers is closer to 50% per $1°C$ decrease in temperature than to the $13\%/°C$ reduction in primary production.

The early version of the tundra ecosystem model used here is more oversimplified and unrealistic than any other of the models tested in this experiment. It is being superseded by an entirely new model. Nevertheless, it is instructive that the simulated system breakdown due to temperature change resulted entirely from differential changes in relationships among system components. The model was initially designed to reproduce an assumed three-year lemming population cycle, but no explicit timing mechanism was incorporated. The emergent property of cyclical behavior resulted from interactions among primary producers, consumers, and predators. The principal state variables and forcing functions in the tundra model were described earlier.

When the simulated mean temperature was increased by 1°C, primary production increased and the peak abundance of lemmings about doubled. The three-year cyclical behavior was retained, and the population dropped between highs to the same very low level as in the control case. When the temperature was reduced by 1°C, primary production decreased proportionately, and the peak lemming population was almost halved. Once again, the three-year cycle remained, but minimum winter population numbers were higher than in the control case. A 2°C temperature decrease, however, completely destroyed the cyclical behavior of the system, primarily because of the loss of significant predation. Net plant production then stabilized at a level considerably lower than would have been predicted from the temperature-induced reduction of 13%/°C in gross primary productivity discussed in the previous section. This, of course, was due to heavy grazing by lemmings.

The significant conclusion from this experiment with the tundra model is not that weasels and jaegers are predicted to be harmed by a moderate decrease in temperature, or even that an apparent system breakdown occurs at a temperature reduction of 2°C but not at a smaller decrease. These quantitative predictions may or may not be vindicated by reality. It is significant, however, that the simulation model predicts an entirely new structural relationship among system components — primary producers, consumers, and predators — as a consequence of a climatic change of relatively modest proportions.

A somewhat similar prediction resulted from the Deciduous Forest Biome simulation. Aphids were the only consumer organisms studied in these model runs. Since no predator was included in implementation of this section of the model, response of aphid populations is dominated by direct temperature effects on aphid metabolism and reproduction and by changes in primary production of the plants upon which they feed.

Aphid populations decreased as temperature was reduced. At moderate reductions, the population stabilized at a lower level commensurate with the reduced primary production. As with the tundra consumers, however, the proportional reduction in aphid population was substantially greater than the reduction in primary production. The most severe perturbation tested in the Deciduous Forest Biome simulation was a 2°C decrease in temperature. This eliminated the aphids entirely in about seven years.

These results definitely do not imply that long-continued temperature reduction is an effective device for controlling noxious insects, including aphids. They do point out the non-linear nature of second- and higher-order ecosystem responses beyond the initial level of primary production, and the need for additional field analyses to explain some of the simulated results.

The complexity of consumer interactions with other ecosystem components is further illustrated by the Grassland Biome simulation. When warm-blooded consumers (birds and mammals) were added to the control ecosystem, primary production, above-ground production, transpiration, and microbial respiration all increased, perhaps as a result of the greater turnover of nutrients brought about by the consumer activity. Essentially similar results were obtained when consumers were added to the simulated ecosystem under a 2°C reduction in temperature. Almost all response variables were increased slightly over the corresponding case without consumers, although production levels were well below those of the base climate. Interestingly, fixation of carbon in consumer biomass was essentially identical at the control and the reduced temperature.

These results are at variance with observations from long-term grazing experiments. Aspects of this portion of the model will require careful scrutiny before its results can be accepted in their entirety.

DECOMPOSITION AND MINERAL CYCLING

Microbial decomposition of dead plant and animal litter, and the consequent release of bound mineral elements, were studied in several of the models. In the Deciduous Forest Biome simulations, the environmental perturbations had little effect on the total annual turnover of fresh and partly decomposed litter, but the seasonal pattern was significantly affected. The quantity of material and of stored chemical energy in the two litter layers was essentially the same in late autumn under all climatic assumptions. Thereafter, however, microbial respiration rates and the rate of degradation of fresh litter to partially decomposed litter were inversely related to temperature. Consequently, detritus passed into the partially decomposed state at a slower rate the lower the temperature. By June there was significantly more material in the partially decomposed layer under higher temperatures than under lower. Although the total change in decomposition indicated by the model was relatively slight, the alteration in seasonal patterns could have a profound effect on remineralization in the forest litter. This change in pattern might be expected to cause alterations in the mineral cycling of the forest which are not reflected in the results of the simulations.

In the drier climate of the grassland region, available moisture might be expected to have a significant effect on decomposition. However, microbial respiration is more sensitive to temperature than to rainfall in the grassland simulations. The temperature response was slightly curvilinear. Soil respiration was about 12% less than in the base case with a drop in mean temperature of 2°C. A temperature reduction of 1°C, coupled with the other climatic changes postulated by CIAP, decreased respiration by 7%. A temperature increase of 0.5°C accelerated soil respiration by 4%. Here also, changes in the pattern of mineralization of plant litter may alter long-term successional trends in relatively complex ways.

The Coniferous Forest Biome modeling group independently concluded that the lack of a mineral-cycling component is perhaps the most glaring deficiency in their ecosystem model. They pointed out that the long-term effect of climatic changes would undoubtedly be manifested in changes in rates of mineral release and uptake, but their simulation provides no means for taking account of this process.

TOTAL ECOSYSTEM STANDING CROP

The Coniferous Forest Biome model was the only one to predict the consequences of climatic change on the total standing crop of unharvested old-growth forest stands. In such stands, primary production is normally in rough balance with mortality. This was the case with the coniferous-forest simulation under normal climatic conditions; the system was nearly in balance, with net primary production nearly (if not over-) compen-

sated for by mortality. This is also true of the actual stands on the H.J. Andrews Experimental Forest from which the data for the model were derived.

Temperature change altered simulated detritus production very little. Detritus was defined as the sum of litter fall, wood fall, stem mortality, and root death. Insect consumption was not included in this figure, but was small in all cases. The maximum simulated change in detritus production under the most extreme climatic change was less than 1%, an amount almost undetectable in the field.

It was difficult to establish a ratio of net ecosystem production under perturbed and natural conditions because the natural condition was nearly zero. Since, however, net primary production decreased as temperature decreased (as discussed in a previous section), and detritus production was not materially affected, total ecosystem standing crop had to decrease as temperature was reduced. The coniferous-forest simulation suggests that this reduction would amount to about 0.85 metric tons of carbon fixed per hectare per year in unharvested Douglas fir stands if all plant material is considered. If wood production alone is measured, the reduction amounts to about 0.09 tons of carbon per hectare per year. Since air-dried wood is about 40% carbon, this represents about 0.2 metric tons of wood per hectare per year. Compare this with the normal standing crop of wood in old-growth Douglas-fir stands on the H.J. Andrews Forest: 400-500 tons per hectare.

The predictions of the tundra and grassland simulations concerning standing crop were somewhat contradictory. In these communities dominated by herbaceous vegetation, turnover of biomass is obviously much faster than in forests, and removal by grazing is more important. In the tundra, the pattern of simulated biomass was generally similar to that of the coniferous forest. Both above- and belowground biomass at the end of the growing season are smaller under lower temperatures than under higher. In the grassland, however, the pattern was different. Although aboveground production was reduced by a lowering of temperature, peak aboveground standing crop was slightly increased. The belowground standing crop was smaller at lower temperatures than at higher, a reversal of the pattern for aboveground biomass. The reasons for this anomaly are not fully understood and are currently being investigated. It is possible that the simulated increase in peak aboveground live biomass may be completely nonbiological, and instead be an artifact of model structure.

TRANSPIRATION, PRECIPITATION, AND WATER STRESS

Water relations are at least as important in natural, noncultivated ecosystems as in agricultural systems. To a considerable extent, natural ecosystems occupy sites unsuitable for cultivation because of deficiencies in either moisture availability or soil properties. Soil deficiencies are as often related to the ability to supply adequate moisture during · the growing season as to inherent fertility. Plant-water relations are therefore particularly crucial in assessment of the impact of man-induced climatic change on natural ecosystems.

The PROSPER model of atmosphere-soil-water relations (Goldstein and Mankin, 1972) was run for four years, using observed environmental data from the Walker Branch Watershed on the AEC reservation at Oak Ridge, Tennessee. The environmental perturbations were assumed to occur uniformly throughout the year and were simply added or subtracted from the actual data for each day. For lack of better information, atmospheric vapor pressure was assumed to remain constant.

In accordance with the CIAP assumptions, precipitation was decreased along with temperature. Evapotranspiration was predicted by the model to decrease proportionately more than precipitation. A 5% decrease in precipitation, coupled with a 2°C decrease in temperature, reduced evapotranspiration by 14.5-11.5%, depending upon weather conditions in individual years. The reduction in transpiration was more closely related to the seasonal distribution of precipitation than to its total annual amount. The decrease of evapotranspiration with reduced temperature and precipitation in the simulation can be attributed in part to there being less available water to evaporate, but more to reduced potential evaporation because of lowered temperature and solar radiation.

The balance between precipitation and transpiration largely determines the effect of soil water on vegetation. The Deciduous Forest Biome team estimated that, as a rough approximation, a plant water potential less than -5 bars represents slight water stress, -10 bars represents moderate stress, and -15 bars represents severe water stress. The model did not predict a water stress as severe as -15 bars for any day during the four-year period with any of the climatic perturbations tested, or for the base or normal climate.

The simulation indicated that coupled temperature and precipitation reductions would bring about a small reduction in the number of days with slight water stress, especially in years when water stress was already low. The climatic change reduced water stress the least in the year 1970, when water stress under the normal weather was greatest. These results indicate that the growth declines predicted to occur due to temperature and radiation decreases might be partially compensated for by decreased water stress, particularly in years of good growing conditions when water stress is only moderate. Put another way, man-induced climatic changes are likely to have the greatest impact on vegetation in years when growing conditions are most unfavorable. Bad years will be made considerably worse than they would otherwise have been, whereas in years of relatively good growing conditions the deterioration due to lowering of temperatures may be less. To the extent that successional patterns, species replacement, and other ecological processes are sensitive to the most unfavorable and stressful years, this effect could be serious. This problem requires substantial additional study; simulation modeling is virtually the only feasible way to address it.

The grassland simulation, like the deciduous-forest model, also indicated a decrease in total transpiration when both temperature and precipitation were reduced. The decrease in transpiration was about 4% for each 1°C decrease in temperature, the latter being accompanied by a 2.5% decrease in precipitation. The grassland simulation did not explicitly report on soil-water changes, but the combined result of a decrease in precipitation and a slightly greater percentage decrease in transpiration is likely to be one of relatively little change in total soil-water balance, although there may well be changes in seasonal pattern.

These results are consistent with those reported by J. Ritchie (1974) for agricultural croplands in the central and southern United States. He states, "The compensating effects of cooler weather and less potential evapotranspiration offset the expected decreases in precipitation, and drought probabilities are not expected to increase in a general sense throughout the United States." This is probably true for natural ecosystems as well, although additional investigations are required.

SENSITIVITY TO DIFFERENT FACTORS OF CLIMATE

All of the simulation models were tested with the parameterized climatic changes stipulated by CIAP. These involved coupled reductions in temperature, total solar radiation, precipitation, and wind movement. Several of the groups also varied the input in additional ways, to determine the relative sensitivity of the systems to changes in the several climatic factors.

Temperature change in all instances yielded a greater simulated response than did a comparable change in any of the other climatic variables tested. Several of the models were run with and without a change in solar radiation accompanying the temperature change. The Coniferous Forest Biome model showed a 12-times-greater response to a temperature change of -1°C than to a radiation change of -1.5%. Specifically, the decrease in net primary production attributable to temperature alone was about 12%/°C. When solar radiation was added to temperature, the simulated decrease was 13%/°C. Qualitatively similar conclusions were obtained by the deciduous-forest and tundra modeling groups.

In addition, however, factors other than light for photosynthesis are commonly limiting in natural ecosystems. This is especially true in open communities such as grasslands where soil water and available nutrients commonly limit growth. Even in forests light may often be present in excess during the growing season. Therefore, reduction in solar radiation would not be expected to have as much effect as reduction in temperature.

Independent and simultaneous variations in temperature and in rainfall were simulated with the grassland model. The results indicated that decreasing both air temperature and rainfall will cause a decrease in all the output variables except peak aboveground standing crop, which shows a slight increase. These results show that simultaneously decreasing rainfall and air temperature involves an interaction different from independently adding the effects of air temperature and rainfall. The sign of the change in peak aboveground standing crop and in transpiration when temperature and rainfall were varied together might have been predicted by superposition of the independent effects of air temperature and rainfall. The same does not hold for peak belowground biomass, or for several of the other variables tested. The nature and explanation of these interactions requires additional investigation.

TREE-RING ANALYSIS

The approach used by the Laboratory of Tree-Ring Research at the University of Arizona is substantially different from the simulation models discussed above in that it is an extensive rather than an intensive approach. The model is simple and can be applied readily to a wide variety of sites and habitats. It involves a response function which is a least-squares regression equation to predict ring widths for a group of trees on a particular site by using orthogonal transforms of monthly temperature and precipitation data as predictors. The only requirements are that there be variation in ring width from year to year, and that a significant part of that variation be attributable to limiting conditions of climate. The effects of serial correlation of the ring widths up to lags of three years are also incorporated in the regression.

A reduction in temperature of 2°C (3°C for sites north of latitude 40°N) for each of 14 months was substituted into the regression equation for 26 tree sites including three different species. Calculations also included precipitation changes of +5% and -5% (± 10% for sites north of latitude 40°N). Some of the calculations that were carried out for five successive years are shown in Figure 4. A growth change of more than 30% was obtained for more than half the sites when temperatures were decreased by 2°C (3°C) and precipitation decreased by 5% (10%). More than half these responses were significantly greater than the calibration error of the model at the 95% confidence level. Reversing the sign of

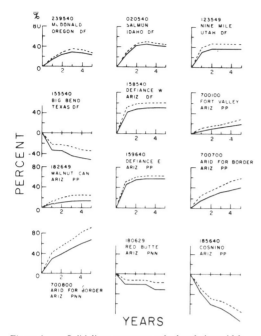

PERCENT

YEARS

Figure 4.   Solid lines represent calculated ring-width response for 1-5 years of persistence of the most extreme climatic conditions postulated by CIAP to result from a fleet of supersonic transport aircraft (temperature decrease of 2°C and 5% precipitation decrease between latitudes 20°N and 40°N, and temperature decrease of 3°C and 10% precipitation decrease between latitudes 40°N and 60°N). Dashed lines show the results obtained when the sign of the precipitation anomaly was changed to positive. Ordinate values are in terms of percentage of present normal ring width at each site. Identification numbers and locations of each site are indicated. Response functions and site data are found in Fritts (in press). The letters DF, PP, and PNN indicate species: Douglas fir (*Pseudotsuga menziesii*), Ponderosa pine (*Pinus ponderosa*), and Pinyon pine (*Pinus edulis*), respectively.

precipitation did not change the sign of, or otherwise substantially alter, the growth response, as the effects of the specified precipitation changes were small compared to those of the specified temperature changes (Figure 4).

The results from 28% of the sites are the same as those obtained from the coniferous-forest and deciduous-forest simulation studies, in that growth declined in response to a reduction in temperature. However, 72% of the sites gave a

response opposite to that of the forest simulations, but resembling the growth response for the shrub simulated in the Desert Biome. Considerable variation among the 26 sites was attributed to differences in microenvironments, differences among species, and differences among regional climates (Fritts, in press).

In most of the selected habitats, high rather than low temperatures are naturally limiting. The lowering of temperature may decrease evapotranspiration, which reduces water loss. Lowering of temperature also decreases respiration and favors more rapid net photosynthesis. The result is an increase in accumulation of stored foods and other growth-promoting substances, which leads to increased growth.

The coniferous-forest simulation was for a site where the water budget was not found to be limiting. The McDonald Forest sample used in the tree-ring analysis, though also in Oregon, represented a stand where low moisture and high temperatures are undoubtedly limiting. Thus, it is not surprising that the calculated growth response for the McDonald site was more like that of arid-site trees in the southwest than that of moist-site trees used for the coniferous-forest simulation.

When values specified for changes in temperature and precipitation were reduced to one-half and one-quarter of those used above, there was a proportional reduction in the computed growth changes. When a temperature increase rather than decrease was substituted, and no change in precipitation was specified, there was a reversal in sign of the growth response. As noted previously, reversing the sign of the precipitation had little effect on the growth-response curves, and never resulted in a change in sign of the growth response. However, the precipitation effect on growth response is always in the same direction as the sign of the specified precipitation change. That is, when the precipitation anomalies are positive the growth-response curves tend to be more positive (less negative). It is also seen from Figure 4 that the growth at some sites begins to level off after a few years as the effects of serial correlation appear.

In this analysis the response of the trees was calculated for a temperature change of the same amount during all seasons of the year. The trees, however, respond differently to temperatures in different seasons (Fritts, in press). If the specified changes in temperature were restricted to one or more particular seasons, such as winter and spring, the nature of the growth responses would be different. Also, it should be noted here that, in terms of the natural climatic variability at a site, the proposed temperature changes were of the order of a standard deviation of monthly temperature, while the proposed precipitation changes were considerably less than that in terms of standard deviation units (particularly for arid regions). This is a likely reason why the effects of the specified temperature changes dominated the effects of the specified precipitation changes in terms of growth response.

CONCLUDING REMARKS

The results reported here have been based on acceptance of the CIAP parameterized climatic changes, with precipitation assumed to decrease with temperature. Although this is presumably true on a global basis, zonal averages may not be applicable to important areas within each latitude band. It may be, for example, that 80% of the area within a given latitude band will undergo changes of 25% in annual precipitation, with the positive changes tending to cancel out the negative ones in such a way that the average anomaly over all longitudes is -5%. Historical evidence, for instance, quite clearly shows that in the Great Plains of the United States, cooler average temperatures have generally been accompanied by higher precipitation, and vice versa. This could materially change some of the predictions made here. In the case of the grassland region, soil-water deficiencies would be appreciably reduced if precipitation were to increase at the same time that mean temperatures were lowered. This question, too, requires additional investigation through simulation modeling and other approaches.

The results of this study point up the need for a variety of models, different approaches, and analyses of many different sites as a basis for large-scale evaluation of environmental impact. The analysis reported here has answered some critical questions, but more work on a greater variety of models, sites, and ecosystem types is needed before these results can be confidently generalized to predict the total ecological impact of man-induced climatic change.

REFERENCES

Fritts, H.C., T.J. Blasing, B.P. Hayden, and J.E. Kutzbach (1971), "Multivariate techniques for specifying tree-growth and climate relationships and for reconstructing anomalies in paleoclimate," J. Appl. Meteor. 10, 845-864.

Fritts, H.C., in press, "Relationships of Rings in Arid-Site Conifers to Variations in Monthly Temperature and Precipitation," Ecological Monographs.

Goldstein, R.A. and J.B. Mankin, Jr. (1972), "PROSPER: A model of atmosphere-soil-plant water flows," in the *Proceedings of the 1972 Summer Computer Simulation Conference* (San Diego, California), pub. Simulation Councils, La Jolla, California, 1176-1181.

Goodall, David W. (1972), "Potential applications of biome modelling," Terre et la Vie 1, 118-138.

Miller, Philip C. and Larry Tieszen (1972), "A preliminary model of processes affecting primary production in the arctic tundra," Arct. Alp. Res. 4, 1-18.

O'Neill, R.V., R.A. Goldstein, H.H. Shugart, and J.B. Mankin, Jr. (1972), "Terrestrial Ecosystem Energy Model," Eastern Deciduous Forest Biome, Memo Report 72-19, Environmental Sciences Division, Oak Ridge National Laboratory, Oak Ridge, Tennessee, 38 pp.

Ritchie, J. (1974), in Section 4.2.2 of Volume V of the CIAP Monograph Series, U.S. Department of Transportation. NOTE: At present, these monographs exist only in an early draft form. They will be publicly available after completion in September 1974.

Sollins, P., D.E. Riechle, and J.S. Olson (1973), "Organic Matter Budget and Model for a Southern Appalachian *Liriodendron* Forest," USAEC Report EDFB-IBP-73-2, Oak Ridge National Laboratories, Oak Ridge, Tennessee.

Timin, Mitchell E., Boyd D. Collier, Jon Zich, and David Walters (1973), "A Computer Simulation of the Arctic Tundra Ecosystem near Barrow, Alaska," U.S. Tundra Biome Rept. 73-1, Tundra Biome Center, Univ. of Alaska, Fairbanks, Alaska, 82 pp.

22

Reprinted from pages 86–108 of Conf. Computer Support of Environmental Science and Analysis, Albuquerque, N. M., Proc., U.S. Energy Research and Development Administration, 1975, 511pp.

THE APPLICATION OF COMPUTERS IN AN ASSESSMENT OF THE ENVIRONMENTAL IMPACT OF POWER PLANTS ON AN AQUATIC ECOSYSTEM*

W. Van Winkle
Research Associate

Environmental Sciences Division
Oak Ridge National Laboratory**
Oak Ridge, Tennessee 37830

ABSTRACT

The application of computers in the area of power plant impact assessment is approached from the standpoint of addressing a representative and important species, the striped bass, which is subjected to numerous stresses by virtue of the operation of five power plants along the lower Hudson River in New York State. Four particular applications of computer usage are discussed: (1) bibliographic information services; (2) data management; (3) data analysis; and (4) simulation modeling.

I am addressing the topic of the application of computers in the area of power plant impact assessment by using a specific example as opposed to a generic discussion. The example chosen is an evaluation of the potential impact of Consolidated Edison Company's Indian Point Nuclear Generating complex, located on the Hudson River at Buchanan, New York (Fig. 1). The use of this region of the Hudson River to produce electricity at the Indian Point complex, plus at four fossil fuel plants (Bowline,

---

*Publication No. 719, Environmental Sciences Division, Oak Ridge National Laboratory.

**Research supported by the U.S. Energy and Research Development Administration and the U.S. Nuclear Regulatory Commission under contract with Union Carbide Corporation.

ES-559

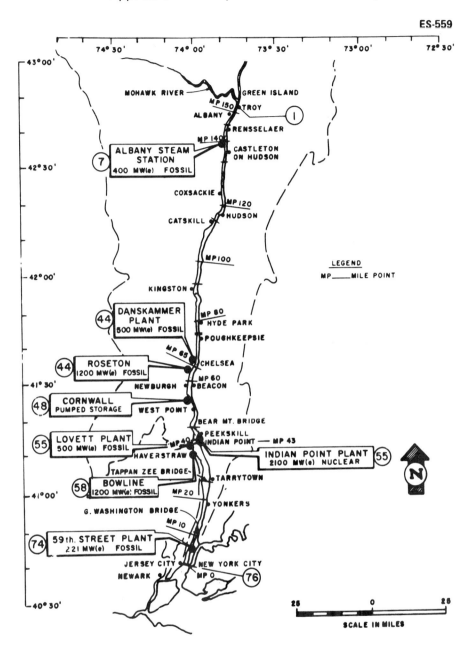

Fig. 1. The Hudson River showing major
existing and planned power generating
plants (U.S. Nuclear Regulatory
Commission, February 1975).

Lovett, Roseton, and Danskammer) and a proposed pump-storage plant (Storm King), is in competition with the use of this region of the Hudson River as an important spawning and nursery area for several species of fish, striped bass in particular. The five existing plants have the capacity to generate a total of 5620 MWe and to withdraw a total of 9150 cfs of water from the Hudson River (U.S. Nuclear Regulatory Commission, 1975). The very real need for additional electrical generating capacity in the New York metropolitan area, coupled with concern for the impact of these plants on the fish populations, has resulted in considerable controversy and lengthy licensing procedures during the past ten years and holds the promise of three or more years of litigation.

During these past ten years, a great deal of research and data collection has been done on the Hudson River ecosystem, supported primarily by Consolidated Edison, and a great deal of analysis and assessment has been performed with these data by Consolidated Edison, the U.S. Nuclear Regulatory Commission (formerly USAEC), the U.S. Environmental Protection Agency, the Federal Power Commission, New York State, and various citizen groups. In the process of these evaluations computer technology has been employed in a number of ways. I will discuss four such applications in this paper: (1) bibliographic information services; (2) data management; (3) data analysis; and (4) simulation modeling.

(1) Bibliographic Information Services

Dr. Ulrikson discusses "Bibliographic Information Services for ERDA Activities" elsewhere in the Conference Proceedings. One of the bibliographies prepared by personnel from his division and the Environmental Sciences Division at Oak Ridge National Laboratory is entitled "Striped Bass - A Selected, Annotated Bibliography" (Pfuderer et al., 1975). This

collection of papers relating to striped bass was initiated in 1971 when
it became apparent that the impact on the Hudson-River striped-bass population of Indian Point and the other power plants was of concern. The
number of striped-bass papers of interest quickly became unmanageable,
and we found it necessary to automate this information file. One of the
two products is a bibliography of 479 references concerning striped bass
and East Coast legislation dealing with state fishing regulations on that
species. These references are currently abstracted and indexed to reflect
the information content of the documents and our particular interests,
namely, information on fecundity, fishing effort, mortality, recruitment,
and the effects of environmental variables such as temperature. Indices
are provided for (1) author, (2) keywords, (3) subject category, (4) geographic location, and (5) title (permuted index of title). This annotated
bibliography saves us considerable time either by directly providing us
with the information we need or by enabling us to quickly select the
references we need to examine for a particular topic. In addition to
this concise encyclopedia to the striped-bass literature, a second product
of developing this automated information file is the capability to further
use the computer in either of two ways. Firstly, the automated information
file can be easily and rapidly corrected and updated as new references
become available. Secondly, one can specify keywords and have printed
those references containing these keywords. For example, if I were interested in all those papers dealing with the combined effects of temperature
and salinity on growth and survival of striped bass, I would specify a
search of the title, abstract and keywords of each reference for the key-
word combinations TEMPERATURE AND SALINITY AND GROWTH and TEMPERATURE AND
SALINITY AND SURVIVAL.

(2)   Data Management

As I mentioned in my introductory remarks, a great deal of research has been done on the Hudson River ecosystem during the past ten years. The published research reports for 1973 from the three major contractors (Lawler, Matusky & Skelly Engineers; Texas Instruments, Inc.; and New York University Medical Center), working for three different utilities (Consolidated Edison Company of New York, Inc., Orange and Rockland Utilities, Inc., and Central Hudson Gas & Electric Corporation), comprise a pile more than three feet high. The contractors' research necessitates exchanging raw data and communicating with each other extensively, Large-scale research involving several different contractors and employers and a consequent necessity to exchange data have made numeric data management an important activity for each of the contractors. The manner in which Lawler, Matusky & Skelly Engineers has integrated data management with its other research activities is illustrated in Fig. 2.

A general framework illustrating the role of numeric data management in a large-scale research program is given in Fig. 3 (Strand et al,, in press). The computer is essential at several points, for example, computerization of data, numeric data storage and retrieval. Data analysis (including data display) and models are considered separately in the following two sections of the present paper.

Texas Instruments' Fisheries Survey of the Hudson River (Texas Instruments, 1974) provides an example of the magnitude of the research program and of the role of numeric data management. Ichthyoplankton samples are collected weekly from March through December from near the dam at Troy, New York to below the Battery at the southern tip of Manhatten, a distance of over 150 miles. The following variables are recorded for

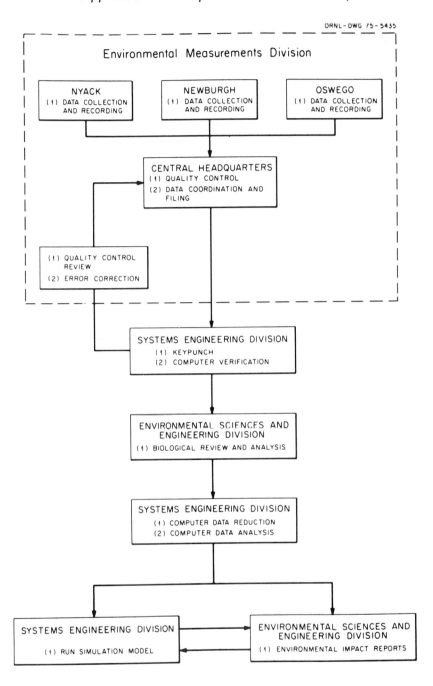

ORNL-DWG 75-5435

Fig. 2. Flow chart illustrating the manner which Lawler, Matusky & Skelly Engineers has integrated data management with its other research activities.

ORNL-DWG 75-3000

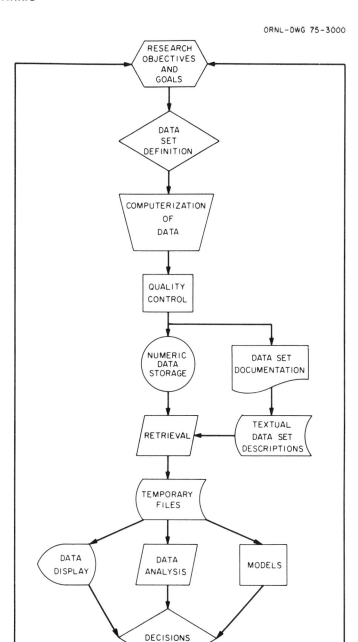

Fig. 3. Flow chart illustrating the role of
numeric data management in a large
scale research program (Strand et al.,
in press).

273

each sample: river mile, site (i.e., west, middle or east side of the river), time, river depth, sample depth, stage of the tide, sampling gear used, mesh size of the sampling net, number of specimens by species and life stage (i.e., egg, yolk-sac larvae, post yolk-sac larvae, and juvenile), initial and final digital flowmeter readings, temperature, dissolved oxygen, pH, salinity and turbidity. To proceed from the field data through data display, data analysis, and simulation models (which is being done not only by Texas Instruments but by other contractors, governmental agencies and interested parties) to the point of providing input to the decision-making process (Fig. 3) is the challenge of numeric data management. Without the aid of the computer it is difficult to envision being able to adequately meet this challenge.

(3) Data Analysis

Data analysis (including data display; see Fig. 3) is one aspect of data management which merits separate consideration. Use of the computer makes possible statistical analyses that otherwise would not be attempted because of the volume of data involved and time limitations. Three examples of data analysis that were greatly facilitated by the use of computers are given below.

a. The length frequency distributions of white perch for May 1973 in beach seine samples and bottom trawl samples are illustrated in Fig. 4. In addition to computer-generated graphics of publication quality, it is a simple matter to have printed frequency distributions for other months, for other fish species, and for alternative length intervals. Furthermore, statistical analysis software packages that include not only routine procedures but also graphical procedures (e.g., histograms as in Fig. 4), are becoming increasingly available (Barr and Goodnight, 1972).

ORNL−DWG 75−5433

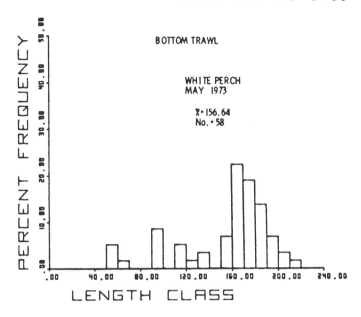

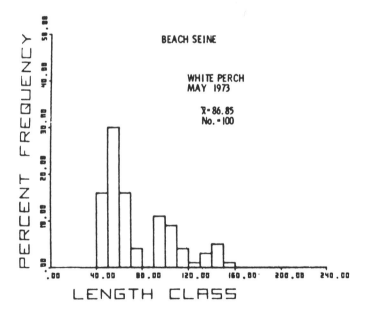

Fig. 4. Length frequency distributions of
white perch for May 1973 in beach
seine samples and bottom trawl samples
as an example of the use of computers
for directly printing figures of publication quality (Texas Instruments,
November 1973).

275

b. A second example (Fig. 5) illustrates the type of statistical analysis which would not be attempted without the aid of a computer except in a special case. A covariance analysis with two covariates and over 400 observations is a full day of tedious work on a calculator. However, covariance analysis is an indispensable tool in attempting to address such concerns as whether there has been a significant reduction in a fish population over a several year period, perhaps due to impingement at power plants. In the absence of such an analysis one is faced with a set of highly variable data involving numerous factors from which it is difficult to reach any sound conclusion.

c. A third example (Fig. 6) deals with the controversial issue of how the average power plant intake concentration of striped-bass ichythyo-plankters differs from the average river concentration in the vicinity of a plant (U.S. Nuclear Regulatory Commission, 1975). Sampling programs were designed and extensive data were collected to address this issue. In analyzing the data, we were concerned with:

i. estimating the ratio of intake to transect concentration at four plants on two or more dates for three life stages;

ii. comparing alternative methods of weighting the middepth and midchannel samples in calculating the average transect concentrations;

iii. calculating a concentration as the ratio of the number of organisms in each sample to the volume of water filtered for that sample, and then calculating the average of the individual concentrations over a 24-hr period (Method 1) versus calculating the 24-hr average number of organisms in all the samples and the 24-hr average volume of water filtered, and then taking the ratio of these two 24-hr averages (Method 2);

iv. testing for normality and homogeneity of variance of the concentration values and of the $\log_{10}$ concentration values; and

ORNL-DWG 75-5434

ANALYSIS OF COVARIANCE OF WHITE-PERCH CATCH-PER-EFFORT BEACH-SEINE DATA FOR 1969, 1970, AND 1972 AT FIVE STATIONS IN INDIAN POINT VICINITY

Source	Sums of Squares	df	Mean Square	F
Year	66.445	2	33.222	4.605*
Month	252.811	8	31.601	4.381**
Station	225.524	4	56.381	7.815**
Covariates	76.843	2	38.421	5.326**
Temperature	62.408	1	62.408	8.651**
Salinity	3.521	1	3.521	0.488
Error	3174.068	440	7.214	

* F statistic significant at 0.05 level

** F statistic significant at 0.01 level

Fig. 5. An example of the type of statistical analysis which would not be attempted in general without the aid of a computer (Texas Instruments, April 1973).

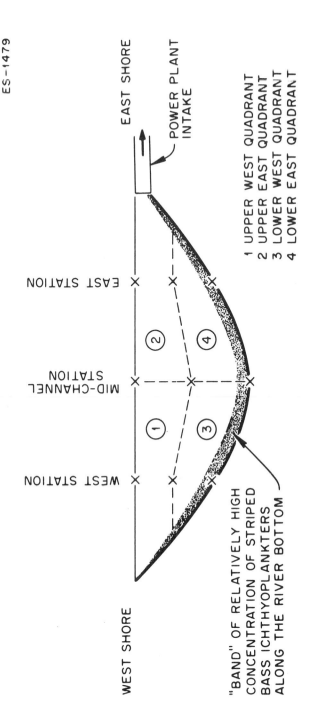

ES-1479

EAST SHORE

POWER PLANT
INTAKE

1 UPPER WEST QUADRANT
2 UPPER EAST QUADRANT
3 LOWER WEST QUADRANT
4 LOWER EAST QUADRANT

EAST STATION

MID-CHANNEL
STATION

WEST STATION

WEST SHORE

"BAND" OF RELATIVELY HIGH
CONCENTRATION OF STRIPED
BASS ICHTHYOPLANKTERS
ALONG THE RIVER BOTTOM

Fig. 6. Hypothetical cross section of the
Hudson River illustrating four quadrants,
the location of the three typical
sampling stations and sampling depths,
and a band of relatively high concentration of striped bass ichthyoplankters
along the river bottom (U. S. Nuclear
Regulatory Commission, February
1975).

278

v.   calculating approximate confidence intervals around the point estimate of the intake-transect concentration ratio.

This example provides a clear indication of how relying on the computer to do the arithmetic and repetitive calculations provided us with the time to do additional analyses beyond the minimum (which was Item i), such as Items ii-v.

(4)   Simulation Modeling

Our major use of the computer in assessing the potential impact of the power plants on the Hudson River striped-bass population has been in the area of simulation modeling. We have developed two types of models: a young-of-the-year population transport model (Eraslan et al., in press) and a life-cycle population model (Van Winkle et al., 1974) (Fig. 7). This research has been supported by both ERDA (Division of Biomedical and Environmental Research) and the Nuclear Regulatory Commission.

The striped-bass young-of-the-year model considers six life stages (egg, yolk-sac larva, post yolk-sac larva, and three juvenile stages); and it includes dependence of spawning rate, mortality rates, growth rates, apparent survival probabilities and maximum swimming speeds on temperature, salinity and population densities. The transport of these life states in the Hudson River is formulated in terms of a daily transient (tidal-averaged), longitudinally one-dimensional (cross-section-averaged) hydrological transport scheme. The major features of the model are represented schematically in Figs. 8 and 9. The validation procedure for this model involves comparing simulated and observed weekly standing crop values in the Hudson River for each of the young-of-the-year life stages (Fig. 10).

From the striped-bass young-of-the-year population model, we obtain forecasts of the percent reduction due to mortality at the power plants in

279

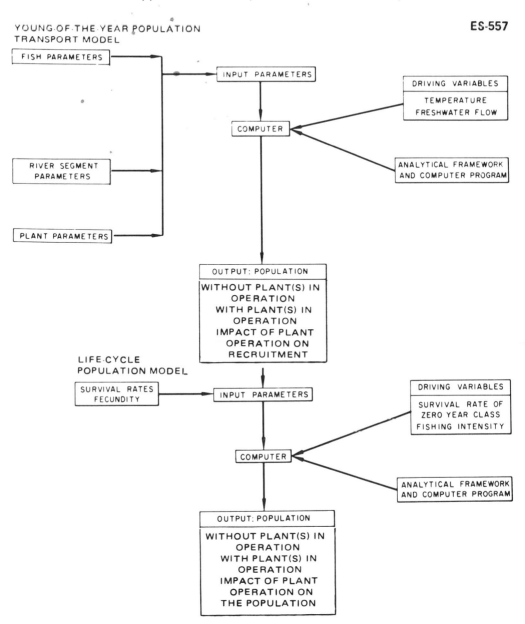

Fig. 7. Overview of the striped bass young-
of-the-year population transport model
and life-cycle population model (Van
Winkle et al., December, 1974).

ES-558

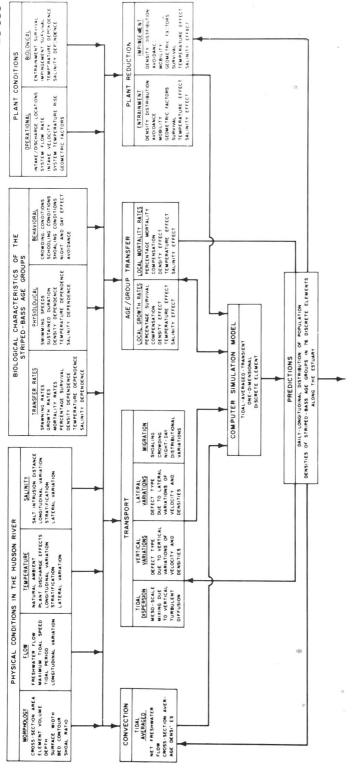

Fig. 8. Schematic representation of the computer simulation model for the striped bass young-of-the-year population in the Hudson River (Eraslan et al., in press).

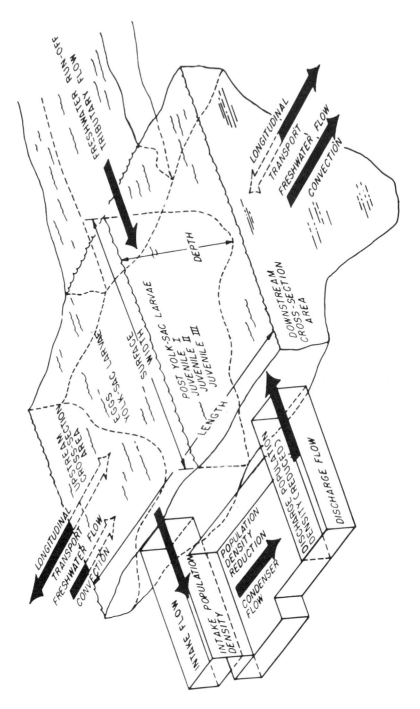

Fig. 9. Schematic representation of a discrete
element as considered in the computer
simulation model for the striped bass
young-of-the-year population in the
Hudson River (U.S. Nuclear Regulatory
Commission, February 1975).

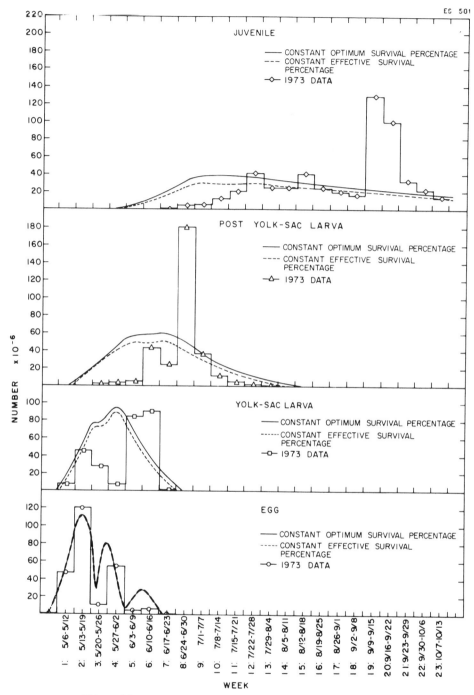

Fig. 10. Validation procedure for the striped
bass young-of-the-year population
transport model of comparing
simulated and observed weekly
standing crop values for each of the
life stages (eraslan et al., in press).

283

the number of striped bass surviving their first year. This percent reduction value provides input to the striped-bass life-cycle population model (Fig. 11). The life-cycle model is designed to evaluate the long-term impact on the striped-bass population of changes in mortality in the youngest age class. The general question concerns what happens to a fishery when density-independent sources of mortality are introduced which act on the young-of-the-year. This model considers all age classes of striped bass from young-of-the-year to fifteen-year olds and older. The model is strictly time-dependent and, unlike the young-of-the-year model, does not include spatial considerations. In the model the striped-bass population is presently assumed to be regulated in the long-term by fishing mortality which varies with the weight of fish available to the fishery in a compensatory manner. Typical but hypothetical results from the life-cycle model are illustrated in Fig. 12. Relative yield is defined as the ratio of the yield to the fishery with a power-plant impact to the yield with no power-plant impact.

The computer program for the young-of-the-year model was too big to run except as a batch job. The life-cycle model, however, was initially programmed for interactive use on a terminal. Subsequently, as the need for production runs arose, a second program was written for batch jobs. The interactive computer program has been extremely valuable for demonstration purposes and for allowing us to efficiently explore the behavior of the model in an intuitive manner prior to doing predetermined and systematic production runs.

In summary, computer technology has played an important role in our assessment of the potential impact of power plants along the lower Hudson River on the striped-bass population in that our efficiency and capabiliti

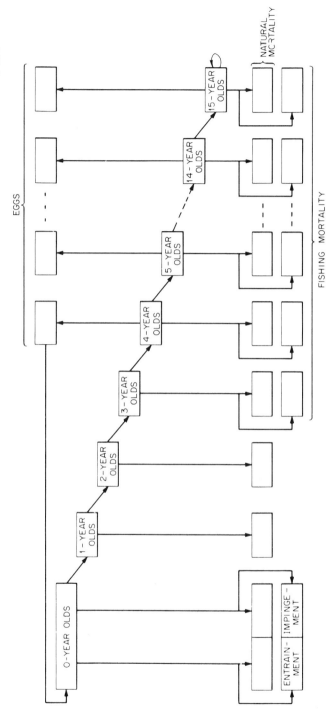

Population Model.

Fig. 11. Box and arrow diagram for the striped bass life-cycle population model illustrating aging transfers, production of eggs by sexually mature females, and losses due to natural mortality, fishing mortality, and entrainment and impingement at power plants (Van Winkle et al., December 1974).

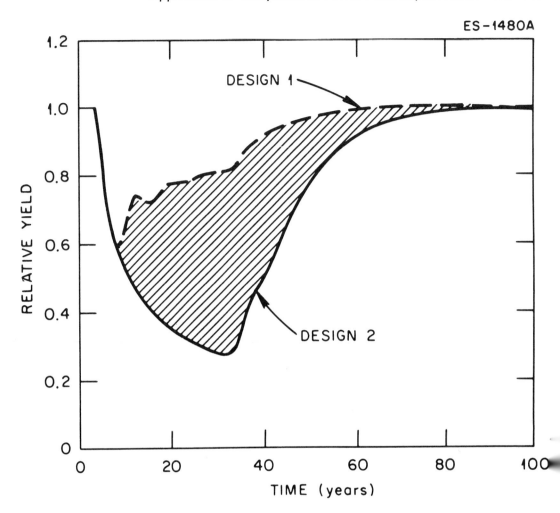

Fig. 12. Typical output from the Calcomp
Plotter for the striped bass life-
cycle population model illustrating
two hypothetical curves of relative
yield versus years.

have been appreciably increased.   Undoubtedly there will be different emphases in other impact assessments, although it seems likely that most applications will fit under one of the four headings of bibiliographic information service, data management, data analysis and simulation modeling.

ERRATA

Page 104, the last line of the figure caption should read: "December 1974)."

REFERENCES

1. Barr, A. J. and J. H. Goodnight. Statistical Analysis System. Student Supply Stores, North Carolina State University, Raleigh, North Carolina, 1972.

2. Eraslan, A. H., W. Van Winkle, R. D. Sharp, S. W. Christensen, C. P. Goodyear, R. M. Rush and W. Fulkerson. A Computer Simulation Model for the Striped Bass Young-of-the-Year Population in the Hudson River. Oak Ridge National Laboratory, Oak Ridge, Tennessee, ORNL-TM-4711, in press.

3. Pfuderer, H. A., S. S. Talmage, B. N. Collier, W. Van Winkle, Jr., C. P. Goodyear. Striped Bass - A Selected, Annotated Bibliography. Oak Ridge National Laboratory, Oak Ridge, Tennessee, ORNL-EIS-74-73, ESD-615, March 1975.

4. Strand, R. H., N. Ferguson, B. S. Ausmus, and D. M. Swift. Numeric Data Management for Ecosystem Analysis. Oak Ridge National Laboratory, Oak Ridge, Tennessee, ORNL-EDFB, IBP, in press.

5. Texas Instruments, Inc., Hudson River Ecological Study, First Annual Report, prepared for Consolidated Edison Company of New York Inc., April 1973.

6. Texas Instruments, Inc., Hudson River Ecological Study, Second Semi-annual Report, prepared for Consolidated Edison Company of New York Inc., November 1973.

7. Texas Instruments, Inc., Fisheries Survey of the Hudson River, March-December 1973, Volume IV, prepared for Consolidated Edison Company of New York Inc., September 1974.

8. United States Nuclear Regulatory Commission, Office of Nuclear Reactor
    Regulation. Final Environmental Statement related to operation of
    Indian Point Nuclear Generating Plant Unit No. 3, Consolidated Edison
    Company of New York., Inc., Docket No. 50-286, February 1975.

9. Van Winkle, W., B. W. Rust, C. P. Goodyear, S. R. Blum, and P. Thall.
    A Striped Bass Population Model and Computer Programs. Oak Ridge
    National Laboratory, Tennessee, ORNL-TM-4578, ESD-643, December 1974.

Editors' Comments
on Papers 23, 24, and 25

23 MILLER
Model Validation Through Sensitivity Analysis

24 CASWELL
The Validation Problem

25 MANKIN et al.
The Importance of Validation in Ecosystem Analysis

VALIDATION

Of the technical problems that still require major research thrust, validation is undoubtedly the most important. Misunderstandings about the nature of validating ecosystem models remain as a deterrant to the expanded use of mathematical models in the ecological community. Several approaches to the validation problem have been taken in other fields (Cohen and Cyert 1961; Cyert 1966; Hermann 1967; Holt 1965; Naylor and Finger 1967) but little progress has been made in discussing the problem in the ecological context (Goodall 1973).

A common misconception about validation is that a model should be able to duplicate the time behavior of an ecosystem over a wide range of conditions. This concept follows logically from early, optimistic viewpoints about the potential importance of systems models in ecology. In actual fact, the model can merely summarize and synthesize our understanding of the system. Obviously, if our understanding of the system is not adequate, the predictive capability of the models is limited. The model is supplied with hypotheses about how ecosystem components interact and the model merely draws out the logical implications of these hypotheses. The model can only be as good as our understanding.

A distinction must be made between models designed for management application and models as an integral part of ongoing research. For application purposes, the accuracy of simulations is of immediate importance since model output is to be used in making decisions. In many cases, this is achieved by calibration, which involves altering model parameters until model output matches available data. Then, these parameter values are fixed for subsequent predicitons. This is a practical form of validation that works well and provides a reasonable solution to the problem of confidence in predictions. However, this is not a rigorous test of the model.

In research, the objective is to advance understanding. Model output is a complex hypothesis to be tested. The stress is on exploration of the model to locate counterintuitive results that can be directly tested by experimentation. Therefore, in a research context, the emphasis is toward invalidating the model. If experimental results merely match model output, little new insight is gained.

The three papers presented in this section represent several viewpoints on validation. Paper 23 by Miller presents a unique perspective in which validation is viewed within the context of a user's needs. A model is valid when it fulfills the user's criteria for adequacy. The final two papers, Paper 24 by Caswell and Paper 25 by Mankin et al. are general discussions of the validation question and provide an up-to-date summary of our understanding of the problem.

REFERENCES

Cohen, K. J., and R. M. Cyert. 1961. Computer models in dynamic economics. *Q. J. Econom.* **75**:112–27.

Cyert, R. M. 1966. A description and evaluation of some firm simulations. *Proc. IBM Sci. Comput. Symp. Simul. Models and Gaming.* IBM Corporation, White Plains, New York, pp. 3–22.

Goodall, D. W. 1972. Building and testing ecosystem models, pp. 173–94. In Jeffers, J. N. R. (ed.), *Mathematical Models in Ecology.* Blackwell Scientific Publications, Oxford.

Hermann, C. 1967. Validation problems in games and simulations. *Behav. Sci.* **12**:216–30.

Holt, C. S. 1965. Validation and application of macroeconomic models using computer simulation. In Duesenberry, J. S., G. Fromm, L. R. Klein, and E. Kuh (eds.), *The Brookings—SSRC Quarterly Econometric Model of the United States.* Rand McNally and North Holland Press, Chicago.

Naylor, T. H., and J. M. Finger. 1967. Verification of computer simulation models. *Manag. Sci.* **14**:92–101.

Reprinted from pages 911–914 of *Summer Computer Simulation Conf., Houston, Proc.*, Houston, Tex., LaJolla, Calif.: Simulation Council, Inc., 1974, 954pp.

MODEL VALIDATION THROUGH SENSITIVITY ANALYSIS
D. R. Miller

Abstract

A new approach to model validation is described which depends only on the magnitude of uncertainties in the data, the sensitivity of the results to errors, and comparison of the accumulated uncertainty to a user-defined limit. Results either suggest that the models are valid or indicate specific areas in which further refinement is necessary. The approach is here applied to models of mosquito population dynamics and the flow of pollutants through an aquatic ecosystem.

1. Introduction

In some areas, simulation has reached a high state of development. A great deal is known about implementation of models, simulation languages, statistical analysis of results of simulation runs, and so forth (1). Unfortunately, an important part of the total simulation problem has been relatively neglected, and it may be the most important part for the ultimate user. Certainly, it is one source of the considerable criticism that has been directed against the simulation business in recent years. This is the question of whether the model used actually does correspond to reality, in the sense that it abstracts from the real system those elements which are of importance for the problem at hand and describes their interactions in a roughly correct way.

In most of the simulation literature (1), one finds this problem dealt with through comparison of model performance with historical data. This approach can be criticized, for such a historical record is itself a highly specific time series, and the comparison depends on an intuitive decision as to how much detail the modeller should expect the run to duplicate. Some other validation methods amount to thoughtful re-consideration and are hard to quantify (2,3). In practice, of course, the situation can be a good deal worse; the time lag between formulation and validation may be so great that formulation is done only once, near the beginning of a project, and no action whatever is taken on the basis of the results of a validation study, if indeed one is made at all (4).

Obviously, results corresponding to observed future history are the ultimate validation of any model when viewed as a predictive device. However, procedures which can be used at early stages of the modelling process are needed. This is particularly true if the results are to be of use in refinement of the model, or in planning of the research study that the model is supposed to support (5). Of necessity, such an approach must depend only on the model itself. The obvious approach is to use the model to predict not only the expected behaviour of the system, but also the level of uncertainty of that prediction as it depends on uncertainty in the input information and in the model structure. The procedure described here is a way of doing this. It is probably not a universal validation method, and depends on features of model performance which must be verified experimentally for each model studied. Nonetheless, the approach has turned out to be of use in several cases, it is easy to implement, and it leads to clear-cut decisions that can be readily accepted by the ultimate user.

The procedure is an extension of traditional sensitivity analysis, on which there is a substantial literature, both theoretical (6) and practical (7,8). We begin by making small changes in each parameter separately, using repeated simulation runs to obtain numerical measures of the effects on model behaviour (the "Sensitivity Coefficients"). The measure used is actually a measure of difference, which we call Deviance, between any two runs. It is defined in consultation with the user, who is also asked to define a tolerance limit such that any two runs differing by an amount smaller than this may be regarded as reasonable approximations of each. Finally, the estimated errors in the actual input data used are combined with the sensitivity coefficients, it having first been verified experimentally that the response is linear, to predict a level of uncertainty, or expected deviance, in model performance. This is compared to the tolerance limit and leads to (a) a judgment that the model is valid, (b) identification of the parameters which may efficiently be further refined, or (c) the conclusion that the simulation will never make sensible predictions at the level of detail specified.

2. Mathematical Formulation

We assume that a finite set of input parameters $p_1, p_2, \ldots, p_m$ completely determines the output variables $x_1(t), x_2(t), \ldots, x_n(t)$. The best available estimate of the "true" parameter values are designated p_i^0, and the corresponding results, which serve as a reference or standard run, are written $x_J^0(t)$, $J = 1, 2, \ldots, n$.

We assume that there has been defined a scalar measure of the total difference between this reference and any other run, say $D(x_i(t), x_i^0(t), T)$, where T is the total time over which the simulation is carried out. This will in fact be a function of the p_i and p_i^0 (as well as T). The p_i^0 are regarded as fixed, while we assume the p_i are perturbations of the reference values. Thus we define

$$p_i = p_i^0 (1 + u_i) \qquad (1)$$

where the u_i are small disturbances centred around zero and having a distribution which must be assumed. This means that D is a function of the m perturbations u_i and the total time T.

The critical assumption is that D is in fact a <u>linear</u> function, at least for small values of u_i, so that it can be approximated by its tangent plane, and there exist numerically calculable first partial derivatives.

$$S_i = \partial D / \partial p_i \qquad (2a)$$

as well as

$$R_i = \partial D / \partial u_i = p_i^0 S_i . \qquad (2b)$$

These are the sensitivity coefficients, equal in number to the number of input parameters p_i^0. (The S_i are the coefficients as usually defined. The R_i, or <u>relative</u> sensitivity coefficients, describe the effects of fractional changes in the p_i and are more meaningful for the user.) This linearity of D in u_i implies that, to a first approximation,

$$D = \sum S_i (p_i - p_i^0) = \sum R_i u_i . \qquad (3)$$

To specify input errors we assume each p_i has a normal distribution with mean p_i^0 and variance σ_i^2. Thus, for u_i we have mean zero and variance $(\sigma_i / p_i^0)^2$ so that

$$\text{Exp}(D) = 0$$
$$\text{Var}(D) = \sum S_i^2 \sigma_i^2 = \sum R_i^2 (\sigma_i / p_i^0)^2 \qquad (4)$$

The expectation that D, or rather its absolute value, have a value greater than some limit can now be expressed precisely.

The actual tolerance limit for D must be determined by, or in consultation with, the user. Once that is done, a statement of how uncertain the predictions are relative to that limit can be made in a way that is fully meaningful to him. Indeed, this is the advantage of making the sweeping assumption of normality as we have.

In those cases where Var(D) is large compared to the limit of acceptance, we can think of individual terms $S_i^2 \sigma_i^2$ as components which identify the contribution to the overall uncertainty of each of the input parameters. Thus we can see immediately what parts of the mechanism are responsible for the largest part of the resultant error, and this can be used as a guide to allocation of future resources. This, in fact, is the application to which Sensitive Analysis has traditionally been put (1,2). In some other cases we may abandon study of a particular component entirely if the corresponding coefficient R_i is extremely small, or perhaps improve the model by combining compartments which individually have large values of $S_i^2 \sigma_i^2$ but may in combination contribute less error. These and other applications are deserving of further study but clearly go beyond the specific purpose of this paper.

3. Assumptions and Intuitive Judgments

The above development contains three aspects which are not subject to precise analysis, and this section describes the author's experience in dealing with them. The problems are those of estimating the input errors, deciding on a measure of deviance, and defining limits of acceptability from this measure.

The first is relatively easy, as users are normally prepared to include upper and lower extremes when supplying estimates of parameter values, in fact are often relieved when asked to do so. Considerable experience in obtaining information of this kind has been accumulated, for example, by practitioners of such techniques as PERT and CPM (10). In addition, users seem quite willing to discuss the variation in experimental determinations and to comment on values to be found in the literature. Although there can never be any final answer about the type of distribution a parameter might arise from, a rough indication of its scatter is not hard to obtain.

The definition of the D-measure, or the measure of deviance as we call it, is both model- and purpose-dependent. Also, the relative magnitudes of the sensitivity coefficients (and perhaps the range over which D remains linear) may change noticeably if a different deviance measure is used. There does not appear to be any better way to proceed than by example, paying strict attention to the purpose for which the study was undertaken. The examples presented below represent cases in which deciding on a definition of D was easy and difficult, respectively.

Finally, one must define a limit of acceptability in D, say D_{crit}. To do this the author has adopted a procedure which seems to have some appeal, namely to use overlays of graphical displays of simulation runs against the reference run. A number of these are prepared corresponding to different values of D over a range of perhaps two orders of magnitude. These are produced by running the model with all the u_i drawn randomly from distributions having different variances. Then, the displays are shown to field experts and they are asked to indicate which they feel would be noticeably different when observed in the field. The spread between, say, the lowest D-value rejected by anyone and the highest D-value accepted by anyone is usually a factor of two or three, and a range within which D_{crit} must be found can be established.

4. Example: Mosquito Population Dynamics

A model designed to simulate the response of an isolated, well-mixed mosquito population during a sterile-male release program has been described elsewhere (8), and may be considered from the standpoint of the above discussion. The model itself is a 29-compartment structure in which the passage of individual mosquitoes is assumed to be synchronized with daily intervals. Ten parameters control the operation of the model; these are shown on Table I with their estimated uncertainties.

The definition of D in this case is reasonably straightforward if one asks the critical question of what an entomologist is concerned about in making field measurements (9). What is normally measured is either the biting population, since this is so important to human observers, or the number of larvae or eggs, rather easier field measurements. However the concern is with total population, and with relative changes in total population. An appropriate D-measure is therefore one which covers all population compartments, concentrating on their changes as time progresses. Thus we may define a measure of change of the population on a given day as

$$WT = \Sigma_j x_j(t)/x_j(o)$$

where $x_j(o)$ is the starting (steady-state) value, and then a measure of D which effectively integrates this over time:

$$DWT = (\Sigma_t WT - \Sigma_t WT^o)/\Sigma_t WT^o .$$

Several other definitions of D-measures were tried, and led to the same general results (9).

Once the formula for DWT was agreed upon, calculating the sensitivity coefficients S_i and testing for linearity presents no problem. Results of plotting $|DWT|$ against $|\Sigma u_i|$ for the case where the u_i were changed simultaneously with relative amounts and scaled by the R_i are shown in Figure 2. Finding an appropriate value for D_{crit} proceeds without incident. A number of runs for different DWT were prepared, and the range of acceptability as given by a number of respondents was between 0.275 and 0.530. The predicted variance of DWT was .177, showing that the model was acceptable to the observers

questioned.

5. Example: Pollutant Kinetics

Models purporting to describe the cycling of various pollutants through the environment seem to be proliferating faster than the pollution itself, and it is natural to ask whether the results of such models have much validity in practical terms. For the sake of discussion, we refer to a particular six-compartment model (Figure 3) developed by the author and co-workers as part of the Ottawa River Project (11). Each of the compartments represents both a mass of water (or sediment, etc.) and a mass of pollutant. Flow of water in and out of the system is used as a driving mechanism, so we have 11 equations. These involve seventeen coefficients.

Modelling of such a system is extremely difficult for a number of reasons, of which the author is particularly concerned with two. These are, first, that the parameters and interactions vary widely with environmental conditions and, second, that many different measures of system performance may be used and lead to quite different results. The solution to the first problem is for the modeller to identify the most critical parameters and then somehow convince laboratory and field biologists that specific experiments to determine the behaviour of that parameter should be undertaken. This is sometimes viewed by the biologist and others as meaning that science is to be directed by a computer, and the reaction can be incredibly hostile. Nonetheless, this is an example of the kind of iterative modelling that must be carried out if such a system is to be understood, and modellers must strive to get it done. Until it is done, estimates of uncertainty in rate constants and other parameters must be large enough to accommodate fluctuations in temperature and whatnot, and this typically causes the results to be so vague that the whole modelling approach is in danger of being rejected.

For the second problem, the author believes that the solution is not the task of the modelling professional at all, but rather of those with the responsibility for formulating policy in the area involved. All the modeller can do is make sure that when a measure is chosen, the technology is developed to identify those parameters to which that measure is particularly sensitive.

Having no such guidelines in the present case, the author chose a measurement more or less at random for illustration, namely the amount of the pollutant found in fish. (The Ottawa River Project is concerned with mercury in addition to certain other pollutants, so perhaps there is some reason in this choice.) Once this choice was made, the rest proceeded quite normally: parameters were estimated, and errors were guessed at, in consultation with biologists; linearity of the D-measure was tested, and found to be satisfactory; sensitivity analysis was performed and overall uncertainty in D was estimated. A summary of the results of this procedure is shown in Table II.

It turns out that the model has certain definite shortcomings; for example, it predicts an unrealistically small value for the invertebrate population. However, over the time scale used, the concentration of pollutant in fish is followed quite well, and when that particular deviance measure is used, the model is valid for a period of months. The sources of error, which must be improved if the system performance is to be followed over a period of a couple of years, are clearly identified. We see that of the seventeen parameters, three contribute eighty percent of the error. Research to better understand their specific interactions is currently underway. Because resources can to some extent be shifted away from the interactions now known to have little effect, the additional effort is not hard to manage.

The essence of the modelling process is precisely the kind of repetitive formulation and validation that such sensitivity analysis makes possible. A model should go through many formulations, each one improved on the basis of the worst faults of its predecessor. Rather than be disappointed at the fact that the model is not accurate when first formulated, we should have been amazed if success were achieved the first time around.

6. Conclusions and Discussion

The procedure described here is attractive in several senses; the steps are not difficult, involvement of the user or field scientist is required at several stages, and the conclusions are clear, if perhaps a bit pessimistic. If validity is not indicated, the direction work must take to improve things is indicated.

The largest limitation would seem to be the linearity requirement. Local linearity is not too much to expect, and one could even quote theorems about continuous dependence on initial conditions and the like. However, we require that the tangent plane approximation be valid for some considerable range of values, and we find that it actually is so. For now, one must think of experimentally verifying linearity of response in each case.

One particular extension of this method seems to be both obvious and desirable for certain problems, and that is to let the quantity T, the maximum time over which the simulation is run, also be variable. Then for given errors in the input parameters one could determine the maximum value of T such that $D(T) \leq D_{crit}$. Presumably, this would provide a time-limit over which results could be expected to be meaningful, and might provide a better way of assessing models purporting to predict the far future than seems to have been used so far.

References

1. Naylor, T.H. Bibliography on Simulation and Gaming. Computing Reviews 10:61-9 (1969).

2. Hermann, C.F. Validation Problems in Games and Simulations. Behavioural Science 12:216-31 (1967).

3. Mihram, G.A. Some Practical Aspects of the Verification and Validation of Simulation Models. Operational Research Quart. 23:17-29 (1973).

4. Thomann, R.V. Models for the Transport of DDT: Verification Analysis. Science 172:84 (1971).

5. Patten, B.C. A Primer for Ecological Modelling and Simulation. In B.C. Patten, ed., Systems Analysis and Simulation in Ecology, Vol. 1, Academic Press, New York, p. 1-22 (1971).

6. Tomovic, R., and Vukobratovic, M. General Sensitivity Theory. New York, American Elsevier (1972).

7. Meyer, C.F. Using Experimental Models to Guide Data Gathering. J. Hydraulic Div. Amer. Soc. Civil Eng. 10:1681-97 (1971).

8. Miller, D.R., Weidhaas, D.E., and Hall, R.C. Parameter Sensitivity in Insect Population Modelling. J. Theoret. Biol., 42:263-74 (1973).

9. Miller, D.R. Sensitivity Analysis and Validation of Computer Simulation Models. J. Theoret. Biol., to appear (1974).

10. Ackoff, R.L., and Sasieni, M.W. Fundamentals of Operations Research. New York, Wiley, p. 275 ff (1968).

11. Miller, D.R., ed. Distribution and Transport of Pollutants in Aquatic Ecosystems. Interim Report No. 2, National Research Council, Ottawa, Canada (1974).

TABLE I. Validation of Mosquito Model

Symbol	Estimate of Parameter	σ_i/p_i^o Estimated	$(\sigma_i/p_i^o)^2$ Estimated	R_i	$R_i(r_i/p_i^o)^2$
P_1	150	.067	44.89	.337	5.09
P_2	.75	.033	10.89	5.09	282.14
P_3	.90	.02	4.00	.865	2.99
P_4	.01	.167	278.89	.111	3.44
P_5	.10	.167	278.89	-0.00677	.01
P_6	.11	.15	225.00	-9.00415	.004
P_7	.70	.047	22.09	-.482	5.13
P_8	.10	.167	278.89	+0.0022	.001
P_9	.50	.067	44.89	+0.0737	.244
P_{10}	.50	.133	176.89	+0.0132	.03

Var (D) = 299.09

Std. Dev. of D = 17.29

Acceptable Range = 27.5 to 75.0

Conclusion: Acceptable

TABLE II. The seventeen parameters of the aquatic ecocystems model. Parameters are designated k_{ijm} where i=1 if the mechanism is a mass movement, i=2 for movement of pollutant. The subscripts j and m designate respectively source and destination of transfer; 1=water, 2=bottom sediment, 3=suspended material, 4=invertebrates, 5=plants, 6=fish. Values shown are for inorganic mercury. For units, see (11). $R_i/10$ is value of deviance for pollutant in fish produced by 10% change in a parameter.

Parameter	Base Value	Estimated σ_i/p_i^o	Measured $R_i/10$	$(R_i\sigma_i/p_i^o)^2$
k_{124}	0.05	0.5	-.073	0.13
k_{126}	0.056	0.2	+.944	3.56
k_{134}	0.28	0.2	+.060	0.01
k_{142}	0.045	0.5	+.006	0.00
k_{146}	0.042	0.5	+.932	21.72
k_{152}	0.005	1.0	+.0046	0.00
k_{155}	0.01	1.0	-.070	0.49
k_{156}	0.042	0.2	+1.760	12.39
k_{162}	0.135	0.1	-2.840	8.07
k_{166}	0.0011	0.2	.016	0.00
k_{212}	0.5	0.3	-.046	0.02
k_{213}	1.6×10^6	>2.0	-.068	1.85
k_{214}	50.0	1.0	-.015	0.02
k_{215}	5000.	0.8	+.736	34.67
k_{216}	21.6	2.0	+.013	0.07
k_{221}	0.91×10^{-3}	0.3	+.250	0.56
k_{231}	1.0×10^{-7}	2.0	10^{-9}	0.00

Square root of Total = 9.14

Estimated Acceptable = 20.0(roughly)

Conclusion: Borderline

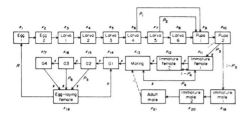

Figure 1.   Twenty-nine compartment model for mosquito
population in the presence of sterile males (8,9)

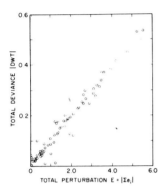

Figure 2.   Linearity of mosquito population model.

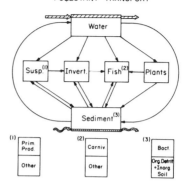

Figure 3.   Six-compartment model for transport of pollutants
in an aquatic ecosystem (11).

295

24

Reprinted from pages 313–325 of *Systems Analysis and Simulation in Ecology,*
Vol. IV, B. C. Patten, ed., N.Y.: Academic Press, 1976, 593pp.

The Validation Problem

HAL CASWELL*

DEPARTMENT OF ZOOLOGY AND DIVISION OF ENGINEERING RESEARCH
MICHIGAN STATE UNIVERSITY, EAST LANSING, MICHIGAN

I. Introduction

In this chapter I want to present some thoughts on model "validation," particularly as it confronts the ecologist constructing a large, complicated ecosystem model. Little of what I will offer would appear new to someone trained in the philosophy of science. My justification for presenting the material is its complete absence from the systems ecology literature. I will not give any dramatic solutions or new techniques; I will be satisfied to direct some attention to a facet of modeling that will become more and more important as more and more ecological modeling projects near completion.

* Present address: Ecology Section U-42, University of Connecticut, Storrs, Connecticut.

II. Normal Concept of Validation

What *is* "validation," as it is used in systems ecology? My dictionary sent me from "validate" to "verify," which was defined as "to establish the truth, accuracy, or reality of." This definition already hints at the roots of the ambiguity in our use of the word. As I will show later, the distinction between truth or reality and accuracy is an important one.

The use of the word in modeling acknowledges the basic scientific practice of comparing one's ideas with reality to see how they measure up. Although there are signs in this book that the situation is changing, the process in systems ecology currently seems to go something like this: A model, consisting of a set of differential or difference equations, is constructed, and the parameters estimated from field data or literature. The initial conditions of the model are set to correspond with the initial point of some sequence of data (hopefully different data from that used to evaluate the parameters), and the resultant solution of the model generated by simulation or analysis. This solution is compared with the data sequence: if the agreement is good, the model is validated; if not, the usual procedure is to modify the model—perhaps by changes in parameter values, perhaps by more basic structural changes—until agreement is obtained.

III. A Model Paradox

I want to discuss briefly an example that highlights one of the problems with this approach to validation. It is taken from an interesting series of papers by von Foerster *et al.* (1960, 1961a,b, 1962) and von Foerster (1966). The problem at hand is to model the growth of the human population of the entire world.

Let $N(t)$ represent the total world population at time t. Then we can write

$$dN(t)/dt = f(N)N(t), \tag{1}$$

where the growth rate function $f(N)$ expresses the effect of density on population growth. This much is ecological old hat; one of the classic derivations of the logistic equation follows by letting $f(N)$ be a decreasing linear function of N. This is a sensible first approximation for a group of organisms whose only interaction is to interfere with one another in a scramble for a limited supply of resources. But, reasons von Foerster, this is not the case with man at all. The logistic model describes a population whose members take part in a multiperson, zero-sum game among themselves for resources. Man's ability to construct societies, civilizations, and technologies turns the situation into a two-person game, man versus nature, and man's success varies directly with population size. Under this hypothesis $f(N)$ is not a decreasing but an increasing function of N, for which von Foerster chooses

$$f(N) = aN^{1/K}, \qquad K \simeq 1, \tag{2}$$

where a is a constant and K is slightly less than one. This is still ecologically reasonable; in fact, von Foerster is not alone in proposing $f(N)$ functions with at least some regions of positive slope (e.g., Slobodkin, 1953; Odum and Allee, 1954; Rosenzweig and MacArthur, 1963; Kilmer, 1972). The resulting equation is

$$dN(t)/dt = aN(t)^{1+1/K}, \tag{3}$$

which has the solution

$$N(t) = N(t_0) \, [(t^* - t_0)/(t^* - t)]^K, \tag{4}$$

where t^* is a constant whose importance will become apparent shortly.

Very well then, here we have a model of an ecological system. A much simpler model, to be sure, than any ecosystem model, but one which should be amenable to the same validation procedure. Von Foerster and his colleagues estimate the parameters in the model from 24 estimates of world population size, carefully screened for mutual independence, over the last 2000 years or so. Figure 1 shows the agreement between the model and the data. Considering the magnitude of the sampling error which must be inherent in this data, the agreement is outstanding. More importantly, there is no sign of any systematic deviation from the straightline relation predicted by the model.

Table I shows another indication of the accuracy of the model. The United Nations' median projection of world population size for the year 2000 A.D. is compared with the prediction based on Eq. (4). This projection is based on far more detailed demographic data than is Eq. (4). Yet, as the U.N. accumulates more and more information, its projections seem to be converging to the prediction made by von Foerster's model.*

TABLE I

	U.N. estimate of population at 2000 A.D., estimate made in:				Estimate from von Foerster's model, made in 1960
1950	1957	1958	1959	1970	
3.20	5.00	5.70	6.20	6.49	6.91 (billions)

*Note added in proof: As of 1975, there is additional evidence of the accuracy of Eq. (3). The Population Reference Bureau's estimate for world population in mid-1975 is 3.97 billion; von Foerster's prediction, made 15 years earlier, is 3.65 billion. We are actually ahead of schedule on the way to Doomsday. See J. Serrin (1975).

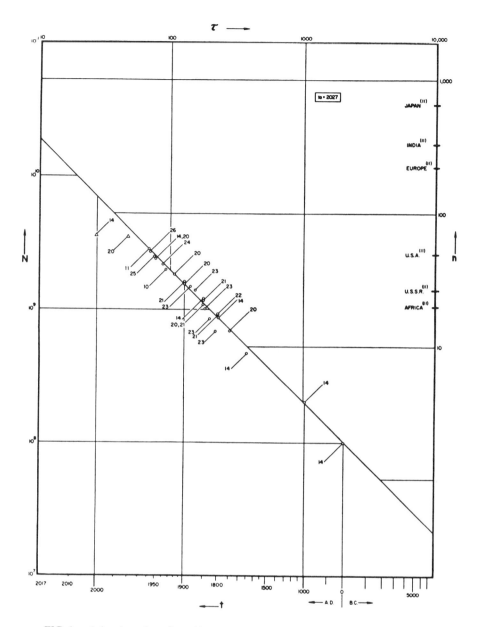

FIG. 1. A log–log plot of world population size N against time t (note the reversed scale for t), compared to the straight line predicted by Eq. (4). The upper abscissa measures time in units of $\tau = t^* - t$; the right-hand ordinate expresses population density (individuals per square mile) n, with density estimates for various countries in 1958 indicated for comparison. The triangular points on the graph are not data points, but projections made by various authors. The numbers indicate references in von Foerster's original paper. Reproduced with permission from von Foerster *et al.* (1960). Copyright 1960 by the American Association for the Advancement of Science.

316

So far, so good. The agreement of the model and available data is satisfactory; certainly it has passed the usual validation procedure. But so far we have examined only its accuracy; what can we say about its truth or reality? A glance at the denominator in Eq. (4) reveals that this model is a member of a class of models which reach, not approach but *reach*, infinite values in finite time. When $t = t^*$, $N(t) = \infty$. The parameter values obtained by von Foerster *et al.* (1960) allow t^*, "Doomsday" as they call it, to be estimated. Their estimate places it at 2026.87 A.D. ± 5.5 years. If their model is correct, the number of people on earth will become infinite in 50 years.

So we are faced with a quandary. This model gives an excellent fit to data, but generates projections that seem unbelievable. (Systems ecology is not devoid of similar examples; consider a donor-controlled compartment model of a food chain which produces solutions that agree with the data, but which also claims that the system is completely insensitive to variations in the top trophic level.) Has the model of the world population been validated or not?

IV. Two Purposes of Modeling

The problem, which lies at the root of this quandary, is one of purpose. The purposes for which a model is constructed have much to say about how important accurate predictions and a good fit to data are in the process of validation.

Models of systems are systems themselves, the interacting components of which are mathematical variables and expressions. They are, moreover, man-made systems, "artificial" systems in the terminology of Simon (1969). The construction of such artificial systems is a design problem, and the process of design is in essence a search for agreement between properties of the artificial system and a set of demands placed on it by the designer (Simon, 1969; Alexander, 1964; Chermayeff and Alexander, 1963). It is impossible to evaluate the success or failure of a design attempt without specification of these demands, the task environment in which the artificial system is to operate. Validation of a model is precisely such an evaluation, so we need to consider the purposes for which models are constructed.

In particular I will distinguish between (i) models that are constructed primarily to provide accurate *prediction* of the behavior of a system, and (ii) models that, as scientific theories, are attempts to gain *insight* into how the system operates. Both the means and the ends of what we call "validation" are different for these two types of models.

A. Models for Understanding

The goal of scientific hypotheses or theories is to increase our ability to explain how nature works. "Theories," says Popper (1959), "are nets cast to catch what we call 'the world'. . . . We endeavor to make the mesh ever finer and finer." In logical form, theories are universal statements, claims that something is always and/or everywhere true. Philosophers of science are fond of the example, "All ravens are black."

Figure 2 outlines the statements involved in a systems theoretical model. Statements are made concerning the identity and behavior of objects in the system and describing the interactions between objects. These statements are embodied in mathematical representations as "free-body models" and "constraint equations," respectively (Caswell et al., 1972). Note that the mathematical representation of a particular set of statements is not necessarily unique. The free-body models and constraint equations are combined to give a mathematical representation of the behavior of the entire system. Conclusions about the system are derived from this representation by deductive logic in the form of mathematical analysis or simulation. As in the example above, these

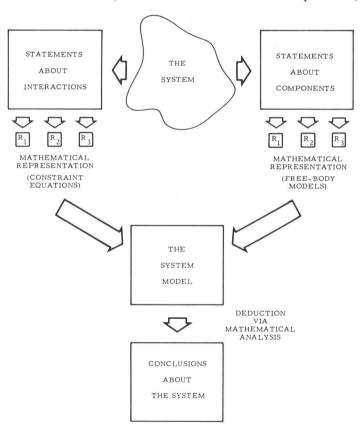

FIG. 2.   The theoretical statements involved in the construction of a system model.

conclusions include not only solutions of the model but more qualitative aspects dealing with limiting behavior, sensitivity, etc.

For these theoretical models the problem of validation is simply the old philosophical problem of induction—how can one logically infer the truth of universal statements from observations of particular phenomena? Popper (1959, 1963) has studied this problem in depth, and concluded that in fact there is *no* way to show a theory or any universal statement to be true via induction. No matter how many black ravens we see, the next one may be an albino. More importantly, he makes a convincing case that the procedure of science should be quite the opposite; the scientist's energies should be devoted to attempts to *invalidate* his theories. This invalidation is logically possible because one can deduce, from the universal statements of the theory and certain singular statements defining the details of the situation ("initial conditions"), sets of conclusions which can disagree with particular observations. If this happens, the universal statements from which the conclusions were derived have been invalidated. Science, Popper maintains, is a sequence of "conjectures and refutations," hypotheses or theories put forward by the imagination of the scientist and tested until they are refuted. A theory inherently incapable of refutation belongs not to science, but to metaphysics.

B. Models for Prediction

The process of evaluating models by attempting to refute them needs to be looked at in more detail, but first I want to examine the validation problem as it applies to another class of models. These I refer to as *predictive* models, models designed primarily to provide accurate quantitative predictions of the behavior of the system.

Predictive models crop up regularly in situations where one's livelihood depends on the ability to make accurate predictions—in applied sciences, in resource management, in engineering. The important point is that the truth or reality of the model is never at issue. You can navigate on the basis of a model of the universe that ignores relativity, survey as if the world were flat, and build a telescope without worrying about whether light is a set of waves, a stream of particles, or something more mystifying. Perhaps the canonical form of purely predictive model building is stepwise multiple regression, where variables are included in or rejected from the model solely on the basis of their contribution to its accuracy. Figure 3 is an outline of the construction of a purely predictive regression model, written with industrial application in mind. Nowhere is there any hint of testing the truth or reality of the model. In fact, as Popper (1963) points out, more often than not such testing is redundant, since the model is known in advance to be false. Instead, our concern is to evaluate its accuracy and the range of conditions over which it is useful.

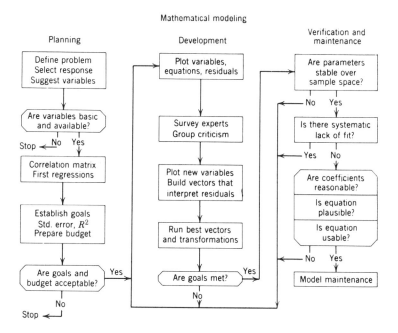

FIG. 3.   A flow chart outlining steps in construction of a predictive model by multiple regression analysis. From Draper and Smith (1966) with permission.

V. Corroboration versus Validation

"Validation" of a predictive model, then, means something very different from "validation" of a scientific theory. In the first case the goal is to determine when and where the model is a useful predictor. In the second case the goal is to attempt to refute the model, and our confidence in its statements increases as it passes more and more severe tests. To distinguish these two approaches I will henceforth use Popper's term "corroboration" to refer to the theoretical case:

> So long as a theory withstands detailed and severe tests and is not superseded by another theory in the course of scientific progress, we may say that it has proved its mettle or that it is *corroborated* (Popper, 1963, p. 33).

I will reserve "validation" for the case of the predictive models.

What then of von Foerster's world population model? If it is interpreted as a predictive model, it seems to be doing reasonably well; it is at least more highly validated than either the exponential or logistic models. If it is a theoretical model, it may be considered refuted, unless we are willing to refute instead some highly corroborated theories, which maintain that there is only a finite amount of matter in the universe. To the extent that it mixes those purposes we are left in confusion. Many ecological modeling efforts do in fact mix the two purposes. This is particularly true of the International Biological Program (IBP) modeling activities, for the IBP contains the mixture in its basic philosophy. This is not necessarily bad, and at any rate it is unavoidable. The distinction should be kept in mind though; allowances should be made for the existence of models (or segments of models) that have been refuted but validated, or corroborated but invalidated.

There is a very important practical difference between corroboration of a theoretical model and validation of a predictive model. As Fig. 3 shows, it is possible to establish, a priori, criteria that allow one to stop and say that a predictive model is satisfactorily validated. These criteria can be stated in terms of the error of prediction, the proportion of the variance explained, the cost of improving the model versus the cost of a given level of error, etc. There is no way this can be done for a theoretical model. The process of testing must continue indefinitely, since no matter how many tests the model has passed, it may fail the next one.

VI. Testing Theoretical Models

I want to return now to the process of testing theoretical models. The usual procedure of "validation" described earlier can be a highly inefficient way of testing a model. Most ecological models, and in particular ecosystem models, are very complex, at least in terms of the number of variables and parameters they contain. This makes it difficult to know when to reject a model by comparing output time series with observed data. Statistical criteria of rejection are usually not applicable, and without such criteria it is difficult to decide just *how much* disagreement between model and data is required for rejection. But the corroboration of a theoretical model increases with the severity of the tests it passes. If we want highly corroborated models, statements about ecosystems in which we can have a high degree of confidence, then we need approaches that will reduce to a minimum the uncertainty about rejection of a false model. Some suggestions follow.

First, don't test entire models; test them in pieces (the component models *and* the interaction constraints). This not only reduces the complexity of the models being tested, it has other advantages that will become apparent shortly. Secondly, theories are not to be tested in a vacuum, but are to be compared to alternative theories (Popper, 1959). Platt (1964) has given the systematic application of this technique the name "strong inference." He outlines the procedure as follows:

1. devising alternative hypotheses,

2. devising a crucial experiment, the outcome of which will exclude at least one of the hypotheses,

3. carrying out the experiment, and

4. returning to 1, refining the problem further and further.

Platt claims that the different rates of progress in different branches of science are due in large measure to the frequency with which their scientists practice strong inference. While it may often be difficult to decide if the behavior of a model is aberrant enough to warrant rejection, it is much easier to compare the results of two and decide which is better. The almost total neglect of the construction of alternative models in systems ecology is no doubt a result of our training. Most ecologists encounter the idea of alternative hypotheses only in statistics. In statistics, the construction of alternative hypotheses demands little or no ingenuity beyond the decision between a one-tailed or two-tailed test. When dealing with theories on the order of complexity of ecological models, developing alternative models demands as much ingenuity as constructing the initial model, but it is effort well-spent.

A third point—don't overlook the role of designed experiments in testing theoretical models. The conclusions of a theory that provide the most severe tests, and hence the highest degree of corroboration, are those that are the most surprising; that are expected on the basis of the theory in question, but *not* on the basis of existing or alternative theories (Popper, 1959, 1963). Such unexpected conclusions are not likely to involve time series of model output. Any pair of alternative models will both generate output, and if the modeling process has proceeded very far the outputs can be expected to be quite similar. The surprising consequences of a model may often be qualitative, or at least testable in qualitative fashion. Conclusions concerning the number, position, and stability properties of equilibria, or the insensitivity of certain variables to others (cf., Blalock, 1971) would seem to be prime candidates for this type of testing. For example, a test comparing a donor-controlled compartment model with a Volterra-style model (using cross-product terms) might be carried out much more rigorously by comparing the sensitivity of the herbivore level to changes in predator abundance than by comparing output time series from the two models with data.

The corroboration of a theoretical model demands that it be subjected to as severe a set of tests as possible. Toward this end we can test component and interaction models rather than entire system models, and devise explicit alternative hypotheses and crucial distinguishing experiments. But after this, what? The process of scientific development outlined by Popper and Platt is cyclical; each refutation must be followed by a new set of conjectures. It is not enough to reject a model and throw it away. This would be a waste of the insight that went into the model in the first place, and would quickly drain the imagination of any scientist. Instead we need to be able to pin the blame, at least tentatively, on some portion of the model. By changing this portion and continuing the testing procedure we can incorporate what was learned from the experiments or observations that led to rejection.

In the case of von Foerster's world population model the blame can be fixed relatively easily. We have enough experience with alternative models of similar form to guess that the growth rate function, $f(N)$, is responsible for the conclusion of infinite population size. We would propose then, as an alternative model, that at some critical value of N, $f(N)$ will change from $aN^{1/K}$ to some other function. This is in fact von Foerster's interpretation. He points out (von Foerster *et al.*, 1961a) that, in physical systems described by such equations, the values do not actually become infinite. Before that happens the system undergoes a radical change—a change of phase, ionization, breakdown, an explosion—in other words, the system breaks, so that it no longer exhibits the same behavior.

The ease with which we can tentatively attribute the refutation of von Foerster's model to a particular statement is due to its small size. It would be impossible to do the same with an ecosystem model in which a behavior is the result of the interaction of literally hundreds of variables, parameters, and functions. This is another reason to conduct tests on portions of the model rather than the whole thing. This may be the strongest reason, for without the ability to recycle our testing procedure we have gained nothing by refuting a model. The use of explicit alternative models is also helpful. If two models differ in only some of their statements, differences in their behavior can be at least tentatively attributed to the differing statements.

I have encountered the suggestion that sensitivity analysis be applied to this problem. The idea is that the parameters to which the misbehaving variable is most sensitive should be prime candidates for adjustment to correct the model. This makes some sense if the model is purely predictive, but none at all in the case of a theoretical model. In the latter situation such alterations represent ad hoc modifications of the statements of the model, and moreover assume that only a small subset (those dealing with parameter values) of the statements are open to correction.

It might be objected that if the corroboration process I have outlined was followed rigorously, systems ecology would be brought to a grinding halt. No one, it might be claimed, can build a model of an ecosystem that cannot be refuted almost immediately. In one sense this is true, but in another important sense it is not. I alluded earlier to the possibility that a model might be invalidated as a predictor but *not* refuted as a theory. The conclusions of a theoretical model that are usually tested are those deduced from one of the mathematical representations of the theoretical statements (see Fig. 2). This may force us to deal with far more detail than our statements intend. Suppose the statement "the variable X varies in the same direction as the variable Y" is embodied in the mathematical representation $X = a + bY$. This representation might be a poor predictor of X, and a test applied to the mathematical representation might result in refutation of the model. But a test of the *statement* itself might corroborate it. Efforts could then be directed to incorporating more detail in the statement. The point is that we must be careful to keep in mind just what we are testing when we attempt to refute a theoretical model.

There is another side to this coin which is worth mentioning in passing. The conclusions from a model might be a result of the particular mathematical representation chosen (see, e.g., May, 1973) rather than of the statements. Because he deals with relatively simple mathematical representations of rather subtle ideas, Levins (1968) has been particularly concerned with this problem. He refers to results which can be obtained from several different mathematical representations as "robust theorems." This is an internal check on the model, not an attempt to refute it by comparison with the real world; hence it is not really a form of corroboration.

One often gets the impression that "validation" is something to be done to a model after it is completed. The outline I have presented suggests that generating alternative models, devising tests, and revising models should be carried out, component by component and interaction by interaction, while the model is being constructed. It also speaks strongly for constructing separate models of the components and interactions that make up the system, a point which has been made before for other reasons (Caswell *et al.*, 1972).

VII. Conclusion

I hope that this discussion of the validation and corroboration of models will be useful to systems ecologists engaged in actual modeling efforts. We need models that do more than accurately describe ecological systems; we need models that represent steps toward a theory of ecosystem dynamics. We won't get them until we begin treating our models like theories and subjecting them to rigorous tests. I would like to see alternative models and the results of comparative tests set out explicitly in modeling publications. At this point we have as much to learn from our failures as our successes.

Acknowledgments

This research was supported by the National Science Foundation through Grant GI-20 and a Graduate Fellowship.

REFERENCES

Alexander, C. M. (1964). "Notes on the Synthesis of Form." Harvard Univ. Press, Cambridge, Massachusetts.

Blalock, H. M. (1971). "Causal Models in the Social Sciences." Aldine-Atherton, Chicago, Illinois.

Caswell, H., Koenig, H., Resh, J., and Ross, Q. (1972). *In* "Systems Analysis and Simulation in Ecology" (B. C. Patten ed.), Vol. II, p. 3. Academic Press, New York.

Chermayeff, S., and Alexander, C. M. (1963). "Community and Privacy." Doubleday, Garden City, New York.

Draper, N. R., and Smith H. (1966). "Applied Regression Analysis." Wiley, New York.

Kilmer, W. L. (1972). *J. Theoret. Biol.* **36**, 9.

Levins, R. (1968). "Evolution in Changing Environments." Princeton Univ. Press, Princeton, New Jersey.

May, R. M. (1973). *Amer. Natur.* **107**, 46.

Odum, H. T., and Allee W. C. (1954). *Ecology* **35**, 95.

Platt, J. R. (1964). *Science* **146**, 347.

Popper, K. (1959). "The Logic of Scientific Discovery." Harper, New York. (Originally published in 1935 as *Logik der Forschung.*)

Popper, K. (1963). "Conjectures and Refutations: The Growth of Scientific Knowledge." Harper, New York.

Rosenzweig, M., and MacArthur, R. (1963). *Amer. Natur.* **97**, 209.

Serrin, J. (1975). *Science* **189**, 86.

Simon, H. A. (1969). "The Sciences of the Artificial." M.I.T. Press, Cambridge, Massachusetts.

Slobodkin, L. B. (1953). *Ecology* **34**, 430.

von Foerster, H. (1966). The Numbers of Man, Rep. #13.0, Biological Comput. Lab., Univ. of Illinois, Urbana, Illinois.

von Foerster, H., Mora, P., and Amiot, L. W. (1960). *Science* **132**, 1291.

von Foerster, H., Mora, P., and Amiot, L. W. (1961a). *Science* **133**, 936.

von Foerster, H., Mora, P., and Amiot, L. W. (1961b). *Science* **133**, 1931.

von Foerster, H., Mora, P., and Amiot, L. W. (1962). *Science* **136**, 173.

25

Reprinted from pages 63–71 of *New Directions in the Analysis of Ecological Systems, Part 1*, G. S. Innis, ed., Simulation Councils Proc. Ser., Vol. 5, No. 1, LaJolla, Calif.: Simulation Councils, Inc., 1975, 132pp.

The importance of validation in ecosystem analysis

by

J. B. Mankin, R. V. O'Neill,
H. H. Shugart, and B. W. Rust
Environmental Sciences Division
and
Computer Sciences Division
Oak Ridge National Laboratory
Oak Ridge, Tennessee 37830

ABOUT THE AUTHORS

It is difficult to write a brief, personalized biographical sketch for four authors, but we shall make the attempt. J. B. MANKIN is a native of Chattanooga, Tennessee, and he has a PhD in electrical engineering from the University of Tennessee. R. V. O'NEILL is a native of Brooklyn, New York, and he has a PhD in zoology from the University of Illinois.

H. H. SHUGART is a native of El Dorado, Arkansas, and he has a PhD in zoology from the University of Georgia. B. W. RUST is a native of Bells, Tennessee, and he has a PhD in astronomy from the University of Illinois. All four authors are employed at Oak Ridge National Laboratory, Oak Ridge, Tennessee, where Rust and Mankin are in the Computer Sciences Division and O'Neill and Shugart are in the Environmental Sciences Division.

We are all members of the Systems Ecology Group sponsored by the Environmental Sciences Division. This group is involved in many diverse areas of ecological modeling such as primary production, consumption and decomposition processes, plant-soil-water interactions, and ecosystem theory and analysis. This group is composed of people from diverse backgrounds in mathematics, engineering, and the biological and physical sciences, and it is structured informally to provide a broad spectrum of expertise to be brought to bear on environmental problems. The group is characterized by team effort and close cooperation among its members during the infrequent periods when we are all speaking to each other. We had some difficulty in choosing a leader, but this problem was overcome when the second author took up karate. At the meeting to decide the ordering of authors, however, the first author brought a baseball bat.

SUMMARY

This paper attempts to develop objective criteria for validating ecosystem models. Application of these criteria requires precise definition of some commonly used terms. A valid model has no behavior which does not correspond to system behavior; a useful model predicts some system behavior correctly. Since no model is perfect, available ecosystem models may be described as invalid but useful models. Model validation thus requires an objective evaluation of the performance of the model. The following performance criteria are proposed. Adequacy is the fraction of system behavior which is duplicated by the model, and reliability is the fraction of model behavior which duplicates system behavior. Some uses of these criteria in model evaluation are suggested, and Bayes' theorem is used to show how a model of adequacy 0.5 or greater may be used to reduce sample size in an experimental program.

INTRODUCTION

Validation of ecosystem models is a challenging problem for the modeler and the difficulty of validation is a major deterrent to acceptance of the modeling approach by experimental ecologists. A significant literature exists on the validation problem in other fields,[3,5,9,10,19] but little has been written about validation of ecological models.[8]

Validation concerns a model's credibility and accuracy. Do model predictions correspond to ecosystem behavior? On its face the question appears simple — though possibly expensive — to answer. For example, we might compare model predictions for some set of conditions to measured values and evaluate the ade-

DEFINITIONS

a   "Adequacy" of a model, the number of data points in Q divided by the number of data points in S (roughly, the fraction of the system that is correctly modeled)

M   Model outputs for a particular experiment under the conditions defined for S

P   The set of properties observable in the system

Q   Model outputs M that match data S, i.e., intersection of M and S

r   "Reliability" of a model, the number of data points in Q divided by the number of data points in M (roughly, the fraction of the model's output that is correct)

S   Data which are gathered under defined conditions and which represent measurable behavior related to a specific problem

X $M - Q$. A model is valid only if $X = \phi$

µ   A measure of volume in hyperspace

φ   The null set

quacy of the simulation, but such a comparison does not generally validate the model.

The problem is more complex than it would first appear to be; the present paper will attempt to clarify the validation issue. We shall not explicitly treat statistical problems of "goodness of fit" between predictions and measurements. Rather, we shall focus on the role of the models in ecosystem research and the evaluation of individual models, with emphasis on dynamic ecosystem models which explicitly consider processes that regulate and control the system.[1,2,15,22,26,28] However, much of the discussion will be relevant to dynamic models for individual populations or processes in the ecosystem.

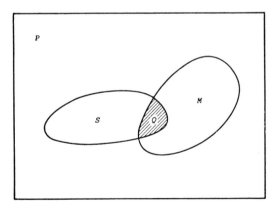

Figure 1 - Venn diagram illustrating the universe P of properties observable on an ecosystem. S is a set whose elements are data from measurements on the ecosystem. M is a set of model responses. The intersection of S and M ($S \cap M$) is denoted Q.

VALIDATION AND MODELING OBJECTIVE

Virtually every aspect of model-building depends on the purposes for which the model was designed. The system variables and the mathematical representation which are used and the data required for quantification will be dictated by the aspects of system behavior to be investigated in the project. One would expect, therefore, that validation would also be affected by objectives.

We may achieve some of the objectives of a model without validating it. For instance, many ecosystem models are developed for purposes other than predicting system behavior. A model can be a rigorous statement of a working hypothesis about system behavior. The primary purpose of such a model is to guide research rather than make predictions. Models have been used in large interdisciplinary studies (such as the biome studies of the U. S. International Biological Program) to identify the interconnections between diverse studies. Indeed, some authors[11,12,23] have maintained that rigorously systematizing a concept in the form of a model is the most valuable contribution which modeling can offer to ecology.

In other cases, ecosystem models summarize large data sets rather than predict behavior. Some excellent examples exist of this application of ecosystem models[17,20,25] which permit an extensive data base to be summarized onto a single printed page. Similarly, a model can use linear regression equations so that slopes and intercepts can be substituted for extensive tabulation of individual data points.

In still other cases, a model is developed because actual field experiments cannot be performed. For example, O'Neill and Styron[21] investigated the effects of radiation on a field population of soil invertebrates. The problem involved high levels of ionizing radiation imposed at numerous periods during the year; such drastic experiments could not actually be performed. Therefore, validation data for a model could not be obtained. Indeed, if validation data had been obtainable, the model might never have been developed in the first place.

For the objectives outlined above, the question of validation is secondary, at best. If the model is a guide for research efforts or a summary of data, then its usefulness is the only matter in question. If the model is designed to "predict the unmeasureable," the model must be examined for reasonableness and completeness without reference to validation data.

The points raised above may still apply even when the primary objective of a model is to predict or to understand (which implies prediction). A large-scale ecosystem model usually has several implicit objectives which may include some of the objectives discussed previously. Therefore, one must be skeptical of evaluations based solely on ability to duplicate system behavior. A model may be very useful even if it can't predict behavior.

Given the present understanding of ecosystem function, no ecosystem model can duplicate ecosystem behavior under all conditions. Models only provide "best possible" simulations. Ecosystem models will be useful for the purposes outlined above for some time to come, and their greatest value may be to indicate where knowledge is lacking.

since the model must omit so much. A classic example of the invalid-useful model is provided by the Lotka-Volterra equations[16] in that some parameter values produce patently unrealistic simulations. There also appears to be system behavior which the model cannot simulate.[27] And yet $Q \neq \emptyset$ since examination and analysis of this model continues to provide insights and suggest new research.[16]

Since ecosystem models can be presumed to be imperfect, the real validation question concerns *how much* the model can predict correctly, i.e., the size of Q relative to M and S. The set of all the results for all possible values of the model parameters is a complicated set in a multidimensional space. The overlap of S and M is the volume of their intersection; this is schematically represented by the area of Q in the Venn diagram of Figure 1. A Venn diagram is a simple method of representing a complicated phenomenon; area in a Venn diagram is the two-dimensional equivalent of volume in higher dimensions. Let us define μ as the measure of volume in a generalized hyperspace. Therefore, $\mu(S)$ would be the volume of the system and $\mu(M)$ would be the volume of the model. We can now introduce two more definitions.

(3) Model reliability r will be defined as $\mu(Q)/\mu(M)$.

Reliability is defined as the fraction of M which is contained in S, the fraction of model simulations which match system behavior. By emphasizing reliability in the choice of a model we minimize the probability of making an error in accepting model predictions.

(4) Model adequacy a will be defined as $\mu(Q)/\mu(S)$.

Adequacy is defined as the fraction of system behavior which can be simulated with the model. Thus, the larger the fraction in Definition (4), the more adequate the model will be for addressing present study objectives.

The concept of volume in three-dimensional space is easy to understand. However, the concept of volume in an abstract hyperspace is less comprehensible. We would like to make this concept simpler. Our knowledge of nature comes from observations and inferences drawn from these observations. If we conduct n_s experiments on the system represented by the set S, our knowledge of S comes totally from these experiments. Similarly, we can observe the model response n_m times. One possible measure of the volume of $Q = S \cap M$ is the number of agreements n_q between model and experiment. Therefore, concrete and practical definitions of adequacy and reliability would be

$$a = \mu(Q)/\mu(S) = n_q/n_s$$

and

$$r = \mu(Q)/\mu(M) = n_q/n_m$$

We can now pose the validation question in terms of reliability and adequacy.

CHOOSING BETWEEN ALTERNATIVE MODELS

Goodall[8] has pointed out that validation may actually involve a choice between alternative models. The question may be not whether a model is *valid*, but which of two models is *better*. Considered simply, we would compare outputs from both models with a single data set S and choose the model with the closest fit.

In some cases, study objectives may require a completely adequate model (i.e., $S \subset M$, all of S is contained within M, but M may be larger than S). For example, to examine the *worst imaginable* case in an environmental assessment problem, the model should simulate all possible behaviors of the system, even though it may also possess unrealistic behavior. Even the unrealistic behavior may be within some tolerable limit (e.g., a legal standard).

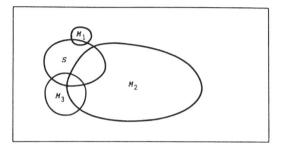

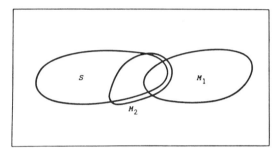

Figure 2 – Venn diagrams representing models of differing reliabilities and adequacies with respect to the set S. In (a) three models which simulate different parts of the system are illustrated. Model 1 (with outputs M_1) is a set disjunct from either of the other two models. Models 2 and 3 are models which have some simulation capabilities in common. In (b) two models are illustrated in which Model 2 is both more adequate and more reliable and also contains all the responses M_1 of Model 1 which intersect the system S.

In other circumstances, one might prefer a completely reliable model ($M \subset S$; i.e., all model responses match system responses, but not necessarily all of the system responses). Such a model cannot produce unrealistic behavior, but might describe very little of the total system behavior. This is frequently the rationale of using a model which is linearized about a system operating point and then considering only small perturbations.

However, in most cases, the comparison of models involves situations such as those depicted in Figure 2. In the upper portion of the figure, we have represented three models. Models 1 and 2 (with responses M_1 and M_2) do not overlap. Model 1 captures less of the system's behavior (is less adequate) but has less unrealistic behavior (is more reliable) than Model 2. Although we might choose between the models based on adequacy or reliability, the models each predict dif-

PREDICTION AS A MODEL OBJECTIVE

Of course, we frequently use ecosystem models to predict system behavior, and we must establish the credibility of these models. We shall proceed by establishing terminology.

Let P be the set of properties observable in an ecosystem. The observations may be direct (i.e., measurements) or indirect (e.g., model predictions). An element of P might be a vector representing the behavior of system components or the behavior of the model in time. Consider a model whose parameters are known only within certain tolerances. Each parameter has a continuum of possible values; hence the model has a continuum of possible solutions (assuming all the other parameters are held constant). Such a complex solution set is not finite or even countable. Therefore, let

$$P = \{p_\lambda : \lambda \in \Lambda\}$$

where Λ is some abstract (not necessarily countable) indexing set. We can then let

$$S = \{s_\gamma : \gamma \in \Gamma \subseteq \Lambda\}$$

(where Γ is a proper subset of Λ) be the subset of P representing outcomes of direct measurements on the system relevant to the objectives of a particular study. An element of S would be the data from measurements taken from a particular experiment. The term *experiment* is used in a general sense to represent a specific set of inputs, initial conditions, system structure, etc., whether or not the researcher imposed these conditions.

The elements of S are data gathered under a defined set of conditions. S is defined relative to study objectives so that S does not represent all measurable behavior, but only that behavior related to a specific problem. Similarly, we can now represent model behavior by the subset $M = \{m_\delta : \delta \in \Delta \subseteq \Lambda\}$. The elements of M represent model output for a particular experiment conducted on the model.

Let $Q = M \cap S = \{q_\nu : (q_\nu \in M) \wedge (q_\nu \in S)\}$

$$= \{q_\nu : \nu \in N = \Gamma \cap \Delta\}$$

be the intersection of the sets M and S. Operationally, the statement that an element belongs to both sets implies some statistical test of "goodness of fit" between predictions made by the model and data. The present study is not concerned with this statistical problem, but any reasonable test would suffice (such as the model's outputs falling within the 95% confidence limits of the measurements). The sets P, M, S, and Q can be represented by the Venn diagram shown in Figure 1.

We are now in a position to propose some simple definitions that will be used in subsequent discussions:

(1) A model is *valid* if and only if $M - Q = X = \emptyset$,

where $\emptyset$ is the null set and $M - Q = X$ is the relative complement of S defined as

$$X = \{x_k : k \in K = \Delta \cap N^C\}$$

A model is valid if its behavior corresponds to system behavior under *all* conditions of interest. A model is considered invalid if we can devise an experiment in which the model's outputs disagree with system measurements within the specified area of interest.

(2) A model is *useful* if and only if $Q \neq \emptyset$.

The second definition indicates that a model is useful if it accurately represents some of the system behavior under consideration and useless if it does not.

These definitions clarify certain ambiguities associated with model validation. A model is invalid if experimental measurements exist which the model cannot duplicate. But this offers little information on predictive capabilities of a model for unmeasured aspects of system behavior. Alternately, if the model does duplicate the outcome of a *single* experiment, we cannot therefore assume the validity of the model for *every* aspect of system behavior. Furthermore, if the model cannot simulate all the ecosystem behavior, it may still be useful for parts of the problem.

Table 1

Possible combinations of the useful-useless and valid-invalid dichotomies in terms of the sets X and Q. ϕ is the null set.

	VALID	INVALID
USEFUL	$X = \phi$ $Q \neq \phi$	$X \neq \phi$ $Q \neq \phi$
USELESS	$X = \phi$ $Q = \phi$	$X \neq \phi$ $Q = \phi$

Table 1 contains the possible combinations of the valid-invalid and useful-useless categories. In the case of the valid-useful model M is a subset of S — ideally, M equals S, in which case M is a perfect model. In the case of the valid-useless model there are no elements of M which are in S and no elements exist which are not in S, implying the nonexistence of the model. This case will be of no further interest. In the case of the invalid-useless model, M and S are disjoint, i.e., all of the elements of M are outside S, and the model does not simulate any aspect of system behavior.

Of greatest interest to our present discussion is the invalid-useful model. In this case (Figure 1) some model behavior duplicates that of the real system and some does not. Since we can presume that any ecosystem model is imperfect, they all fall in the invalid-useful category.

Levins[14] has proposed a similar concept for models of animal populations. He points out that any model is useful ($Q \neq \emptyset$) since almost any plausible relationship probably exists. Yet every model is invalid ($X \neq \emptyset$)

ferent things about the system. If we want information about all the system's behavior, we should retain both models. We may then use each model in the specific area where it provides the best results. Models 2 and 3 share some common behavior. Model 3 is more reliable, while Model 2 is more adequate. Although we might choose one model over the other in some particular case, there is no obvious criterion for rejecting either model.

The only situation we can identify in which we clearly choose one model over another is represented in the lower portion of Figure 2. Model 2 is more reliable, more adequate, duplicates all realistic behavior of Model 1 and simulates some regions of S which Model 1 cannot. Apparently three conditions must be satisfied to justify rejection of one model in favor of another:

$$\text{a)} \quad a_1 < a_2$$

$$\text{b)} \quad r_1 < r_2$$

$$\text{c)} \quad Q_1 \subset Q_2$$

If these conditions are satisfied, Model 2 qualifies as a *better* model. But ascertaining that all three conditions are satisfied may be difficult or impractical. Therefore, the best decision is often to use more than one model.

IS THIS MODEL VALID?

We have attempted to pose the validation question in a new context, but the original problem still remains: how to judge the reliability and adequacy of a model? Answering this question requires consideration of efficient experimental designs.

We could design a single experiment to monitor all state variables for comparison with an ecosystem model. By reference to Figure 1, there are two possible outcomes of this design. Model output and system behavior may coincide in some statistical sense (i.e., $Q \neq \emptyset$, and the model is useful). On the other hand, model output and observed behavior may differ (i.e., $X \neq \emptyset$, and the model is invalid). Thus, the single large experiment can establish usefulness or invalidity, but cannot measure reliability or adequacy. Previous discussion has already established that a model can be assumed to be useful and invalid from *a priori* reasoning.

Ordinarily, we have some data for the system we are modeling. Although many parameters may be estimated from these data, model output corresponding to this original data set does demonstrate usefulness in the sense of Definition (2). In addition, a rigorous search of model behavior will ordinarily reveal patently unrealistic behavior for some parameter values, thus establishing the invalidity of the model. Such preliminary testing checks the logical consistency of the model itself and is usually referred to as *verification*.[6,18]

Choice of new experiments to test the model should attempt to delineate the boundaries of Q. An efficient procedure is to monitor a few variables for numerous experiments and thus to locate boundary points where information deviations between output and measurements become significant. If we assume that exploration with the model is far less expensive than the corresponding field measurements (which it

often is), then we should use the model to maximize the information gained by field experimentation. In particular, the model can be used to identify unlikely but reasonable model predictions. Such predictions will probably lie on the boundaries of Q. Testing these predictions will yield more information than testing predictions which are either clearly correct or clearly unreasonable.

In addition to the foregoing use of the model, it can be used to locate optimal monitoring points in the system. For example, under a simulated nutrient enrichment experiment, the model might predict a 10% increase in net primary production which would be difficult to verify in the field. But the model might also predict that consumer biomass should decrease rather than increase as might be expected intuitively. We can test this prediction by a relatively simple (and inexpensive) experiment which detects increase or decrease in consumer biomass.

Systematic exploration of the model's responses can help greatly in identifying those predictions which can be tested with the simplest possible experiments. It is generally not feasible to monitor all variables in a model representing a complex hypothesis since that is too laborious and expensive. Rather, the model should be explored to locate interesting implications of the hypothesis, which should then be tested. This process is directly analogous to "strong inference"[24] as practiced so successfully in molecular biology.

Viewed in this context, model validation is a process, not an event. Over a period of time, a series of carefully devised experiments makes it possible to designate the regions of system behavior which can be successfully simulated. Whatever manpower and budget levels are available, the search for simple significant experiments permits the maximum number of independent tests to be performed and maximum information to be gained about reliability and adequacy.

TESTING HYPOTHESES AND MODELS

As previously stated, a model often is an unambiguous statement of current knowledge about a system. A model reveals those areas which require more experimentation to fill gaps in our knowledge. This section considers the case in which we have a model and some prior knowledge about its adequacy and reliability and are interested in examining some new situation which the model describes. The object is to determine if we can use the model to reduce the number of experiments required to examine the new situation.

We shall use the symbol $\rightarrow$ to indicate agreement in some statistical sense; $\bar{x}$ indicates the outcome of a experiment (i.e., data), and $\hat{x}$ indicates a model result. Therefore, $\hat{x}_i \rightarrow \bar{x}_i$ and $\hat{x}_j \nrightarrow \bar{x}_j$ indicates that the ith model run (simulation) agrees with the ith experiment and the jth model run disagrees with the jth experiment.

We shall use the form of statistical inference known as *hypothesis testing*. More detailed discussions may be found in many texts, including Fisz,[7] Kirk,[13] and Conover.[4] The hypothesis to be tested is called the *null hypothesis* and is denoted by H_0, and the *alternative hypothesis*, denoted by H_1, is the negation (converse) of the null hypothesis. There are two ways of making an incorrect decision in hypothesis testing. A *type I error* is the error of rejecting a true null hypothesis. A *type II error* is the error of accepting a false null hypothesis. The possible outcome of a hypothesis test is illustrated by the following:

	Accept H_0	Reject H_0
H_0 is true	Correct decision Pr = 1 - α	Type I error Pr = α
H_0 is false	Type II error Pr = β	Correct decision Pr = 1 - β

The quantity α is known as the *level of significance;* we often want to obtain a desired level of significance or confidence (e.g., 95% confidence level corresponds to significance α = 0.05).

We shall consider the probability of making a *type I error*. Although failure to reject the null hypothesis means that we cannot distinguish the true situation from the null hypothesis, we shall speak of the acceptance or rejection of a true null hypothesis for clarity. Let A be the event that the null hypothesis is accepted when it is true, and A^C the event that the null hypothesis is rejected when it is true. Therefore,

$$P(A) = P(\bar{x} \to H_0 | H_0 \text{ is true}) = 1 - \alpha \quad (1)$$

and

$$P(A^C) = P(\bar{x} \not\to H_0 | H_0 \text{ is true}) = \alpha \quad (2)$$

Let B be the event that *the model results indicate that the null hypothesis should be accepted*, and let B^C be the event that the model results indicate that *the null hypothesis should be rejected*. In order to use the model to strengthen the hypothesis testing procedure, we must have some knowledge of the probability that the model will agree with the experimental evidence indicating acceptance or rejection of the null hypothesis. We have this from our *a priori* information about the adequacy and reliability of the model. Agreement can occur in two ways: the model can indicate acceptance of the null hypothesis when the experimental evidence indicates acceptance, and the model can indicate rejection of the null hypothesis when the experimental evidence indicates rejection. We will assume that these occurrences have equal probability. Therefore,

$$P(B|A) = P(\hat{x} \to H_0 | \bar{x} \to H_0) = a \quad (3)$$

where a is the adequacy as previously defined. We also need to determine $P(B|A^C)$. This is the probability of disagreement between model and system given that the result is an element of the system. Therefore,

$$P(B|A^C) = P(\hat{x} \to H_0 | \bar{x} \not\to H_0) = 1 - a \quad (4)$$

Can we use this *a priori* information to increase the probability of making a correct decision? Conversely, can we decrease the probability of making a *type I error*?

The answer is *yes, if the model is sufficiently adequate*. We need an estimate of the *a posterior* probability of accepting the null hypothesis when it is true, given that our model indicates that it should be accepted. By applying Bayes' theorem,[7] we have

$$P(\bar{x} \to H_0 | \hat{x} \to H_0) = P(A|B) = \frac{P(B|A)P(A)}{P(B|A)P(A) + P(B|A^C)P(A^C)}$$

Using equations 1, 2, 3, and 4, we have

$$P(\bar{x} \to H_0 | \hat{x} \to H_0) = \frac{a(1 - \alpha)}{a(1 - \alpha) + (1 - a)\alpha}$$

$$= \frac{1 - \alpha}{1 - \alpha + \frac{(1-a)\alpha}{a}}$$

$$= \frac{1 - \alpha}{1 - \left(2 - \frac{1}{a}\right)\alpha}$$

To improve our prior estimate we must satisfy the following conditions:

$$1 - \alpha \leq \frac{1 - \alpha}{1 - \left(2 - \frac{1}{a}\alpha\right)} \leq 1$$

This is equivalent to

$$\tfrac{1}{2} \leq a \leq 1$$

This result gives an estimate of the degree of adequacy a model must have in order to aid in experimental design, that is, $a \geq 0.5$. Note that the error involved in the estimate of $\hat{a}$ has been neglected in this section.

For example, the number of samples needed to estimate the mean of some ecosystem response to within a confidence limit d (following the classic technique of Stein[29]) is

$$n = t_1^2 \, s^2 / d^2$$

where

n = the number of samples needed

t_1 = the tabulated value of the t-statistic for the desired confidence level and for the degrees of freedom of the initial sample

d = the half-width of the desired confidence limit

s^2 = the sample variance.

s^2 has been estimated by some preliminary data on the ecological system; the size of this sample determines the degrees of freedom for the t-statistic.

Now, consider a case in which a model predicts a mean response within the range $\pm d$ of the sample mean obtained in the preliminary sample mentioned above. Then

$$n^* = t^* \, s^2 / d^2$$

where

n^* = the number of samples needed, given the agreement between the preliminary data and the model

t^* = the tabulated value of the t-statistic for the desired confidence region, for the degrees of freedom of the initial sample, conditioned by the prior model agreement.

Now, if $a \geq 0.5$, then $1 - n^*/n$ equals the reduction in sample size to estimate a given confidence inter-

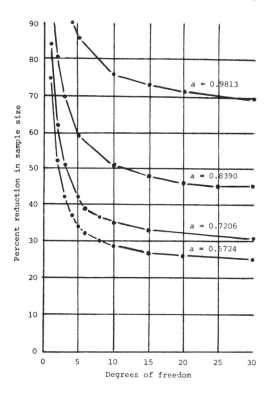

Figure 3 - Degrees of freedom in a preliminary sample versus percent reduction in sample size to obtain an estimate of a mean with some confidence range for models of different adequacies *a*. Sample size reduction calculated from Stein.[29]

val on a mean. Figure 3 plots this reduction in sample size against varying degrees of freedom in the initial calibration sample for four models of different adequacy. As one might expect, (a) the more adequate a model, the greater the reduction of sample size, and (b) the more degrees of freedom in the calibration sample, the less the reduction. The hyperbolic form of the curves indicates that the use of a model (even one of rather low adequacy) in cases in which the number of preliminary data samples is small can greatly reduce the number of subsequent samples needed. Using a graph such as Figure 3, we can (a) compare the cost of collecting a sample to the cost of developing a model with some given adequacy and (b) calculate the cost effectiveness of modeling. Figure 3 is in reality a rather special case of model application in hypothesis testing; at present the adequacies of ecosystem models are not known. Figure 3 shows that a model of adequacy *a* = 0.6724 gives about a 30% reduction in a sample of 15 or greater observations. Such a model could be expected to predict some ecosystem response incorrectly twice in a row about 10% of the time. Ecosystem analysis is approaching a stage in which models are being tested against reality in what are frequently called "validations." Testing these models against some single ecosystem response is very wasteful.

DISCUSSION

Thus far, we have emphasized the model as a finished product. In fact, modeling and validation are continuous processes in time. When we perform preliminary simulations with a computer-implemented model, which at the present state of the art is necessarily invalid, we find errors and correct them by redesigning the model or limiting the parameter values. Concurrently, new experimental data may indicate new and interesting regions of system behavior not included in the model as it is formulated, which calls for revising the model. The whole process of studying ecosystems can be viewed as a continuous attempt (never fully realized) to alter models and achieve $M = S$.

The critical question for the advancement of ecosystem analysis is not whether present models are valid. We can assume with a high degree of certainty that they are not. The critical question is whether or not model exploration linked with continuing field research is useful, i.e., whether or not it is a more efficient methodology than experimentation alone. The question is not whether our present models are valid, but whether we can utilize our invalid models to increase the efficiency with which we study ecosystems.

In many cases, the validation process is focused too directly on the mathematical or computer-coded entity called the "model." In reality the model merely formalizes unambiguously our insight into system behavior, and so the validation question can be divided into two parts. First, we may ask whether the model faithfully represents our insights. This is the traditional verification question and relates to the completeness and logical consistency of the formulation. Secondly, we ask whether our insights are adequate. This must be answered by comparing model output to system behavior. If there is less than perfect correlation, we cannot judge the model formulation as inadequate; rather, we must judge the present level of understanding as inadequate.

These considerations have led the authors to the conclusion that too much time and energy has already been expended on the "validation problem." The scientifically important question is not whether the present generation of models can simulate everything of interest about ecosystems. Our present level of understanding simply does not permit this. Let us admit that the models are invalid and ask whether the models can be applied in the sense of "strong inference" to provide a more rigorous and systematic approach to research. Let us dismiss the question: *Have you proven that your model is* valid? with a quick *No.* Then let us take up the more rewarding and far more challenging question: *Have you proven that your model is* useful *for learning more about the ecosystem?*

ACKNOWLEDGEMENT

This research was supported in part by the Eastern Deciduous Forest Biome, US-IBP, funded by the National Science Foundation under Interagency Agreement AF 199-40-193-69 with the Energy Research and Development Administration Oak Ridge National Laboratory, and in part by the U. S. Energy Research and Development Administration under contract with Union Carbide Corporation. Contribution No. 221 from the EDFB-IBP. Publication No. 749, Environmental Sciences Division, Oak Ridge National Laboratory.

REFERENCES

1   ANWAY, J.C.   BRITTAIN, E.G.   HUNT, H.W.
    INNIS, G.S.   PARTON, W.J.   RODELL, C.P.
    SAUER, R.H.
    Elm: Version 1.0
    Grasslands Biome Technical Report 156   Colorado
    State University   Fort Collins   1972

2   BLEDSOE, L.J.   FRANCIS, R.C.   SWARTZMAN, G.L.
    GUSTAFSON, J.D.
    PWNEE, a Grassland Ecosystem Model
    Grasslands Biome Technical Report 64   Colorado
    State University   Fort Collins   1970

3   COHEN, K.J.   CYERT, R.M.
    Computer Models in Dynamic Economics
    Quarterly Journal of Economics   vol. 75   1961
    pp. 112-127

4   CONOVER, W.J.
    Practical Nonparametric Statistics
    Wiley   New York   1971

5   CYERT, R.M.
    *A Description and Evaluation of Some Firm
    Simulations*
    *Proceedings of the IBM Scientific Computing
    Symposium on Simulation Models and Gaming*
    IBM Corporation   White Plains, New York   1966
    pp. 3-22

6   FISHMANN, G.S.   KIVIAT, P.J.
    The Statistics of Discrete-Event Simulation
    Simulation   vol. 10   1968
    pp. 185-191

7   FISZ, M.
    Probability Theory and Mathematical Statistics
    Wiley   New York   1963

8   GOODALL, D.W.
    Building and Testing Ecosystem Models
    In J.N.R. Jeffers, editor, *Mathematical Models
    in Ecology* (Blackwell Scientific Publications,
    Oxford, 1972), pp. 173-194

9   HERMANN, C.
    Validation Problems in Games and Simulations
    Behavioral Science   vol. 12   1967
    pp. 216-230

10 HOLT, C.S.
    *Validation and Application of Macroeconomic
    Models Using Computer Simulation*
    In J.S. Duesenberry, G. Fromm, L.R. Klein, and
    E. Kuh, editors, *The Brookings-SSRC Quarterly
    Econometric Model of the United States* (Rand
    McNally and North Holland Press, Chicago, 1965)

11 INNIS, G.
    *An Experimental Undergraduate Course in Systems
    Ecology*
    Bioscience   vol. 21   1971
    pp. 283-284

12 JEFFERS, J.N.R.
    *The Challenge of Modern Mathematics to the
    Ecologist*
    In J.N.R. Jeffers, editor, *Mathematical Models
    in Ecology* (Blackwell Scientific Publications,
    Oxford, 1972), pp. 1-11

13 KIRK, R.E.
    *Experimental Design: Procedures for the Behavioral
    Sciences*
    Brooks/Cole Publishing Company   Belmont, Califor-
    nia   1968

14 LEVINS, R.
    *The Strategy of Model Building in Population
    Biology*
    American Scientist   vol. 54   1966   pp. 421-431

15 MacCORMICK, A.J.A.   LOUCKS, O.L.   KOONCE, J.F.
    KITCHELL, J.F.   WEILER, P.R.
    *An Ecosystem Model for the Pelagic Zone of Lake
    Wingra*
    Eastern Deciduous Forest Biome memo report
    72-122   University of Wisconsin   Madison   1972

16 MAY, R.M.
    Stability and Complexity in Model Ecosystems
    Princeton University Press   Princeton, New Jersey
    1973   235 pp.

17 McGINNIS, J.T.   GOLLEY, F.B.   CLEMENTS, R.G.
    CHILDS, G.I.   DUEVE, M.J.
    *Elemental and Hydrologic Budgets of the Panamanian
    Tropical Moist Forest*
    Bioscience   vol. 19   1968   pp. 697-700

18 MIHRAM, G.A.
    *Some Practical Aspects of the Verification and
    Validation of Simulation Models*
    Operational Research Quarterly   vol. 23   1971
    pp. 17-29

19 NAYLOR. T.H.   FINGER, J.M.
    Verificaiton of Computer Simulation Models
    Management Science   vol. 14   1967   pp. 92-101

20 ODUM, H.T.
    *Summary: an Emerging View of the Ecological Sys-
    tem at El Verde*
    In H.T. Odum and R.F. Pigeon, editors, *A Tropical
    Rain Forest* (U.S. Atomic Energy Commission
    Division of Technical Information, Oak Ridge,
    Tennessee, 1970), pp. I-191-I-298

21 O'NEILL, R.V.   STYRON, C.E.
    *Applications of Compartment Modeling Techniques
    to Collembola Population Studies*
    American Midland Naturalist   vol. 83   1969
    pp. 489-495

22 PARK, R.A. *et al.*
    A Generalized Model for Simulating Lake Ecosystems
    Simulation   vol. 23   August 1974   pp. 33-50

23 PATTEN, B.C., editor
    Systems Analysis and Simulation in Ecology
    Academic Press   New York   1971   vol. 1   607 pp

24 PLATT, J.R.
    Strong Inference
    Science   vol. 146   1964   pp. 347-353

25 REICHLE, D.E.   O'NEILL, R.V.   OLSON, J.S.,
    editors
    Modeling Forest Ecosystems
    EDFB-IBP-73-7   Oak Ridge National Laboratory
    Oak Ridge, Tennessee   1973   339 pp.

26 SHUGART, H.H.   GOLDSTEIN, R.A.   O'NEILL, R.V.
    MANKIN, J.B.
    TEEM: a Terrestrial Ecosystem Energy Model for Forests
    Oecologica Plantarum   vol. 9   1974
    pp. 231-264

27 SMITH, F.E.
    Experimental Methods in Population Dynamics: a Critique
    Ecology   vol. 33   1952
    pp. 441-450

28 SOLLINS, P.   REICHLE, D.E.   OLSON, J.S.
    Organic Matter Budget and Model for a Southern Liriodendron Forest
    EDFB-IBP-72-2
    Oak Ridge National Laboratory   Oak Ridge, Tennessee   1973   150 pp.

29 STEIN, D.
    A Two-Sample Test for a Linear Hypothesis Whose Power is Independent of the Variance
    American Mathematical Statistics   vol. 16   1945
    pp. 243-258

DISCUSSION

Kickert: Your approach is only helpful when the logical sequence of research design has been followed, i.e., modeling, then field data collection in an experimentation context. What suggestions do you have when the modelers find themselves in a program in which the data collection has been begun by the program director from the start?

O'Neill: In a well-designed research project, systems analysis must play a role in all aspects of the study, including initial research design. However, a great number of variables enter into the initial decision process, and the resulting design may, therefore, be suboptimal from the analyst's viewpoint. The only advice I can offer is to continuously provide input and attempt to influence the design of the research.

Mauriello: What assurance is there that model output which might seem unlikely lies on boundaries of Q?

O'Neill: It seems reasonable to us that model exploration would reveal (1) clearly unreasonable output, (2) output that appears to represent known behavior of the system, (3) output which represents reasonable but questionably realistic behavior. We believe that experimentation on the third class of phenomena which are not clearly in Q or X represent the most information-rich and efficient approach.

Tracy: What use does one make of a useful model? Don't people build models to see if the set M never intersects S? That being the case, a useless model has been useful.

O'Neill: While it is true that the process of constructing, testing, and rejecting a model yields information, most commonly a model developed by competent scientists will not result in $Q = \emptyset$. In our opinion, the greatest utility of the model comes with the use of a tested model in conjunction with ongoing research to provide a more efficient methodology for studying ecosystems.

Innis: You seem to have used "useful" in both a technical and a nontechnical sense.

O'Neill: I apologize for any semantic ambiguities in my oral presentation. I hope the ambiguity can be eliminated in the written text.

Innis: The heuristic beauty of your structure is encouraging. The space P and the subsets S, M, and Q are finite, and thus Q is almost certainly void. The establishment of measures on P might also be difficult in practical situations.

O'Neill: The problem with definition of the sets appears to arise when we attempt to define practical measures of r and a. At this point, a finite approximation (number of successful simulations/total simulations) is offered as a measure of the complex volumes. The adequacy of this approximation might well be questionable at the present stage of concept development. Since the sets are now approximated by finite sets, the probability that a single model's output will be identical to a single measurement set approaches zero. However, it still seems possible to define a reasonable criterion for inclusion into Q, such as model output falls within 20% of the measured values.

Patten: I think this paper is significant as a beginning model clarifying the validation problem. The set theoretic presentation is heuristically very useful. I think further clarifications may follow this initial effort. I could only say that time series fits are not the only criterion of validation, but for additional criteria the same general approach to analysis follows upon renaming the P universe. In fact, analysis in $P_1 \times P_2 \times \ldots \times P_n$ can be foreseen.

Editors' Comments
on Papers 26 and 27

26 BOTKIN, JANAK, and WALLIS
Some Ecological Consequences of a Computer Model of Forest Growth

27 LEVINS
The Qualitative Analysis of Partially Specified Systems

NEW APPROACHES

The last two papers are included in this volume because they represent significant new directions being taken in systems ecology. Paper 26 by Botkin et al. considers a model for stand growth dynamics. Individual trees grow, die, and are replaced on a predetermined stand. The result is a mechanistic model of forest growth and succession that has the ability to duplicate actual stand dynamics. The model is unique in that the number of state variables (i.e., individual trees) changes during the simulation as trees die and are replaced.

Paper 27 by Levins considers an original approach to understanding ecological dynamics when only minimal data are available. Ordinarily, an ecological model requires considerable data for parameterization. In the approach taken by Levins, an understanding of the interactions among system components, even without data on actual rates, can be utilized to predict system behavior and stability.

26

Reprinted from *J. Ecol.* **60**:849–872 (1972)

SOME ECOLOGICAL CONSEQUENCES OF A COMPUTER MODEL OF FOREST GROWTH

By DANIEL B. BOTKIN*, JAMES F. JANAK† AND JAMES R. WALLIS†

* School of Forestry and Environmental Sciences, Yale University, New Haven, Connecticut 06511, U.S.A. and
† IBM Thomas J. Watson Research Center, Yorktown Heights, New York 10598, U.S.A.

INTRODUCTION

The complexity of a forest ecosystem makes difficult any attempt to synthesize knowledge about forest dynamics or to perceive the implications of information and assumptions regarding forest growth. Although digital computer simulation seems to offer a potential for creating a complete model of forest growth, little progress has been reported. Computer simulation has been carried out for the growth of trees in even-aged stands of a single species (Mitchell 1969), and for meteorological energy exchange in a forest canopy (Waggoner & Reifsnyder 1968). A specific simulation built directly from Hubbard Brook data has been reported (Siccama *et al.* 1969). Successional change in northern hardwood forests has been predicted from observed birth and death rates (Leak 1970). A conceptual model has been created for the growth of individual tree seedlings from rates of photosynthesis and the distribution of photosynthates (Ledig 1969). Computer simulation has been carried out for some aspects in a few other terrestrial ecosystems, such as productivity in a corn crop (Duncan *et al.* 1967); but apparently no one has successfully reproduced the major characteristics of a mixed-species, mixed-aged forest from a conceptual basis.

A computer simulation of forest growth is now developed that successfully reproduces the population dynamics of the trees in a mixed-species forest of north-east North America. The simulator is designed to be used in the Hubbard Brook Ecosystem Study and to provide output in the same form as the original vegetation survey of that study (Bormann *et al.* 1970). However, the underlying concepts of the simulation are general. The properties of each species are derived from its entire geographic range and in theory any non-hydrophytic species whose relevant characteristics are known can be entered into the simulation. In the present version of the program, the description of the environment is restricted to those features that have been recorded for the Hubbard Brook Forest, but the relative importance attached to each environmental factor has been influenced by the environmental characteristics of the north-eastern United States. It is hoped that a wide dissemination of this simulator will encourage others to test this version with their data and hence lead to later versions of wider usefulness and applicability.

The basic goal was to produce a dynamic model of forest growth, a model in which changes in the state of the forest are a function of the present state and random components. This approach has two advantages over the curve-fitting approach to forest growth: first, the simulator can be regarded as a repository for an integrated knowledge of the ecosystem; second, additional hypotheses can be formulated and tested using Monte Carlo samples of simulator runs and comparing the results with observed data. For

319

example, it would be comparatively simple to operate the simulator to estimate over what range and under what conditions it agrees with the Bartlett Forest birth–death probabilities (Leak 1970).

The simulation was built step by step, beginning with optimum growth for single trees, the effect of less than optimum light and moisture levels on growth, and the allocation of the growth resources among competing trees. The aim was to introduce a minimal number of assumptions and to find the simplest mathematical expression for each factor that was consistent with observation. New factors were introduced only when it was clear that the results of the simulation were not consistent with observation. A primary difficulty has been finding usable data regarding the relationships between tree growth and environmental variables. Where information was lacking, simple yet reasonable relationships were chosen.

In the original vegetation survey of the Hubbard Brook Forest, the species and diameters of all trees with dbh (diameter at breast height) greater than or equal to 2 cm were recorded on each of 208 10 × 10 m plots distributed uniformly over a small watershed. Environmental characteristics recorded for each plot were elevation, aspect, slope, percentage of the plot surface in rock, till depth and an index of soil moisture. The simulator was designed to capitalize on these data, to 'grow' the trees, and allow for manipulation of both stands and site characteristics.

The program is written entirely in FORTRAN IV using only standard library routines and a good uniform random-number generator. A complete listing of the source deck can be obtained from the authors and a flow chart for the main program, called JABOWA, is given in Fig. 1. The program has been successfully operated under the IBM time sharing system (TSS) release 7 and is designed for use with remote terminals and PCS. The remote terminal was an IBM 2741 and PCS is TSS's program checkout system, a command language which allows for interrupting the program during execution, displaying and altering parameter and/or variable values, and dynamically altering the program logic by means of the branch command. A slightly modified version has also been operated successfully under CMS, The Cambridge Monitoring System. Prospective users with similar facilities should have no trouble using the simulator, while those with only batch mode operation at their computer centre should be able to convert the program fairly easily providing that 50K bytes of core storage initialized to zero are available.

It can be seen in Fig. 1 that the innermost loop of JABOWA contains calls to the three work-horse subroutines (GROW, KILL, and BIRTH). The other subroutines of JABOWA are subservient to these three and discussion of these can be found elsewhere (Botkin, Janak & Wallis 1972).

RELATION TO DATA

The Hubbard Brook Forest under investigation contains thirteen tree species: sugar maple (*Acer saccharum**), beech (*Fagus grandifolia*), yellow birch (*Betula alleghaniensis*), white ash (*Fraxinus americana*), mountain maple (*Acer spicatum*), striped maple (*A. pensylvanicum*), pin cherry (*Prunus pensylvanica*), chokecherry (*P. virginiana*), balsam fir (*Abies balsamea*), red spruce (*Picea rubens*), white birch (*Betula papyrifera*), mountain ash (*Sorbus americana*), and red maple (*Acer rubrum*). The present forest, which was cut approximately 60 years ago, is dominated by sugar maple, beech and yellow birch below

* Nomenclature is according to Gleason (1968).

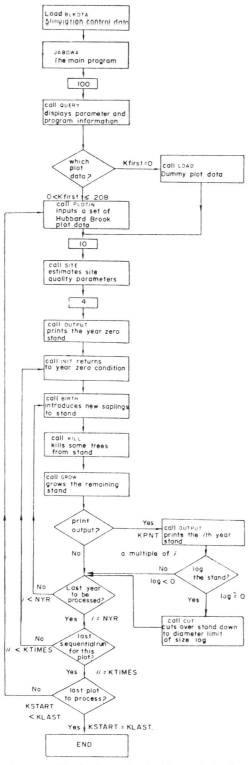

Fig. 1. Flow chart of JABOWA, Version 1 (reprinted with permission from the IBM Journal of Research and Development).

762 m, and red spruce, balsam fir and white birch at higher elevations, although the last three species are also found at lower elevations.

Studies of succession in this area indicate that, in the year following a large clearcutting, pin cherry germinates in great abundance to form dense even-aged stands. Subsequent germination of this species does not occur (Marks 1971). Under some conditions, white or yellow birch may germinate with the pin cherry and continue to enter the stand in subsequent years until light levels become low. It appears that the birches germinate and survive under some shading and can germinate in openings or partial clearings that would not provide sufficient light for the cherry species (Marquis 1969a).

Under heavy shade on the better sites, with the single exception of beech, the existing plot data for the Hubbard Brook Forest do not yield strong correlations between the number of trees, saplings and seedlings of a given species that are present on a plot. At the higher elevations of the Hubbard Brook watershed, the soils tend to be shallower and rockier, and the non-boreal species grow poorly and are subjected to a differentially severe mortality. Observations of the growth of seedlings and saplings of the major shade-tolerant species indicate that under shaded conditions a number on the order of 0–2 individual saplings may be added in any one year to a 10×10 m plot (L. Forcier, personal communication, 1969). The occurrence of saplings of the tolerant species is not clearly related to light intensity at the forest floor, while seed germination, growth and survival of young stems undergo very great yearly variations suggesting cycles of 3- to 5-year duration.

DESCRIPTION OF THE MODEL

In the present, simulator tree species are defined by a few general characteristics: a maximum age; maximum diameter; maximum height; a relation between height and diameter; between total leaf weight and diameter; between rate of photosynthesis and available light; between relative growth and a measure of climate; a range of soil moisture conditions within which the species can grow; and the number of saplings which can enter the stand under shaded, open, or very open conditions. The abiotic environment is defined by an elevation, a soil depth, soil moisture-holding capacity, percentage rock in the soil, a set of average monthly temperatures and precipitation records from a nearby weather station, and a value for the annual insolation above the forest canopy.

Direct competition among individuals is restricted to competition for light (taller trees shade smaller ones, and species with more leaves for a given diameter shade smaller competitors more than other species; under shaded conditions photosynthesis is higher for shade-tolerant species than intolerant ones, and *vice versa*). Species strategy is also invoked by species-specific survival probabilities and by differential addition of new saplings in relation to light at the forest floor. Because the annual probability of survival of an individual is related to the maximum known lifetime of its species, individuals of long-lived species have a better chance of survival in any one year than those with short maximum lives. For each plot year of simulation three major subroutines are called: subroutine GROW, which deterministically provides the annual growth increment for each tree; subroutine BIRTH, which stochastically adds new saplings; and subroutine KILL, which stochastically decides which trees die.

Subroutine GROW

This subroutine uses a tree growth model to augment the dbh's of all trees on a 10×10 m plot by an amount representing 1 year's growth. The model consists of a basic growth rate equation for each species that may be taken to represent the rate of growth of a tree with optimum site quality and no competition from other trees. For each plot-year, this growth rate is decreased by factors taking into account shading and shade tolerance, soil quality, and average climate as measured by the number of growing degree days.

All growth curves tend to be sigmoid and the final growth equation used exhibits this overall property. Some may think that the equations which follow are occasionally based upon rather arbitrary assumptions, but it is probable that there is no unique solution to forest growth simulation, and that many equations based upon different assumptions could yield quite similar results.

A tree growing in the open collects an amount of radiant energy roughly proportional to its leaf area. The JABOWA growth rate equation for a tree growing under optimum conditions has the form:

$$\delta(D^2H) = \text{R} \cdot \text{LA} \cdot \left(1 - \frac{DH}{D_{max}H_{max}}\right) \qquad (1)$$

in which D is the dbh of the tree, H its height, with D_{max} and H_{max} being maximum values of these quantities for a given species, LA is the leaf area, and R is a constant. The equation states that the change in volume (D^2H) of a tree over a period of 1 year is proportional to the amount of sunlight which the tree receives, derated by a factor $(1 - DH/D_{max}H_{max})$ which takes some account of the energy required to maintain the living tissue. The right-hand side of eqn (1) is later multiplied by additional factors to take shading, climate, etc. into account. Values used in JABOWA version 1 for D_{max}, H_{max} and other parameters are given in Table 1.

The height H (in cm) of a tree with dbh D (in cm) is assumed to be given by the following expression (Ker & Smith 1955):

$$H = 137 + b_2 D - b_3 D^2. \qquad (2)$$

The number 137 represents breast height (in cm), and the constants b_2 and b_3 are chosen for each species so that $H = H_{max}$ and $dH/dD = 0$ when $D = D_{max}$. One finds:

$$b_2 = 2(H_{max} - 137)/D_{max}; \quad b_3 = (H_{max} - 137)/D_{max}^2. \qquad (3)$$

Rate of change of height decreases with increasing diameter (eqn 2), and actual change in height becomes negligible for large diameters.

From the above discussion, it can readily be seen that JABOWA currently makes no adjustment for the forester's concept of 'form factor'. For example, it would be comparatively simple to modify eqn (2) of sub-routine GROW to reflect differences in site quality, if it was known how the constants H_{max} and D_{max} varied with measures of growing degree days, and the other requirements for autotrophic plant growth. Field checks would be necessary to quantify such relationships reliably and this was therefore not attempted.

The leaf weight for a tree of species i is taken to be

$$\text{WEIGHT} = C_i D^2 \qquad (4)$$

where C_i is a constant. This equation states that the ratio of leaf weight to stem basal area is a constant from sapling age to death. Data from Hubbard Brook (R. H. Whittaker, personal communication, 1970) and elsewhere (Perry, Sellers & Blanchard 1969; Kittredge 1948; Baskerville 1965) indicate that the actual exponent ranges from 1·5 to about 3. The error associated with these estimates is unknown. Increasing the exponent

Table 1. *Basic parameters*

G = growth constant, C = leaf area constant, AGEMX = maximum age (years); D_{max} = maximum known diameter (cm), H_{max} = maximum known height (cm); b_2 and b_3 are constants in the equation $H = 137 + b_2 D - b_3 D^2$ relating height to diameters; $DEGD_{min}$ and $DEGD_{max}$ are minimum and maximum degree-days; WMIN and WMAX are minimum and maximum values for the index of evapotranspiration (mm of water per year available for evapotranspiration).

	G[5]	C[6]	AGEMX[3]	Type[9]	D_{max}/H_{max}	b_2	b_3	$DEGD_{min}$[4]	$DEGD_{max}$[4]	WMIN	WMAX
Sugar maple	170	1·57[1]	200[3]	2	152·5[2]/4011[3]	50·9	0·167	2000	6300	300	
Beech	150	2·20[1]	300[3]	2	122[3]/3660[3]	57·8	0·237	2100	6000[7]	300	
Yellow birch	100	0·486[1]	300[3]	1	122[3]/3050[3]	47·8	0·196	2000	5300	250	
White ash	130	1·75	100	2	50/2160[3]	80·2	0·802	2100	10700	320	
Mt. maple	150	1·13[1]	25	2	13·5/500	53·8	2·00	2000	6300	320	
Striped maple	150	1·75	30	2	22·5/1000	76·6	1·70	2000	6300	320	
Pin cherry	200	2·45[2]	30	1	28[2]/1126[2]	70·6	1·26	1100	8000	190	
Chokecherry	150	2·45	20	1	10/500	72·6	3·63	600[3]	10000	155	
Balsam fir	200	2·5	80[3]	2	50/1830[3]	67·9	0·679	1100	3700	190	
Spruce	50	2·5	350[3]	2	50/1830[3]	67·9	0·679	600	3700	190	
White birch	140	0·486	80	1	46/1830[3]	73·6	0·800	1100	3700	190	600[8]
Mt. ash	150	1·75	30	2	10/500	72·6	3·63	2000	4000	300	
Red maple	240	1·75	150[3]	2	152·5[3]/3660[3]	46·3	0·152	2000	12400	300	

[0] Values not otherwise referenced were developed during the course of the study.

[1] R. H. Whittaker (1970, personal communication).

[2] Marks (1971).

[3] Harlow & Harrar (1941).

[4] Climatological ranges in growing degree-days, obtained by matching northern and southern limits of range maps (Fowells 1965) to January and July mean world isotherms.

[5] Growth constants adjusted for reasonable growth of individual tree in full sun with climate and soil factors equal to 1 (values of G will give ~2/3 of maximum diameter at 1/2 maximum age starting from an 0·5 cm stem).

[6] Actual leaf area in square metres is ~ $CD^2/15$ for D in cm.

[7] Northern strain.

[8] Calculated for New York City.

[9] Type 1 is shade-intolerant; type 2 is shade-tolerant.

in the leaf weight–diameter relationship (eqn 4) has the effect of steepening this curve for intermediate-aged trees. The exponent could be as small as 1 or as large as 3 without drastically altering the overall shape of the final growth curve.

If it is assumed that leaf area is proportional to leaf weight, and defining $G = RC_i$, eqn (1) can be written in the form:

$$\delta D = \frac{GD\left[1 - DH/D_{max}H_{max}\right]}{\left[274 + 3b_2D - 4b_3D^2\right]}.$$ (5)

A curve of diameter *versus* time resulting from this equation corresponds to an unusually large tree, and reflects the fact that the simulator should be capable of producing trees of any species as large as have ever been observed. Because of the way trees are killed by the simulator, however, the presence of such large trees would be an extremely rare event, and the usual dominant trees produced by the simulator are considerably smaller than the maximum values given in Table 1.

The constant G in eqn (5) sets the initial rate of growth of young trees of species i (the solution $D(t)$ of the equation is asymptotic to D_{max} as $t \to \infty$). Given a maximum observed age for each species, AGEMX, the constant G was arbitrarily chosen so that D/D_{max} was 2/3 for a tree of half the maximum age; this choice of G gives reasonable growth rates for most species (see Appendix).

The growth rate equation actually used in subroutine GROW is obtained by multiplying the right-hand side of eqn (5) by additional factors $r(AL)$, representing the effects of shading, shade tolerance and actual site insolation, T(DEGD), representing climatological effects, and S(BAR), taking some account of soil quality:

$$\delta D = \frac{GD\left[1 - DH/D_{max}H_{max}\right]}{\left[274 + 3b_2D - 4b_3D^2\right]} \cdot r(AL) \cdot T(DEGD) \cdot S(BAR).$$ (6)

These three additional factors are now discussed in detail.

Assuming that a layer of leaves is a uniform absorber of light, it can be shown that the light intensity at a height h is related to the light intensity Q_0 above the top of the canopy by (Kasanaga & Monsi 1954; Loomis, Williams & Duncan 1967; Perry *et al.* 1969)

$$Q(h) = Q_0 \exp{-k \int_h^\infty LA(h')dh'}$$

where LA(h') is the distribution with height of leaf area per unit plot area and k is a constant. In subroutine GROW, this equation is replaced by

$$AL = PHIe^{-k \cdot SLA}$$ (7)

where AL is the available light for a given tree, SLA is the 'shading leaf area', defined as the sum of the leaf areas (obtained from eqn (4)) of all higher trees on the plot (with the heights obtained from eqn (2)), and PHI is the annual insolation in appropriate units. Currently, JABOWA uses a default value of 1 for PHI; a desirable improvement would consist of a subroutine for generating a value of PHI based on latitude and aspect. The constant k in eqn (7) is adjusted for reasonable shading beneath a dense canopy, and it has been found that $k = 1/6000$ (in units reciprocal to those of eqn (4)) gives good results for 10×10 m plots.

Version 1 of JABOWA recognizes two types of trees: shade-tolerant and shade-intolerant. For these two degrees of tolerance, the quantity $r(AL)$ appearing in eqn (6) is

$$r(AL) = \begin{cases} 1 - e^{-4 \cdot 64(AL - 0 \cdot 05)}, \text{ shade-tolerant} \\[2ex] 2 \cdot 24\,(1 - e^{-1 \cdot 136(AL - 0 \cdot 08)}), \text{ shade-intolerant.} \end{cases} \tag{8}$$

In each case, AL is to be obtained from eqn (7). The function r, which may be thought of as a representation of photosynthetic rates, for the two degrees of shade tolerance, contains constants chosen to give reasonable fits to measured photosynthesis curves (Kramer & Kozlowski 1960). Note that the annual insolation PHI can be expressed in any units if appropriate changes are made in the constants appearing in eqn (8).

The function T(DEGD) in eqn (6) represents an attempt to take account of the effect of temperature on photosynthetic rates. It is assumed that each species will have an optimum temperature, and photosynthesis will decrease symmetrically above and below this optimum. A rough index of these thermal effects is obtained from the number of growing degree-days per year (40° F base) for the site. This quantity is defined as the sum of $(T - 40)$ over all days of the year for which the average temperature T exceeds 40 F. Inasmuch as such detailed temperature profiles do not exist for most forest sites, an approximation is used involving only the January and July average temperatures. If one assumes that the annual temperature profile is sinusoidal, it is easy to compute the number of degree-days using the average annual temperature as a base. If this average is not too far from 40° F, a correction to the 40° F base can be made by approximating the temperature curve by straight lines near the average annual temperature. In this way, one obtains the following approximate expression for the number of growing degree-days:

$$\text{DEGD} = \frac{365}{2\pi}(T_{\text{July}} - T_{\text{Jan}}) - \frac{365}{2}\left(40 - \frac{T_{\text{July}} + T_{\text{Jan}}}{2}\right) + \frac{365}{\pi}\frac{\left(40 - \dfrac{T_{\text{July}} + T_{\text{Jan}}}{2}\right)^2}{T_{\text{July}} - T_{\text{Jan}}} \tag{9}$$

in which all temperatures are in degrees Fahrenheit.

For each species, is now set:

$$\text{T(DEGD)} = \frac{4(\text{DEGD} - \text{DEGD}_{\min})(\text{DEGD}_{\max} - \text{DEGD})}{(\text{DEGD}_{\max} - \text{DEGD}_{\min})^2}. \tag{10}$$

This function is a parabola (see Hellmers 1962, p. 284) having the value zero at minimum and maximum values of DEGD, and a value between zero and one for any value of DEGD between the extremes. Values of $\text{DEGD}_{\min}$ and $\text{DEGD}_{\max}$, representing the extremes for which each species will grow, can be obtained reasonably accurately by comparing species range maps (Fowells 1965) to lines of constant DEGD estimated from maps of the January and July world isotherms (Trewartha 1968; U.S. Dep. of Commerce 1968). There are admittedly many micro-environmental effects, such as exposure to wind and available nutrients, that are completely neglected in this approximation; however, the number of growing degree-days proves to be a useful measure of gross thermal effects upon plant growth.

Eqn (6) contains the factor S(BAR), which is simply

$$S(\text{BAR}) = 1 - \frac{\text{BAR}}{\text{SOILQ}} \tag{11}$$

where BAR is the total basal area on the plot. SOILQ is the maximum basal area of a stand of trees under optimum growing conditions on the plot, and the function S(BAR) is a crude expression of the competition for soil moisture and nutrients on the plot.

Subroutine *BIRTH*

In each year, new saplings of each species enter a plot on the basis of their relative tolerance to shade and whether the degree-day and soil moisture conditions allow growth of that species. The model is designed in this regard to mimic the specific behaviour of each species and the process of succession described previously. It assumes a seed source available for each species, but only those species that can grow are added to the stand. The available growing degree-days DEGD at the site is compared with the species vectors of minimum and maximum values of growing degree-days $DEGD_{min}(i)$, $DEGD_{max}(i)$, and a similar comparison of growing season evapotranspiration SOILM is made against the vector of species requirements WMIN(i) to produce a list of allowable species. For all but the birches and cherries a random choice is made from the allowable species list and a random choice of either zero, one, or two new trees of the single selected species are added. The diameter of the added trees depends upon the parameter SIZE (default value 0·5 cm) and a small random addition.

The idea of a random selection of which shade-tolerant tree species to add is not the desperate expedient that it first appears to be. The Hubbard Brook plots show many inexplicable differences in species composition between plots within the elevational bands discussed by Bormann *et al.* (1970). In addition, upon the better sites, with the single exception of beech, the existing plot data do not yield strong correlations between the number of trees, saplings, and seedlings of a given species that are present on a plot. At the higher elevations of the Hubbard Brook watershed, the soils tend to be shallower and rockier, and the non-boreal species grow poorly and are subjected to a differentially severe killing (see discussion of subroutines GROW and KILL).

A range of 0–2 individual saplings is a reasonable rate of annual introduction for the shade-tolerant species on a 10 × 10 m plot (L. Forcier, personal communication, 1969). Seed germination, growth and survival of young stems undergo great yearly variations suggesting cycles of 3- to 5-year duration, and the occurrence of saplings is not clearly related to light intensity at the forest floor; therefore a random introduction approximates current knowledge well. In practice, JABOWA Version 1 produces stands that are similar to those observed at Hubbard Brook, although the above described procedure may produce stands that underemphasize the major species, in particular beech, while over-emphasizing the importance of the minor stand components (e.g. red maple). However, biasing the probabilities for species entry should await extensive testing of the present random choice algorithm. Further field study is also clearly necessary to determine the conditions that promote survival of saplings of the shade-tolerant species.

The four intolerant successional species are handled quite differently. If the total leaf area on a plot (variable WEIGHT) is below a first threshold (variable CHERRY), between 60 and 75 new cherry trees are added to the stand. The number of cherries added by the simulator is far below the dense thickets observed in the field. Cherries are short-lived trees and more realistic modelling may result in an excess of largely unnecessary computation without markedly improving the overall simulation of stand dynamics. In future versions of the program, when the simulation will be used to calculate turnover of water, minerals, and energy, the initial number of cherry stems will be made to agree with observations. As the program now functions, it starts with too few cherry stems and gives them a higher probability of survival than is observed, but the number of cherries 10 years following a clearcut is realistic. If the leaf area is greater than the cherry species cutoff but less than a second threshold value (variable BIRCH) then the two birch species can enter the plot. Between zero and thirteen trees are added as a random choice weighted by stand density, so that the shadier the plot the fewer the trees that are added.

Both the two cherry species and two birch species have overlapping ranges. Choke-cherry has a wider geographical range than pin cherry (Harlow & Harrar 1941), but the latter species is more prevalent near its optimum than pin cherry. Although the two birch species have overlapping ranges, yellow birch goes much farther south than white birch, while white birch survives farther north and on much shallower and rocky sites. In the simulator the choice between cherry or birch species is weighted by site and species constants to reflect the observed ranges.

Subroutine KILL

Few data are available regarding survival rates of trees. Simple assumptions have therefore been made regarding tree death. It is assumed that from sapling age to maturity there are some causes of tree mortality that are age-independent. For actively growing trees, it is assumed that no more than 2% of the saplings of a species should reach the maximum age for that species. This gives a probability, that a tree will die in any year, of

$$p = 1 - (1 - \varepsilon)^n.$$

If $p = 0.98$ when $n = $ AGEMX, the maximum age of the species, and ε is the death probability, then

$$(1 - \varepsilon)^{\text{AGEMX}} = 2 \times 10^{-2}$$

which gives approximately

$$\varepsilon = 4.0/\text{AGEMX}.$$

Trees whose annual increment is below a certain value (0·01 cm in the present version of the program) are subjected to a second death mechanism which assures that such a tree would have only a 1% chance of surviving 10 years in the forest with its annual increment remaining below the minimum (the probability that such a tree will die in any one year is 0·368).

The basic rationale is that a tree which cannot maintain a certain minimum growth rate cannot survive for long in the forest, but that in addition there are numerous other events, such as severe wind, lightning, parasitism, and defoliation, which may result in the death of any tree at any time. While in reality some of these events may be truly random, others are not. The assumption of this model is that even the non-random events have only a certain probability of affecting any one tree and that the sum of the effects of all such events approaches a random probability of killing any tree in any year. That such biotic and abiotic factors affect each species differently is represented by the use of the maximum known age of each species to determine the survival probability for any year.

Subroutine SITE

This subroutine produces indices of the quality of the site for growing trees. It is customary for forest site indices to be based upon the observed height/age relationship of dominant trees of certain key species that hopefully appear on the plot. Such data are time-consuming to accumulate and often difficult to obtain for disturbed stands. JABOWA Version 1 involved a different philosophy; it was desired to index site quality not by what was currently growing on a plot, but by estimates of the exogenous influences, growing degree-days and an index of actual evapotranspiration. This approach also raises interesting questions about the minimum amount of light and moisture that are necessary as well as the optimum levels needed for tree and stand growth. However, in the present version of the program, the soil moisture information is used only to bias the input of new stems.

The growing degree-days parameter, DEGD, has already been described in the discussion of subroutine GROW. To estimate the value of DEGD for an individual site, long-term estimates are needed of the mean January and July temperatures. It is axiomatic that this information is not available for most sites where JABOWA might be used, and so subterfuge may be necessary. Conversion of mean monthly temperatures, BASET, from a nearby U.S. Weather Bureau first order weather station is done by subroutine SITE using an average lapse rate for the difference in elevation between the plot height, IELEV, and that of the base station, BASEH. For the growing season months the lapse rate used is 3·6° F/1000 ft, while the January minimum uses 2·2° F/1000 ft (Sellers 1965). Note that subroutine SITE expects elevation in feet, rainfall in inches per month and temperatures in degrees Fahrenheit; conversion to metric units is carried out in the subroutine. No attempt was made to account for differences in monthly temperatures which result from differing aspects. All of the Hubbard Brook plots have more or less southerly exposure, and differences in mean monthly temperature associated with aspect differences were assumed small; for other areas this may not be a reasonable approximation.

Latitudinal as well as coastal influences on climate are pronounced in New England, and it is hoped that, by using the closest long-term weather station and adjusting for elevation difference, reasonable microclimatic indices can be developed for nearby sites. Elevational transference should be kept less than about 1000 m, as the growing degree-day calculation is sensitive to large elevational changes and is not completely linear.

For deep well-drained forest soils in New England, it is assumed that soil moisture stress is generally not sufficient to restrict stand growth. However, for shallow, rocky soils many species may have difficulty becoming established and, accordingly, an index of actual evapotranspiration, SOILM, an index in millimetres of water per year available for evaporation, was incorporated. The index is developed as a modified Thornthwaite water balance calculation (Sellers 1965). The maximum available moisture storage, STRMAX, is the soil depth, TILL, or 10 m, whichever is the smaller, multiplied by the moisture storage per unit depth of the fine soil fraction, TEXT (Thornthwaite & Mather 1957), discounted by the percentage rock in the soil mantle. The percentage rock in the soil mantle was not estimated for the Hubbard Brook plots, although the percentage of surface area covered by boulders or rock outcrops, IROCK, was observed, and this was used as a substitute.

Monthly precipitation values are needed for the water balance calculation, and lacking other information these are assumed to be the same as for the base station, BASEP. The only other parameter needed for the calculation of SOILM is an estimate of the proportion of the current month's precipitation to be added to SOILM if the potential evapotranspiration is in excess of current storage. The default value for this parameter, EXCESS, is 0·25. No provision was made to adjust EXCESS as a function of STRMAX, although such adjustment may later be found to be necessary.

The initial assumption of no moisture restriction on stand growth for New England forest stands is corroborated by the values of SOILM that were obtained. With the 30-year Woodstock, New Hampshire weather station data, assuming IROCK = 0, and IELEV ≃ 300 m, then TILL has to fall below 0·45 m before SOILM starts to contract. At higher elevations there is even less heat and hence an even smaller requirement for soil moisture storage, so that at 1500 m TILL can fall to 0·33 before SOILM starts to contract.

Something similar to a 'wind-chill factor' may accelerate tree mortality as timber line is approached, but the difficulty of quantifying the wind-chill concept with the available sparse data presented any direct assessment of the chill factor within subroutine SITE. An indirect assessment of the chill factor has probably been built into the simulator in that only those species which can be grown can be added to the stand. (Tests of $DEGD_{min}(i)$ versus DEGD and WMIN(i) versus SOILM are made by subroutine BIRTH.) In particular for shallow, rocky, high elevation sites the value of SOILM falls precipitously; such localities probably form a set that largely intersects the set of wind-chill sites. Later efforts should be directed towards determining whether or not this indirect assessment of the chill factor is indeed adequate.

The parameter SOILQ measures the maximum basal area of a stand of trees under optimum growing conditions for a 10×10 m plot. If rock outcrops reduce the available plot area, the SOILQ should be reduced accordingly. The parameter IROCK is used by subroutine SITE in this capacity; later efforts may show that this approximation needs to be strengthened.

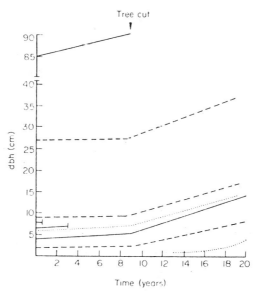

FIG. 2. Simulated diameter growth of individual trees on a single plot. Each line represents a single tree and the end of a line that tree's death. At year 9 the large sugar maple was removed. The remaining trees, no longer suppressed by the large maple, show increased growth rates. ——, Sugar maple; · · · ·, beech; – – –, yellow birch. (Reprinted with permission from the IBM Journal of Research and Development.)

RESULTS AND DISCUSSION

Competition, succession and changes in species composition as a function of time and climate are reproduced successfully by the model. A simulation of competition is illustrated in Fig. 2, where the curves are smoother than in nature because the growth algorithm is deterministic. A single large sugar maple, which grows approximately 0·5 cm in diameter per year, suppresses the growth of smaller trees, none of which grows as much as 2 cm in 10 years. When this large tree is removed growth increases in the suppressed trees. The largest remaining tree, a 28-cm yellow birch, grows 10 cm in the subsequent 10 years. Smaller trees, suppressed only by this yellow birch, show increases in growth rates consistent with the amount of shading each experiences from larger remaining trees. For example, a 9-cm yellow birch, which grew 1 cm before the elimination of the large sugar maple, grows approximately 8 cm by year 20. Species differences in growth and response to competition for light are also indicated. For example, a beech grows more slowly than a sugar maple of similar size.

Simulation of secondary succession at 610 m elevation is illustrated in Fig. 3. This elevation is well within the hardwood forest, and the simulation shows that 15 years after clearcut pin cherry (species 7) provides the most basal area while yellow birch (species 3) is the only other major contributor to basal area of the plot. All the pin cherries disappear by year 30, and afterwards the importance of shade-tolerant species increases at each interval although yellow birch dominates until year 60. These results are consistent with field observations of succession below 760 m elevation in northern New Hampshire (Bormann *et al.* 1970; Siccama 1968).

The model predicts a general increase in density, or stems per unit area (Fig. 4), and a decrease in basal area (Fig. 5) with elevation. Interestingly, it predicts an anomalous point at 762 m with total basal area lowest at this elevation. During succession density fluctuates more at this elevation than at others. The transition from hardwood to conifer forest actually occurs at this elevation in northern New Hampshire and a study also verifies the model's predictions for Vermont (Siccama 1968). It is important to realize that the model is not artificially constrained to produce this anomaly. The model assumes that the same factors which account for the latitudinal distribution of a species also account for its elevational distribution, and the anomaly at 762 m then occurs as a result of competition and species response to growing degree-days, which is non-optimal for all species at this elevation and latitude.

That the degree-day function could account for so much of the known change in vegetation with change in elevation is a surprising result of the model. The degree-day function was introduced with the expectation that other functions, such as a wind-chill factor to represent the desiccating effects of winter winds, would be necessary. Such additional factors may be needed to predict changes in the volume of trees, but they are apparently not necessary for prediction of stem density and basal area.

The behaviour of individual species as a function of time and elevation is also successfully reproduced. The average basal area per plot for 100 identical plots is shown as a function of time for each of the six major species in Fig. 6. This average decreases with elevation at all time intervals for yellow birch (Fig. 6a), beech (Fig. 6b), and sugar maple (Fig. 6c). The average basal area of white birch tends to increase with elevation (Fig. 6d), while the average basal area of red spruce increases with elevation to 914 m, and then remains constant (Fig. 6e). Balsam fir grows best at 914 m but grows better at 1067 m than at 762 m or lower elevations (Fig. 6f). In basal area, birches dominate the early stages of succession: yellow birch at lower and white birch at higher elevations.

The model predicts an old-age forest that is far from uniform (Figs. 7 and 8); its 'climax' state is best viewed as the most probable state at a point in any area which has remained in a constant climate without major catastrophe for a long time. Simulations carried out at 610 m for 2000 years following clearcut result in a rich forest mosaic, generally dominated by beech and sugar maple, but including plots with a wide range of cover. In two different simulations with identical initial conditions, pin cherry occurred on 1% of

Fig. 3. Output of JABOWA, Version 1, as it appears on the user's computer terminal, showing 60 years of secondary succession at 15-year intervals on the test plot (Plot 0). In this case the user specified the plot's altitude and soil depth, the minimum size tree to be cut, the number of years of the simulation and the interval between print-outs. Soil depth is in metres. Index of Actual ET is an index of growing season evapotranspiration in mm. SPEC. is the species number, and refers to the list above see (p. 850). NUM. is the number of trees of each species; BASAR. is the basal area contributed by each species. Numbers under DBH are the diameters of each tree in cm. Leaf area is an index of the total leaf area on the 10 × 10 m plot obtained by dividing total leaf weight of all species on the plot by 45. Dividing this number by 15 gives a rough approximation of leaf area index on the plot. IX is the number initiating the pseudo-random number sequence. Major parameters listed are those which the user will most commonly modify.

IX = 1065786486

PLOT NUMBER	ELEVATION (METRES)	SOIL DEPTH	PERCENT ROCK	GROWING DEGREE DAYS	INDEX OF ACTUAL ET
0	610	10.0	0	2549.4	423.1

YEAR	SPEC.	NUM.	BASAR.	DBH					
0									
	1	4	21.991	2.000	2.000	2.000	4.000		
	2	4	62.832	2.000	2.000	6.000	6.000		
	3	6	130.376	2.000	2.000	8.000	9.000	3.000	2.000
		14	215.200	LEAF AREA = 6.681					

DIAMETER LIMITS FOR LOGGING BY SPECIES

1	2	3	4	5	6	7	8	9	10	11	12	13
0.0	0.0	0.0	0.0	0.0	0.0	0.0	0.0	0.0	0.0	0.0	0.0	0.0

ALL TREES WERE CUT

YEAR 0 NO TREES LIVING

YEAR	SPEC.	NUM.	BASAR.	DBH										
15														
	1	4	7.234	1.890	1.591	1.434	1.024							
	2	3	9.991	2.574	1.873	1.608								
	3	13	139.496	3.722	3.731	3.722	4.417	3.713	3.717	3.814	4.202	3.368	3.889	3.392
				3.022	3.094									
	4	2	1.347	0.970	0.879									
	5	2	0.554	0.590	0.598									
	6	2	1.034	0.843	0.779									
	7	4	283.334	9.625	9.264	9.907	9.172							
		30	442.988	LEAF AREA = 22.638										

YEAR 30

SPEC.	NUM.	BASAR.	DBH
1	5	118.857	6.675, 6.175, 5.894, 5.088, 2.833
2	3	40.943	5.261, 4.860, 0.913
3	13	795.008	9.746, 9.737, 9.717, 10.689, 9.673, 9.799, 9.179, 9.891, 9.206, 8.674, 8.771, 1.611, 1.609
4	3	3.825	1.789, 1.058, 0.744
5	1	1.157	1.213
10	4	4.781	1.564, 1.476, 0.868, 0.842
12	1	0.688	0.936
13	1	1.487	1.376
	31	966.745	

LEAF AREA = 19.433

YEAR 45

SPEC.	NUM.	BASAR.	DBH
1	5	363.035	12.373, 11.789, 10.492, 7.643, 1.298
2	4	142.066	9.172, 8.704, 3.406, 3.066
3	11	1926.324	16.474, 16.476, 16.334, 17.710, 16.210, 16.341, 16.433, 15.584, 15.001, 5.555, 5.547
4	1	5.828	2.724
6	1	2.609	1.822
9	3	74.750	6.713, 6.723, 2.215
10	7	29.539	3.370, 3.493, 2.442, 2.399, 1.192, 0.733, 0.609
13	4	15.453	3.829, 1.846, 0.766, 1.010
	36	2559.603	

LEAF AREA = 60.018

YEAR 60

SPEC.	NUM.	BASAR.	DBH
1	3	604.325	17.833, 17.154, 12.536
2	4	167.922	12.846, 6.395, 2.382, 1.486
3	9	3200.747	22.939, 22.829, 22.485, 24.378, 22.038, 22.166, 21.186, 20.438, 9.194
4	2	10.893	3.604, 0.938
5	1	0.736	0.968
9	3	325.418	17.080, 11.037, 0.887
10	6	31.617	5.329, 2.622, 1.690, 1.134, 0.576, 0.710
12	3	11.352	2.551, 1.994, 1.903
13	3	50.946	6.576, 4.047, 2.291
	34	4403.949	

LEAF AREA = 110.214

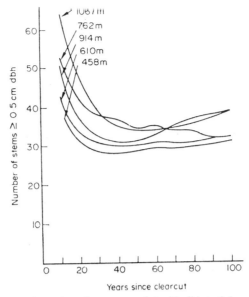

FIG. 4. The average total number of stems per plot with dbh $\geqslant$ 0·5 cm for 100 plots with identical site conditions for a range in elevation from 458 m to 1067 m and a deep, well-drained soil.

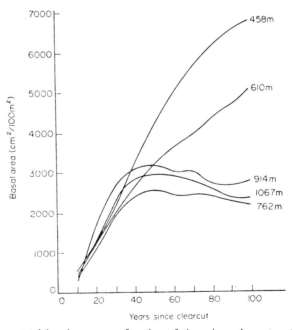

FIG. 5. Average total basal area as a function of time since clearcut and elevation in metres. Each line represents the average for 100 plots at a single elevation with identical site conditions including a deep, well-drained soil.

334

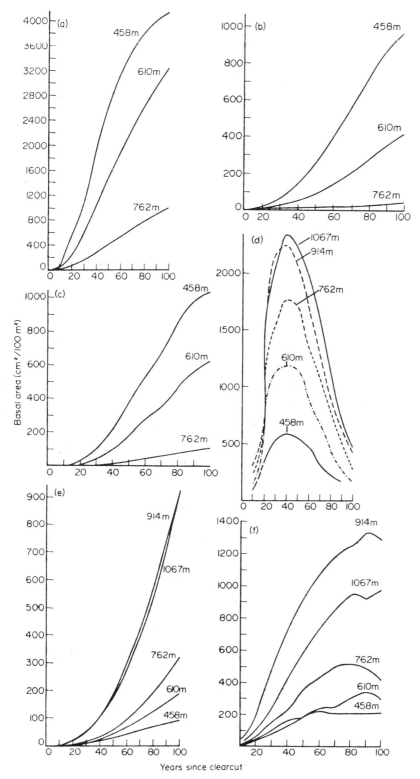

Fig. 6. Average basal area as a function of time since clearcut and elevation for each of the six major species: (a) yellow birch; (b) beech; (c) sugar maple; (d) white birch; (e) red spruce; and (f) balsam fir. Each line represents the average for 100 plots at a single elevation with identical site conditions including a deep, well-drained soil.

335

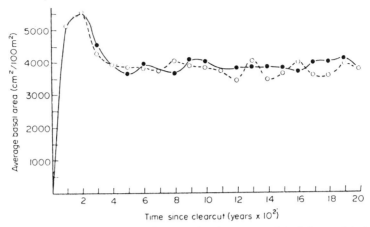

FIG. 7. Two long-term predictions of average basal area per plot of the model at 610 m elevation. Each line represents the average of 100 plots with identical site conditions including a deep, well-drained soil and constant climate, but starting with different pseudo-random numbers.

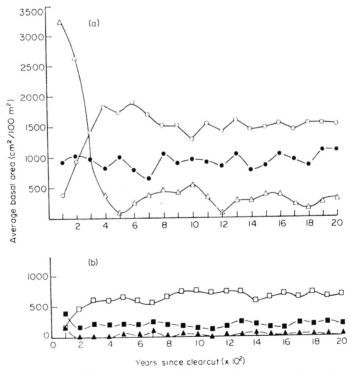

FIG. 8. Long-term predictions of average basal area for six species under constant climate and good, well-drained soil at 610 m elevation for basal area of six major species. (a) Includes those species more typical of the lower elevations: ●, sugar maple; ○, beech; △, yellow birch. (b) Includes those species more typical of higher elevations: ■, balsam fir; □, red spruce; ▲, white birch. Each line represents the average of 100 plots.

the plots after 2000 years. Even in a stable climate openings sufficiently large to permit the growth of pin cherry, solely the result of the 'natural' and nearly simultaneous death of neighbouring large trees, are sufficiently probable to allow a continuing role for pin cherry and other opportunistic species in an old-age forest. Thus, the model suggests a resolution to the long-standing controversy about whether relatively intolerant but

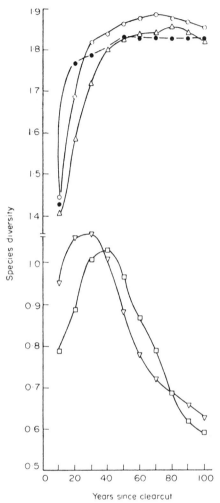

Fig. 9. Species diversity as a function of time since clearcut and elevation. Each line is the mean for 100 plots with identical site conditions including a deep, well-drained soil. ●, 458 m; ○, 610 m; △, 762 m; ▽, 914 m; □, 1067 m.

relatively long-lived species such as yellow birch or white pine could be members of long-undisturbed forests.

The model predicts that an old-age forest will have a smaller standing crop and species diversity* than some earlier stages (Fig. 9). Diversity increases early in succession and reaches a maximum during the first century at all elevations; of course it is much greater below 762 m than above, because sugar maple, beech and yellow birch disappear above

* The Shannon index of species diversity, $H' = \Sigma - p_i \log p_i$, where p_i is the probability of selecting an individual of species i by a random selection process, is used (Pielou 1966).

337

this elevation, but spruce, fir and white birch, which dominate at the higher elevations, also occur at lower ones.

These predictions are consistent with observed patterns in similar hardwood forests of Wisconsin (Loucks 1970), where peak standing crop occurs approximately 200 years after clearcut and both standing crop and species diversity decrease in older stands. Casual observers of forest ecosystems sometimes suggest that a final or climax state would be reached much earlier. However, it is clear from a conceptual point of view that in any ecosystem the effects of the initial conditions on species composition would be present for a period comparable to a small multiple of the lifetime of the longest-lived major species. (In northern New Hampshire, this lifetime is on the order of 400 years.) The model predicts that the final state could not be said to occur prior to year 400 (Fig. 7) and the final order of importance of the major species is not reached until year 600 (Fig. 8). In light of the high probability of a major natural catastrophe during such a long period, and the variability of site conditions over even small areas in New England, the model suggests that the concept of a climax forest as defined earlier in the twentieth century (a uniform pattern of species composition with wide areal and temporal extent) is untenable.

The long-term predictions of the model must be treated cautiously. The simulator currently uses constant climate, and considers all species that can grow at a site as part of the available reservoir for entry into the stand. Presumably during long periods of excessively hot or cold weather certain species would be removed from the active reservoir, and would be reinstated only when the climate returned to a more normal range. That long-term climatic fluctuations probably exist for New England can be deduced from Landsberg's reconstructed 230-year climatic record of Philadelphia (Landsberg, Yu & Huang 1968). It appears that annual growing degree-days could be adequately modelled using a filtered fractional noise (with $h = 0.8$ and $\rho = 0.4$ (Matalas & Wallis 1972)), but it also appears that the model would then have to be modified to allow for species-specific migration in response to long-term climatic fluctuation.

There are three ways to verify models such as this. The first comprises the reactions of experienced field observers to the predictions. The second involves detailed gathering of all information relevant to the model, and the third is a comparison, where possible, with existing records. That the model satisfies the first kind of verification has already been made clear. Observations regarding the second kind have been carried out, but have not yet been analysed and will be reported elsewhere.

Direct quantitative comparison of the average characteristics of real plots in the Hubbard Brook Forest with the simulated plots described here is difficult because the real plots vary considerably in soil depth and elevation and because this forest has a somewhat complicated and poorly documented logging history. The forest was apparently logged twice, once about 1909 when large spruce trees were removed and again about 1916–17 when hardwoods and the remaining spruce were cut. Analysis of growth rings indicates that the loggers left a forest of scattered large hardwoods of beech, birch and maple and smaller hardwood stems. Although the area was probably not completely clearcut, cutting was sufficient at least in some areas to allow establishment of cherry and birch (Bormann *et al.* 1970).

In spite of the many sources of variation, the mean value for the simulated plots for both years 50 and 60 is within one standard deviation of the mean of the observed forest. Furthermore, some of the Hubbard Brook plots show characteristics very similar to those of the optimal plots of the simulator. For example, a real plot at 553 m elevation, till depth of 1·5 m and 1% rock had a total of eighteen stems and total basal area of 3658 cm². Five yellow birch trees contributed 2376 cm², six beech 1005 cm², two spruce 89 cm², two sugar maple 16 cm², with the remaining contributed by understorey species. The average total basal area predicted by the model on 100 plots with identical initial conditions at 610 m is 3250 cm² at year 50 and 3660 cm² at year 60. By species, the predictions at years 50 and 60 are: sugar maple 200 and 300 cm²; beech 100 and 150; yellow birch 1400 and 2000.

It is interesting to compare the predictions of the simulator with present conditions of forests in New England and with the meagre accounts available regarding presettlement forests of this area. In its present version, the model predicts that white birch would be present at elevations below 762 m, during early stages of succession and red spruce would be an important member of old-age stands. Although spruce, fir and white birch did not occur in the Hubbard Brook plots below 709 m, these are found at lower elevations in nearby locations, and were apparently more important in the past. An early report (Chittenden 1905) on conditions of undisturbed forests in Northern New Hampshire stated that white birch was found at all elevations but was much more important at higher ones and that red spruce was 'the characteristic tree' of the northern part of this area. Forests between 550 m and 1070 m were 'characterized by the prevalence of spruce in mixture with balsam and yellow and paper birch'. This report attributed an increase in hardwoods to the effect of lumbering. Thus the long-term predictions of the model are consistent with what is known about the original forest cover of New Hampshire.

The one difference found between the predictions of the model and reports of real forests concerns the distribution of stems by size class in stands 50–60 years old. Three reports, one of an old-age stand in Vermont at 488 m elevation (Bormann & Buell 1964), one of a 25-year-old stand in New Hampshire (Marquis 1969b) and the previously mentioned report of undisturbed forests in New Hampshire (Chittenden 1905) agree on the following: stems of size 2–10 cm are more numerous than stems of 10–20 or greater than 20 cm, and the number of 10–20 cm stems is not greatly different from the number of larger ones.

On the other hand, at year 50 and 60, the model predicts the number of stems 10–20 cm to be 4–5 times the number of larger stems although the predictions for stands older than 70 years agree with observations quoted above. If the few observations available are in fact representative of the New England forests, then perhaps the assumptions regarding mortality rates of trees need modification. Further verification is necessary to test fully this possible deviation, and simulations involving modified mortality functions are planned. The model could be easily modified to make it reproduce more closely the observations of these studies, but this would not constitute verification. To verify predictions of this model, records are needed of the actual growth of trees in forests with known histories and clearly specified environmental conditions, but data of this kind are lacking.

CONCLUSIONS

By defining tree species by nine characteristics, and their abiotic environment by seven, the general dynamic features of a forest ecosystem are reproduced. A number of simplifying assumptions were employed regarding tree growth, and detail was added to the model only as required to account for differences between observation and prediction. The existing discrepancies between the present version of the model and observation seem to relate mainly to the problem of precisely specifying the parameters for the nine species characteristics now in use rather than to the need for entirely new characteristics.

That the general behaviour of an ecosystem as complex as a forest can be reproduced from a few characteristics is itself an interesting and non-obvious result of the simulation. The existing version of the computer program reproduces competition, secondary succession and changes in vegetation that accompany changes in elevation. The existing model appears very flexible in that adjustment of these parameters allows forests to be reproduced with a wide range of characteristics. The model is still experimental and it is hoped that others will use it and help determine its limitations. Detailed field verification of the model is in progress. Further expansion of the model is being made to allow calculation of changes in volume, and therefore biomass and mineral status of the simulated forest. These will require additional species characteristics to relate tree shape to competitive and environmental conditions.

ACKNOWLEDGMENT

This is contribution number 55 in the Hubbard Brook Ecosystem Study. Financial support for this work was partially provided by NSF grant No. 14239.

SUMMARY

Competition, secondary succession, and changes in vegetation accompanying changes in elevation have been successfully reproduced by a computer simulation of a mixed-species, uneven-aged forest ecosystem of north-eastern United States. As much as the data and understanding allow, this simulation has a conceptual basis. Changes in the state of the simulated forest are a function of the present state plus random components. The program predicts a peak standing crop for the forest approximately 200 years following clearcutting, with a relatively stable, but fluctuating, species composition at subsequent periods. Designed to be used in the Hubbard Brook Ecosystem Study, the program provides output in a form compatible with the original vegetation survey of tnat study and allows a flexible interaction between the user and the simulator.

REFERENCES

Baskerville, G. L. (1965). Dry matter production in immature balsam fir stands. *Forest Sci. Monogr.* **9**, 42 pp.

Bormann, F. H. & Buell, M. F. (1964). Old-age stand of hemlock-northern hardwood forest in central Vermont. *Bull. Torrey bot. Club*, **91**, 451–65.

Bormann, F. H., Siccama, T. G., Likens, G. E. & Whittaker, R. H. (1970). The Hubbard Brook ecosystem study: composition and dynamics of the tree stratum. *Ecol. Monogr.* **40**, 373–88.

Botkin, D. B., Janak, J. F. & Wallis, J. R. (1972). The rationale, limitations and assumptions of a northeast forest growth simulator. *IBM J. Res. Development*, **16**, 101–16.

Chittenden, A. K. (1905). Forest conditions of Northern New Hampshire. *Bull. Bur. For. U.S. Dep. Agric.* **55**, 100 pp.

Duncan, W. G., Loomis, R. S., Williams, W. A. & Hanaer, R. (1967). A model for simulating photosynthesis in plant communities. *Hilgardia*, **38**, 181–205.

Fowells, H. A. (1965). *Silvics of Forest Trees of the United States.* Agric. Handbook, **271**. U.S. Government Printing Office, Washington, D.C.

Gleason, H. A. (1968). *The New Britton and Brown Illustrated Flora of the Northeastern United States and Adjacent Canada.* Hafner, New York.

Harlow, W. M. & Harrar, E. S. (1941). *Textbook of Dendrology.* McGraw-Hill, New York.

Hellmers, H. (1962). Temperature effect on optimum tree growth. *Tree Growth* (Ed. by T. T. Kozlowski), pp. 275–87. Ronald Press, New York.

Kasanaga, H. & Monsi, M. (1954). On the light-transmission of leaves, and its meaning for the production of matter in plant communities. *Jap. J. Bot.* **14**, 304–24.

Ker, J. W. & Smith, J. H. G. (1955). Advantages of the parabolic expression of height–diameter relationships. *For. Chron.* **31**, 235–46.

Kittredge, J. (1948). *Forest Influences.* McGraw-Hill, New York.

Kramer, P. J. & Kozlowski, T. T. (1960). *Physiology of Trees.* McGraw-Hill, New York.

Landsberg, H. E., Yu, C. S. & Huang, L. (1968). Preliminary reconstruction of a long time series of climatic data for the Eastern United States. *Univ. Maryland, Inst. Fluid Dyn., Appl. Math. Tech. Note* 13N–571.

Leak, W. B. (1970). Successional change in northern hardwoods predicted by birth and death simulation. *Ecology,* **51**, 794–801.

Ledig, F. T. (1969). A growth model for tree seedlings based upon the rate of photosynthesis and the distribution of photosynthate. *Photosynthetica,* **3**, 263–75.

Loomis, R. S., Williams, W. A. & Duncan, W. G. (1967). Community architecture and the productivity of terrestrial plant communities. *Harvesting the Sun* (Ed. by A. San Pietro, F. A. Greer, & T. J. Army), pp. 291–308. Academic Press, New York.

Loucks, O. L. (1970). Evolution of diversity, efficiency, and community stability. *Am. Zoologist,* **10**, 17–25.

Marks, P. (1971). *The role of* Prunus pensylvanica *L. in the rapid revegetation of disturbed sites.* Ph.D. thesis, Yale University.

Marquis, D. A. (1969a). Silvical requirements for natural birch regeneration. *Birch Symposium Proceedings,* pp. 40–9. Northeastern Forest Experiment Station, Upper Darby, Pennsylvania.

Marquis, D. A. (1969b). *Thinning in young northern hardwoods: 5-year results.* U.S.D.A. Forest Service Res. paper NE-139, Upper Darby, Pennsylvania, 22 pp.

Matalas, N. C. & Wallis, J. R. (1972). Statistical properties of multivariate fractional noise processes. *Wat. Resour. Res.* **7**, 1460–8.

Mitchell, K. J. (1969). Simulation of the growth of even-aged stands of white spruce. *Yale Univ. Sch. For. Bull.* **75**, 48 pp.

Perry, T. O., Sellers, H. E. & Blanchard, C. O. (1969). Estimation of photosynthetically active radiation under a forest canopy with chlorophyll extracts and from basal area measurements. *Ecology,* **50**, 39–44.

Pielou, E. C. (1966). Species-diversity and pattern-diversity in the study of ecological succession. *J Theor. Biol.* **10**, 370–83.

Sellers, W. D. (1965). *Physical Climatology.* University of Chicago Press, Chicago.

Siccama, T. G. (1968). *Altitudinal distribution of forest vegetation in relation to soil and climate on the slopes of the Green Mountains,* Ph.D. thesis, University of Vermont, Burlington, Vermont.

Siccama, T. G., Botkin, D. B., Bormann, F. H. & Likens, G. E. (1969). Computer simulation of a northern hardwood forest. *Bull. ecol. Soc. Am.* **50**, 93.

Thornthwaite, C. W. & Mather, J. R. (1957). Instructions and tables for computing potential evapotranspiration and the water balance. *Publs Clim. Drexel Inst. Technol.* **10**, 180–311.

Trewartha, G. T. (1968). *An Introduction to Climate.* McGraw-Hill, New York.

U.S. Dep. of Commerce (1968). *Climatic Atlas of the United States.*

Waggoner, P. E. & Reifsnyder, W. E. (1968). Simulation of the temperature, humidity and evaporation profiles in a leaf canopy. *J. appl. Meteor.* **7**, 400–9.

(*Received* 24 *January* 1972)

APPENDIX

Integration of growth equation

If, by definition, $x = D/D_m$ and $a = 1 - 137/H_m$, the growth equation, eqn (5), is

$$\frac{dx}{dt} = \frac{G}{2H_m} \frac{x(1-x)(1+ax(1-x))}{[1-a(1-x)(1-2x)]}. \tag{A1}$$

The value of G for each species has been arbitrarily chosen in the text so that x is approximately $2/3$ when $t = \text{AGEMX}/2$. To simplify determination of numerical values of G, and to study other methods of fixing the value of G, it is convenient to have the integral $x(t)$ of this equation. If x_0 is the value of x when $t = 0$, one has from eqn (A1)

$$\frac{Gt}{2H_{\mathrm{m}}} = \int_{x_0}^{x(t)} \frac{dx}{x(1-x)} \left[1 - \frac{a(1-x)^2}{1+ax(1-x)} \right] \tag{A2}$$

or, from tables of integrals,

$$\ln\left(\frac{x}{1-x}\right) + \frac{a}{2}\ln\left(\frac{1+ax-ax^2}{x^2}\right) - \frac{(a+a^2/2)}{\sqrt{a^2+4a}}\ln\left(\frac{2-(\sqrt{a^2+4a}-a)x}{2+(\sqrt{a^2+4a}+a)x}\right) = \frac{Gt}{2H_{\mathrm{m}}} + C \tag{A3}$$

where C is the value of the left-hand side for $x = x_0$. If $x_0 = 1/2D_{\mathrm{m}}$ (i.e., $D(0) = 0.5$ cm), the value of G giving $x = 2/3$ when $t = \mathrm{AGEMX}/2$ is

$$G = \frac{4H_{\mathrm{m}}}{\mathrm{AGEMX}} \left\{ \ln(2(2D_{\mathrm{m}}-1)) + \frac{a}{2}\ln\left(\frac{\frac{9}{4}+\frac{a}{2}}{4D_{\mathrm{m}}^2 + 2aD_{\mathrm{m}} - a}\right) \right.$$
$$\left. - \frac{a+a^2/2}{\sqrt{a^2+4a}}\ln\left[\frac{(3+a-\sqrt{a^2+4a})(4D_{\mathrm{m}}+a+\sqrt{a^2+4a})}{(3+a+\sqrt{a^2+4a})(4D_{\mathrm{m}}+a-\sqrt{a^2+4a})}\right] \right\} \tag{A4}$$

in which D_{m} and H_{m} are in centimetres, and AGEMX is in years.

In some cases, the values of G obtained from eqn (A4) give unreasonable growth rates. This is particularly true of the short-lived species for which eqn (A4) gives growth rates which are too large, and of beech, for which eqn (A4) gives too small a growth rate. The values of G for these species have been adjusted in the simulator to give more reasonable growth rates.

Probably a much better way of determining a value of G for each species lies in demanding that the maximum possible annual diameter increment given by eqn (5) be equal to some value δD_{max}, which could be determined from field observations. One finds that the required value of G is such that

$$\delta D_{\mathrm{max}} \simeq 0.2G\, D_{\mathrm{m}}/H_{\mathrm{m}}. \tag{A5}$$

The value of G for beech used in the simulator corresponds to $\delta D_{\mathrm{max}} = 1.0$ cm, whereas the value of G implied by eqn (A4) for beech would lead to $\delta D_{\mathrm{max}} = 0.7$ cm. The latter value is almost certainly too small; this merely means that the arbitrary assumption that $D/D_{\mathrm{max}} \sim 2/3$ when $t = \mathrm{AGEM\check{X}}/2$ is not correct for all species.

27

Copyright © 1974 by The New York Academy of Sciences
Reprinted from N.Y. Acad. Sci. Ann. **231**:123–138 (1974)

THE QUALITATIVE ANALYSIS OF PARTIALLY
SPECIFIED SYSTEMS

Richard Levins

Department of Biology
University of Chicago
Chicago, Illinois 60637

The most difficult general problem of contemporary science is how to deal with complex systems as wholes. Most of the training of scientists, especially in the United States and Great Britain, is in the opposite direction. We are taught to isolate parts of a problem and to answer the question "What is this system?" by telling what it is made of. The dramatic advances in science in our generation have almost all been in areas where such an approach is practicable. The notable stagnations have been in areas of complex systems approached in pieces.

It is now a commonplace, at least in ecology, that systems are complex and that the one-step linear causality is a poor predictor of ultimate outcome. Consider, for example, the problem of providing more food for hungry people. Since insects destroy a significant portion of the world's crops, and since insecticides can be shown in the laboratory to kill insect pests, it is a plausible inference that the use of insecticides will control insects and increase food available to the hungry. Furthermore, to avoid side effects, laboratory tests may show that insecticides such as heptachlor are relatively nontoxic to mammals. Therefore, it is reasonable to expect that the use of such insecticides would reduce insect pests, increase yields, and alleviate hunger.

But often it does not work that way. First, the application of insecticide does not necessarily control the insect pest, for at least three reasons:

1. Any insect killed by insecticide is that much less food for the predators of the pest. This in itself reduces the predator population, so that the end result is a shift in the cause of death of the pest—more are poisoned, fewer are eaten—but not in the numbers.

2. The insecticide directly reduces the predators of the pest.

3. Natural selection in the target population rapidly builds up resistance to the insecticide. In general, an insecticide is physiologically effective for two to ten years.

The side effects may also behave in unexpected ways. The relatively safe heptachlor may be transformed into highly toxic substances under field conditions where the action of sunlight in the presence of a vast ensemble of organic and inorganic substances promotes reactions that do not occur in the simple laboratory test.

Finally, even the obvious expectation that increased food production alleviates hunger proves false. The whole domain of agricultural economics, grain prices, trade agreements, credits for farmers, land concentration, and speculation intervenes between the harvesting of a crop and its consumption.

Similar problems of complex interactions have arisen in ecology, medicine, economics, administration, and other disciplines. This has led to an interest in complexity *per se* and the exploration of strategies to deal with complexity. Three general approaches have emerged.

1. Statistical-biometrical. Here the system is treated as a black box, and its

behavior is described in terms of the patterns of variation and covariation of the component variables. In conjunction with other approaches its value is in the testing of hypotheses about a system's structure or in posing subproblems. Taken alone, it helps in the prediction of the behavior of very similar systems but not in understanding.

2. In affluent technological societies with a generally reductionist philosophy, the dominant approach is through engineering models. Derived from either the cybernetic school or "systems analysis," these are attempts to measure all the links in a system, write all the equations, measure all the parameters, and either solve the equations analytically or simulate the process on a computer to obtain numerical results.

3. Standing outside of academic science through most of its history, the Marxist tradition has always emphasized complexity itself as an object of study and has stressed interconnection, wholeness, qualitative relations, multiple causality, the unity of structure and process, and the frequently contra intuitive results of contradictory processes. It is the major source of my own research.

The complex networks of biological systems are only partially specified for various reasons:

1. When the number of links is large and each one difficult to measure, the complete description of a given system might be the work of several lifetimes. Thus, while we know that a predator eats its prey, the rate of change of predators as a function of the number of prey depends on its hunting strategy; the effect of hunger on feeding rate; the possibilities of learning to recognize types of prey; food preferences when several prey species are available; reproductive physiology, which establishes the relation between food intake, numbers of offspring, and physical parameters of the environment; and the genetic heterogeneity of predator and prey species with respect to all of these components.

2. While some variables of a system are readily measurable, others are too vaguely defined and yet very real. Thus, in case 3 the levels of blood glucose, insulin, and adrenalin and the kinetic constants describing their rates of breakdown are definable and measurable, but the subjective symptoms of hypoglycemia, adrenalin release, and external stress are not. To insist upon a precise description would therefore mean to exclude these aspects from the system. Yet they are important in its dynamics and often the central issue. We will show later than despite the vagueness of some of the psychological components a qualitative analysis permits a close integration of these with physiology.

3. Often the question at issue is not the interpretation of a particular system but of a class of systems. We may want to know such things as: how many species of birds, more or less as similar to each other, like the warblers of New England, can coexist in a stable community? How does an ecosystem stratified into discrete trophic levels of predators and prey differ in its behavior from one in which a herbivore and its predator have predators in common? How is the stability of a system affected when it grows in the number of variables but not the rules of construction? For problems of this sort, numerical specification is not very helpful.

4. The partial specification may sometimes be closer to the biological reality than a complete one. For instance, it is unlikely that the genotype specifies the complete circuit diagram of the central nervous system. Rather, it seems to specify certain general rules of construction such as connectivity, range of an individual neuron, distribution of thresholds, and ratio of excitatory to inhibitory neurons, for whole regions. We want to know which rules of construction would permit what kinds of behavior.

5. In a deeper sense, no system is ever completely specified. Thus we may observe that a species' rate of increase is reduced when population density increases. This suggests an equation in which the rate of growth, $1/N \, dN/dt$, is a decreasing function of N. But in fact, population size rarely affects the rate of increase directly. There is an indirect pathway, often of many links, from population size to the accumulation of waste products in the medium, to the physiological state of the organism. Between any two variables it is always possible to insert intermediate variables, while each variable itself is the lumping of a heterogeneous ensemble of genotypes, ages, and so on. The qualitative analysis should be capable of indicating when variables may be lumped and when to insert intermediate variables. The network to which we apply our mathematics is already the result of conscious or unconscious qualitative decisions; that is, theoretical interpretation.

A formal mathematical development of the qualitative theory is in preparation for publication elsewhere. Here my purpose is to outline the argument, support it with a heuristic framework, and present the results in a form that can be easily applied.

The first step is to show the partial equivalence of a system of differential equations, a matrix, and a graph. Consider a set of variables x_i and the equations for the rate of change of each:

$$\frac{dx_i}{dt} = f_i(x_1, x_2, x_3 \cdots x_n). \tag{1}$$

At equilibrium, $f_i = 0$ for all i. Then, differentiating (1) with respect to each variable we get

$$\frac{\partial}{\partial x_j} \frac{dx_i}{dt} = \frac{\partial f_i}{\partial x_j} \tag{2}$$

evaluated at equilibrium. If x_j does not appear at all in f_i, this will be zero. These $\partial f_i/\partial x_j$ terms are the elements a_{ij} if the matrix A, an $n \times n$ square array in which $_{ij}$ is the element in the ith row and j^{th} column:

$$A = \begin{matrix} a_{11} \, a_{12} \, a_{13} \cdots a_{1n} \\ a_{21} \, a_{22} \, a_{23} \cdots a_{2n} \\ \vdots \\ a_{n1} \, a_{n2} \, a_{n3} \cdots a_{nn} \end{matrix} \tag{3}$$

The diagonal elements a_{ii} are of special importance. Many biological variables are self-reproducing, and may be represented by equations of the form

$$\frac{dx_i}{dt} = x_i g_i(x_1, x_2, x_h \cdots x_n) \tag{4}$$

in which g_i does not contain x_i explicitly. Then

$$\frac{\partial}{\partial x_i} \frac{dx_i}{dt} = x_i \frac{\partial g_i}{\partial x_i} + g_i \tag{5}$$

But since g_i does not include x_i at all, the first term on the right side is zero, and since we are evaluating these terms at equilibrium, g_i is also zero. Hence for this kind of variable the diagonal element of the matrix,

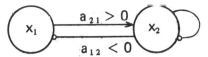

FIGURE 1. The symbols of loop analysis. X_1 and X_2 are variables. The arrow from X_1 to X_2 indicates a positive effect ($a_{21} > 0$), and the line ending in a circle indicates negative effect ($a_{12} < 0$). X_2 also has a negative effect on itself ($a_{22} < 0$).

$$a_{ii} = 0. \tag{6}$$

But when a variable is not self-reproducing, when it is produced or introduced into a system at a rate that does not vanish when the variable itself is absent, while the rate of removal of the variable depends on its own concentration, this is no longer the case. Thus, the concentration of usable phosphate in a lake may follow the equation

$$\frac{dP}{dt} = I - bP \tag{7}$$

where I is the input from streams and b is the rate of phosphate uptake by algae, removal by outflow, or transformation into unusable forms by inorganic processes. Then

$$\frac{\partial}{\partial p} \frac{dP}{dt} = -b \tag{8}$$

and the diagonal element is negative. The same thing happens in a chain of chemical transformations where the rate of formation depends on the concentration of the precursor.

Thus the matrix is derived from the set of differential equations, but represents it only locally, near an equilibrium point.

The next step is to draw the graph of the matrix. Let each variable x_i be represented by a vertex of the graph, and let the line from x_j and x_i be equivalent to a_{ij}. If $a_{ij} = 0$ there is no line from x_j to x_i. Since a_{ij} need not equal a_{ji}, each pair of points may be connected by zero, one or two oriented lines. Further, since we often know only the sign of a_{ij} but not its value, we distinguish positive and negative lines by the symbols in FIGURE 1.

A variable that is self-damped is represented by a loop of length 1 as shown in FIGURE 1.

Corresponding to every square matrix there is also a determinant, which has a numerical value formed by a sum of products of the elements of the matrix. Thus, for example, the 2 × 2 matrix has the determinant

$$\begin{vmatrix} a_{11} & a_{12} \\ a_{21} & a_{22} \end{vmatrix} = a_{11}a_{22} - a_{21}a_{12}. \tag{9}$$

Note that the value of the determinant is equal to the product of the loops of length 1 minus the loop of length 2. This result can be generalized to all determinants: a determinant of order K has the numerical value

$$D^{(k)} = \Sigma(-1)^{k-m}L(m, k) \tag{10}$$

where $L(m, k)$ is a product of m disjunct loops totaling k elements. Since the principal diagonal yields a product of k disjunct loops, its sign is always positive, while a single loop of length k has the coefficient $(-1)^{k-1}$.*

It will be convenient to transform this value into the measure of the feedback of a matrix, which will be defined as

$$F_k = (-1)^{k+1} D^{(k)} \tag{11}$$

or

$$F_k = \Sigma (-1)^{m+1} L(m, k). \tag{12}$$

This measure has the following properties: if in a given product of m loops all loops are negative, then the m minus signs times $(-1)^{m+1}$ contributes a negative term to F_k, and if all loops in a graph are negative F_k is negative. A single positive loop results in positive feedback, two positive loops in a product restore the negative feedback, and so on. Thus, when combining loops into product to get feedback, positive and negative behave opposite to the way they do in ordinary multiplication.

THE LOCAL STABILITY OF SYSTEMS

In order to understand the conditions for local stability we first take a detour to a simple discrete-time, single variable cybernetic mechanism. Suppose that x is a variable that changes at discrete intervals Δt according to the rule

$$x(t + \Delta t) = x(t) + a\Delta t x(t). \tag{13}$$

This system has an equilibrium point at $x = 0$. But if it is not at equilibrium its behavior depends on the value of $a\Delta t$. If $a\Delta t$ is zero (that is, $a = 0$), the system never moves, and it can be described as having neutral or passive equilibrium. If a is positive, we have a system with positive feedback. If $x(t)$ is initially positive it will increase, if initially negative it will decrease, and in both cases it moves away from zero, which is therefore an unstable equilibrium. If $a\Delta t$ is a small negative number (<1), $x(t)$ will approach zero asymptotically from either side. Thus if $a\Delta t = \frac{1}{2}$ and we start at $x(t) = 1$, the successive values are $1, \frac{1}{2}, \frac{1}{4}, \frac{1}{8} \cdots (-\frac{1}{2})^{n-1}$. If the $-a\Delta t = 1\frac{1}{2}$, the successive values starting at $x(+) = 1$ are $1, -\frac{1}{2}, +\frac{1}{4}, -\frac{1}{8} \cdots (-\frac{1}{2})^{n-1}$. As the $-a\Delta t$ term increases beyond 2, we get unstable oscillations: for $-a\Delta t = 3$, the process would go from 1 to $-2, +4, -8, +16 \cdots$. Norbert Weiner therefore defined as a measure of instability the product of the reaction rate, a, times the time lag, Δt.

This simple model could be complicated in many ways. We could make a a function of x, the equation could depend on $x(t - \Delta t)$ and other past values, we could have it a system of many variables. But already the essential features of interest to our argument have appeared:

There are two kinds of instability in systems. When positive feedback predominates, the system moves out from equilibrium in the same direction as its

---

* This relation between determinants and loops has apparently been discovered independently a number of times but does not appear in the generally available literature. While lecturing on these results I was directed to the earlier work of Mason.[5] His work was directed toward developing algorithms for the easier computation of electrical circuits, rather than as an instrument for qualitative understanding. Therefore, even if I had followed the engineering literature I would have missed its significance. Similarly, Sewell Wright used the technique of "path coefficients" to find inbreeding measures.

initial displacement, while if there is excessive negative feedback with long enough time lags there is oscillatory instability.

In systems of differential equations there are no explicit time lags. However, time lags sneak in indirectly through the effects of a variable on itself by way of other variables; that is, through the loops in the system of various lengths.

The analogy to the discrete-time cybernetic unit does not enter into the proof of the conditions for local stability but instead makes them more understandable.

The first necessary condition for stability is that in a system of n variables

$$F_k < 0 \qquad (14)$$

for all $k \leq n$. That is, at each level k negative feedback must outweigh positive feedback. It follows immediately from this that the following systems shown in FIGURE 2 are unstable:

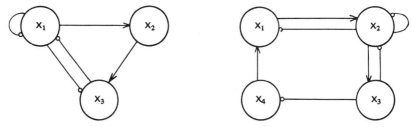

FIGURE 2. Unstable systems with nonnegative feedback. *Left:* the feedback at level 2, F_2, is positive, since it comes entirely from the X_1, X_3 loop. *Right:* $F_3 = 0$. If X_4 were self-damped, instead of X_2, then we would have $F_3 < 0$, since F_3 is a product of disjunct loops of lengths 1 and 2.

Each F_k term is the sum of the feedbacks from all the subsystems of k variables formed by fixing n-k variables at their equilibrium levels and considering only the remaining k as variables. It therefore follows that if $F_k < 0$ at least one subsystem of order k has negative feedback. Another conclusion is that if a system is unstable because feedback is positive at some level k, no changes in loops of length greater than k can stabilize the system.

The second condition for stability, that negative feedback with long-time lags cannot be too large compared to the shorter-loop negative feedback, is more difficult to describe. It is a reinterpretation of the Routh-Hurwitz algorithm (see Gantmacher[1]). This condition is expressed as a sequence of expressions all of which must be positive. For our purposes the first of these is sufficient:

$$F_1 F_2 + F_3 > 0. \qquad (15)$$

If the previous condition is satisfied, F_1, F_2, and F_3 are all negative. Then the condition requires the feedback at level three to be less than the product of the negative feedbacks at the lower levels. But we can carry the analysis one step further: F_3 contains loops of length 3 and also the products of disjunct loops of shorter length. The product $F_1 F_2$ does not involve the loops of length 3. But it includes all products of loops of length 1 with loops of length 2 (disjunct or not) and of length 1 with products of pairs of disjunct loops of length 1. The final result can be expressed in (16), where the subscripts refer to the elements included in a loop and different subscripts are distinct:

$$\Sigma - L_i^2 L_j + L_i L_{ij} + L_{ijk} > 0. \qquad (16)$$

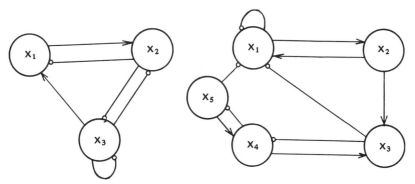

FIGURE 3. Unstable systems with excessive negative feedback. In both cases, the negative loop of length three is not balanced by negative feedback from conjunct loops of lengths 1 and 2.

Thus the following systems are unstable (see FIGURE 3).

There has been much discussion in recent years as to whether complex systems are more stable or less so than simple ones. In 1955, Robert MacArthur proposed that complexity promotes stability, and this generalization has been accepted within ecology and also in discussions of the environmental crisis. However, in 1968 Levins showed that as a community of competitors increases in the number of species it eventually becomes unstable, and this point is reached when the number of species is roughly the reciprocal of the average value $a_{ij}a_{ji}$. Gardner and Ashby[2] used computer simulation to show that if the interaction coefficients a_{ij} are chosen from random distributions, the probability of stability decreases from nearly one to almost zero when n increases through values of roughly $1/\text{var}(a_{ij})$. And Robert May[6] also argued analytically that large systems are usually unstable.

In the argument that follows we use (16) to show what happens as a system increases in size or connectedness. Suppose that a system has n variables all of which are self-damped with $a_{ij} = -1$. Suppose further that the probability of a link between any pair of variables is p, and that if they are connected the magnitude of the interaction averages $-a$. Finally, we assume that all connections are independent of each other. Then the first term in equation 16, $-\Sigma L_i^2 L_j$, is equal to the number of ordered pairs of elements or $n(n-1)$. The second term is $-n(n-1)p^2a^2$, while the number of loops of length three is $\frac{1}{3}n(n-1)(n-2)p^3$ and the whole stability requirement becomes

$$n(n-1) - n(n-1)p^2a^2 - \tfrac{1}{3}n(n-1)(n-2)p^3a^3 > 0. \qquad (17)$$

Since $n(n-1)$ factors out, we are left with

$$1 - p^2a^2 - (n-2)p^3a^3 > 0. \qquad (18)$$

But for n large enough, this will certainly be violated. This is a robust result. If we change the model so that whenever x_i is connected to x_j there is a reciprocal reaction of opposite sign but otherwise links are independent, we get

$$1 + pb^2 - (n-2)p^3a^3 > 0. \qquad (19)$$

The point is, the loop of length three involves three distinct elements and the number of such combinations is of the order n^3, whereas the terms that involve lower order feedbacks have conjunct loops involving only n^2 combinations. Similar stability criteria involving feedbacks of greater length behave in the same way.

Consider now a crude model of the central nervous system in which a region of many neurons consists of excitatory and inhibitory links in such a way that on the average, inhibitory links exceed excitatory links. The self-damping term depends on the physiology of the separate neurons. The interaction intensity $-a$ is a sigmoid function of the general level of sensory input. Then for a given number of neurons in the region, n, and probability of connection, p, there is a threshold input intensity below which the mean firing rate of neurons is stable and above which the system will oscillate. (Wilson and Cowan,[8] using simulation and a different mathematical formulation, showed that intensity of stimulus may be coded by the frequency of oscillation. Their approach is more useful for a detailed investigation of neural nets, while the method described here gives a more intuitive over-all understanding.)

Instead of holding probability p of connection fixed, we define the mean connectivity $c = p(n - 1)$, equation 3 becomes approximately, for large n,

$$n(n - 1) - c^2a^2 - c^3a^3 > 0. \tag{20}$$

If c is held fixed, the system is eventually (for large enough n) stable. But for any given n, a high enough connectivity makes the system unstable.

The general conclusion that systems of many variables with high connectivity are likely to be unstable (and that the higher the connectivity the greater the preponderance of long loops and of very low frequency, long-term oscillations) is rich in implications for a number of fields.

1. Instability must not be confused with lack of persistence. In cellular metabolism and in the central nervous system it is rather interpretable as spontaneous activity. In cells, where the connectivity of biochemical networks is relatively low, this results in periodic behavior, cell cycles. In the central nervous system it is possible that regions of high connectivity yield cycles so long that they are never completed. This simulates randomness and results in a certain unpredictability of neural phenomena.

2. Only where variables are self-reproducing does instability suggest extinction. This relation has made it possible to interpret the number of coexisting species in communities of different structures (Levins and colleagues,[4] Vandermeer[7] . . .).

3. A system involves variables the reactions of which proceed at commensurate rates. Those which are much slower are held as constant parameters, whereas the variables of interest equilibrate; those which are much faster are treated as already at their moving equilibria. They are therefore replaced by functions of the main variables. This sheds a new light on the significance of enzymes. It is not merely that they increase the rates of reactions by many orders of magnitude and make new reactions possible. By so doing, they simplify the biochemical network of cells. The network of enzymatic reactions at high velocity is very much less interconnected than the network of slow interactions among thousands of molecular types. The relative simplicity results in either stable or rapidly oscillating systems upon which natural selection may act. But once the enzymes are killed the slow reactions creep back into the system; the very high connectivity results in nonperiodic instability; dust returns to dust not by assuming the same chemical composition as its surroundings but by having its own kinetics merge in a sea of commensurate reaction rates.

The Response of Equilibrium Level to Parameter Change

A system may be at equilibrium and yet change. If the parameters of a system that appear as constants in the equations in fact change slowly compared to the changes of the variables themselves, we can have a system that follows a moving equilibrium. We therefore want to know what happens to the equilibrium levels in a system when the parameters of that system—temperature, input rate, genetic properties, kinetic constants—are altered.

Consider the system of equations we examined before, but with the parameters explicitly included:

$$\frac{dx_i}{dt} = f_i(x_1, x_2 \cdots x_n; c_1, c_2 \cdots c_m) \tag{21}$$

where the c_n are constant parameters. At equilibrium the right hand side of f_i is zero and we can differentiate the system with respect to a single parameter, say c_h, to obtain the new system of linear equations in $\partial x_j/\partial c_h$:

$$\Sigma \frac{\partial f_i}{\partial x_j} \frac{\partial x_j}{\partial c_h} + \frac{\partial f_i}{\partial c_h} = 0 \tag{22}$$

The $\partial f_i/\partial x_j$ are the familiar a_{ij} of the previous sections. The $\partial x_j/\partial c_h$ are new variables, while $\partial f_i/\partial c_h$ is a constant in each equation. Therefore, this system can be solved directly by determinants. For instance,

$$\frac{\partial x_1}{\partial c_h} = \begin{vmatrix} -\partial f_1/\partial c_h & a_{12}a_{13}\cdots a_{1n} \\ -\partial f_2/\partial c_h & a_{22}a_{23}\cdots a_{2n} \\ \vdots & \vdots \vdots \quad \vdots \\ -\partial f_n/\partial c_h & a_{n2}a_{n3}\cdots a_{nn} \end{vmatrix} \Bigg/ \begin{vmatrix} a_{11}a_{12}\cdots a_{1n} \\ a_{21}a_{22}\cdots a_{2n} \\ \vdots \vdots \quad \vdots \\ a_{n1}a_{n2}\cdots a_{nn} \end{vmatrix} \tag{23}$$

That is, the solution for $\partial x_i/\partial c_h$ is found by substituting the column $-\partial f_i/\partial c_h$ for the ith column of the determinant of the system and dividing this determinant by the determinant of the whole. These results can be reexpressed in terms of feedback and pathways in the system, as follows: Let $p_{ij}^{(k)}$ be a simple open path (one that does not cross itself) from variable j to variable i in the system which includes k elements (and therefore $k - 1$ links). Define the complementary subsystem to $p_{ij}^{(k)}$ as the subsystem formed by only those elements which do not appear in $p_{ij}^{(k)}$. Finally, let $F_{n-k}^{(p-)}$ be the feedback of the complementary subsystem. For formal reasons we must define a subsystem of zero elements as having feedback

$$F_0 = -1 \tag{24}$$

and an open path from an element to itself

$$p_{ii}^{(1)} = 1. \tag{25}$$

Then we can express our final result as follows:

$$\frac{\partial x_i}{\partial c_h} = \sum_{j, k} \frac{\partial f_j}{\partial c_h} p_{ij}^{(k)} F_{n-k}^{(p-)}/F_n. \tag{26}$$

That is, $\partial x_k/\partial c_h$ is the sum of the products of $\partial f_j/\partial c_h$ times the products along each path from x_j to x_i, each multiplied by the feedback of the complementary subsystem and all divided by the feedback of the whole. Of special interest are those cases where c_h is simply the input to variable h. Then only $\partial F_h/\partial c_h$ is different from zero (in fact, equals 1). Since F_n is the feedback of the whole system, which is taken to be stable, it is negative. If the complementary subsystem has zero feedback then a path has no effect. Finally, if the complementary subsystem has net positive feedback, then $\partial x_i / \partial c_h$ has a sign opposite to that of the path.

We illustrate these principles with a few simple examples in FIGURE 4.

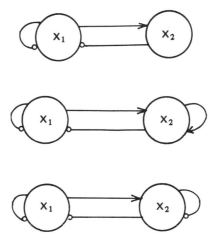

FIGURE 4. Equilibrium changes in two-variable systems. *Top:* X_1 has zero complement, and is therefore unchanged by direct input. *Center:* the complement is positive. Therefore, input to X_1 affects the equilibrium in the direction opposite to its own sign. Only 4c shows the common-sense effect that adding X_1 increases its level.

Here an input to x_1 has as its complement the system $[x_2]$ with zero feedback. Therefore, adding x_1 does not change the level of x_1 at all. But adding x_1 has a positive effect on x_2, which adsorbs the whole increase. An input to x_2 will increase x_2 (the complementary subsystem $[x_1]$ has negative feedback) and reduces x_1. In FIGURE 4 (center), x_2 has positive feedback. Now increasing the input to x_1 actually reduces the level of x_1. This comes about because the increase in x_1 increases x_2, but x_2 has positive feedback (i.e., is autocatalytic) and increases even more. Finally, the greatly increased x_2 reduces x_1. In FIGURE 4 (bottom), x_1 and x_2 are both self-damped.

Therefore, increasing the input to either one increases its own equilibrium level. But if x_1 is increased, then so is x_2, and the two variables will show a positive correlation in time or space, whereas increasing the input to x_2 reduces x_1 and results in a negative correlation. Suppose now that x_2 is sensitive to temperature, whereas x_1 is sensitive to soil pH. Variation in soil pH results in positive correlation between x_1 and x_2, temperature variation in negative correlation, and a combination of the two could give very confusing results.

We now proceed to consider a few semirealistic cases. It must be emphasized that these are not being offered as full models of the situations in question but only as illustrations of how qualitative analysis may be applied.

Case 1. Control of Insect Pests of Crops with Insecticides. Here P_1 is the crop plant that is self-damping because of crowding; H_1 is the herbivore, the pest species that eats the plant; P_a is the specialized parasitoid, usually a wasp, which kills only H_1; P_r is a generalized predatory insect or spider that eats H_1 and also H_2, a herbivore that feeds on other plants. This scheme is shown in FIGURE 5, where I = insecticide.

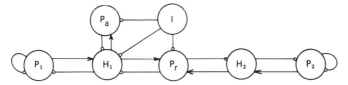

FIGURE 5. A model of a cultivated field community with Plant (P_1), herbivores (H_1 and H_2), predators (P_r and P_2), a parasite (P_A), and insecticide (I).

An increase in the level of insecticide use, I, has a direct negative effect on P_a, P_r, and H_1. But the direct path $[I, H_1]$ has as its complement the subsystem

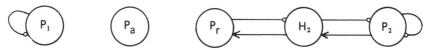

which has zero feedback because the parasitoid is isolated. Therefore this path has no effect. The other two paths, $[I, P_a, H_1]$ and $[I, P_r, H_1]$ are both positive (a product of two negative links) but only the first of these has a nonvanishing complement. Therefore, the final result of adding insecticide at a new, constant dosage is to increase the herbivore species. Any events further along the path (P_r, H_2, P_2) likewise leave H_1 unaltered in numbers. For instance, if H_2 increases this increases P_r, which eats more H_1, reduces the food supply for P_a, and results finally in a shift in the cause of death of H_1 but not its numbers. The only way to affect H_1 is through P_a. Thus, if a species is partly controlled by a specialized predator, its equilibrium level is very stable, and control through its specialized predator is more effective than by way of generalized insectivores or insecticides.

Of course, this argument holds only if the system remains at equilibrium. High enough dosages of insecticide can destroy the equilibrium and wipe out the species.

Case 2. Contamination of a Lake. Here we consider a system with two nutrients that are washed into the lake from outside, nitrate (N) and phosphate (P). We distinguish two kinds of algae: the green algae use phosphate and nitrate and are sensitive to a toxin released by the blue-greens.

The blue-greens release nitrate into the lake but depend on phosphate (FIGURE 6).

If additional nitrate is washed into the lake, the effect on nitrate level depends

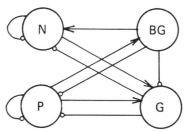

FIGURE 6. Nutrients and organisms in a lake. A two-nutrient, two-alga community, with N = nitrate, P = phosphate, G = green algae and BG = bluegreen algae.

on the feedback of the complementary system P, BG, G. This consists of a single positive loop of length three. Therefore, adding nitrate results in a decrease in the nitrate level, an increase in green algae at the expense of blue-greens, and an unaltered level of phosphate. The addition of phosphate increases the blue-green algae, reduces the green algae, and increases the level of nitrate. The phosphate level itself is unchanged. If we now add a herbivore that can eat green but not blue-green algae, the graph becomes as shown in FIGURE 7.

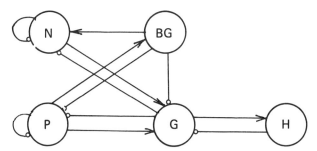

FIGURE 7. Here we have added the herbivore H, which eats only the green algae. (See FIGURE 6.)

Now the complement to N is a system with negative feedback:

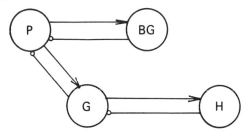

Thus now an increase in nitrate input raises the nitrate level at equilibrium, increases the herbivore population, and has no effect on the other variables. An increase in phosphate input increases nitrate and blue-green levels as before and leaves phosphate unchanged. But instead of decreasing the level of green algae, it reduces the herbivore population and leaves the green algae unchanged. Finally, the addition of toxic substances that kill the herbivores (a negative input to H) will increase green algae, reduce the blue-green and nitrate levels, and leave phosphate unaltered.

These cases are simplified models modified from a large-scale modeling of aquatic systems that I am working on in collaboration with Patricia Lane of the Kellogg Biological Laboratory.

Case 3. This model is not intended as a realistic description of hypoglycemic diabetic mechanisms. Instead, it is intended to illustrate the possibility of integrating physiological and psychological processes. It would be impossible at present if we insisted on full quantification.

In this model we include the variables insulin (I), blood glucose (G), and adrenalin (E), and we lump the subjective states associated with stress, hypoglycemia,

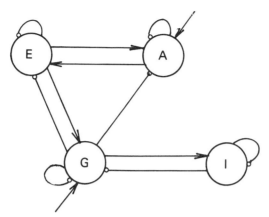

FIGURE 8. A model of part of carbohydrate metabolism. E = epinephrine concentration; G = blood glucose; I = insulin; A = level of anxiety.

and high adrenalin levels into the single variable A. This is not necessary; a more serious analysis would make more distinctions among subjective states.

All of the causal links are the familiar ones. Glucose increases insulin while insulin reduces glucose; the stressed subjective state can come from external stimuli, high levels of adrenalin, or hypoglycemia; stress calls out adrenalin; and all of the variables are self-damped. The self-damping of adrenalin and insulin depends on the rate of metabolic breakdown of these substances and are therefore physiological parameters of the individual. The self-damping of the glucose depends on the equilibrium level of insulin and the effectiveness of the insulin. The self-damping of the subjective state A is more difficult to describe. It is the recovery from anxiety, the "pulling oneself together" that depends on the individual psychology and also can be influenced by tranquilizers and other drugs. The model is shown in FIGURE 8.

First, consider the consequences of increasing the rate of input of glucose either by eating or by increased conversion of glycogen. Then, following the rule of Equation 6, the level of anxiety will decrease (both by the direct path and through E). The level of adrenalin will decrease. But the effect on the level of glucose itself and on insulin will be positive only if the subsystem $[E, A]$ has a net negative feedback. The loop of length two is positive, but may be outweighed by the product of the self-damping terms for E and A. The self-damping term of A is especially variable from person to person. In an anxious person for whom recovery from stress is slow, the self-damping of A may be weak enough so that the E, A subsystem has net positive feedback. Then we have the anomalous physiological results that increasing sugar input reduces the equilibrium level of blood sugar and reduces the insulin level, and that increasing insulin input results in an increase in equilibrium blood sugar and a reduction of insulin. However, the role of insulin in increasing adrenalin and anxiety levels is unaffected.

This particular response, due to the positive feedback of the E, A subsystem, may be reversed if either tranquilizers or changes in psychological mechanisms strengthen the self-damping of A. But it need not be necessarily of psychogenic origin. Any mechanism that reduces the rate of breakdown of adrenalin, increases its release for a given hypothalamic stimulus, or increases the sensitivity of the subjective state to adrenalin would also increase the positive feedback of the subsystem.

But if a reduction in glucose release increases anxiety in a typical hypoglycemic bout, laboratory tests during the anxiety state should show low blood sugar. If the primary event, however, is an increase in external stress, the increase in A results, via adrenalin, in an increase in blood sugar. Now laboratory tests would show a positive correlation between anxiety states and blood sugar. Since external stress and reduced glucose intake may themselves be positively correlated (for instance, coming into the hospital for a fasting glucose test itself causes anxiety), the opposite effects are confounded, resulting in normal or irregular, ambiguous laboratory findings.

If persons with primary pancreatic insufficiency also suffer from slow damping of anxiety, the typical diabetic symptoms may fail to appear; the weakening of the pancreas will be misread by a secular increase in blood insulin and fall in blood sugar, while, on the other hand, pancreatic tumors can result in falling insulin levels, rising blood sugar, and greater anxiety.

So far we have examined changes in the equilibrium level. Two other properties of the system are worth noting: first, the E, A system may be unstable because of positive feedback exceeding negative feedback when the self-damping of A is weak. However, it is embedded in a larger system including blood glucose, insulin, and other variables, and the whole may be stable. What happens is that when the A, E component is disturbed, variations in sugar level are part of the mechanism that reestablishes equilibrium. Therefore, if a major effort to stabilize blood sugar succeeds in preventing variation, blood sugar ceases to be a variable in the system and the A, E subsystem becomes an isolated whole system that is unstable. Hence, fluctuations in blood sugar act as a cushion for variations in psychological stage, and an overzealous stabilization can have serious psychological consequences.

But if the A, E system may have either positive or negative feedback it can also have values close to zero. When it is near neutral (or passive) stability, it itself stabilizes the equilibrium level of blood sugar.

A second problem relates to the overall stability of the whole system. Oscillatory instability may arise if the negative loop of length 3, AG, A, E, is strong enough. Equation 16 states this condition more precisely. From our model, the condition for stability suggests that an unstable, oscillatory system may occur if the emotional state A is especially sensitive to fluctuations in blood sugar or adrenalin, if the self-damping (rate of breakdown) of adrenalin or insulin is slow, or if the response of blood sugar to these substances is weak.

Since the stability condition involves inequality relations among all the parameters, the locus of an abnormal response need not be a single lesion. Instead, diffuse small differences in the constants of the system can result in pathologies that are impossible to locate at a specific point. In the end we will have to recognize "positive feedback at level two" or "excess negative feedback of length three" as the primary pathologies, and develop the techniques for their measurement and management.

Since the last example given refers to a medical condition, it is necessary to repeat that it is not presented as a model of diabetes but rather as an approach to a complex problem that I hope to be able to follow up some day in collaboration with medical researchers.

EVOLUTIONARY PROCESSES

Although systems that represent very different physical realities may be represented by graphs in the same way, they will evolve very differently. In the systems of interacting physiological processes within the organism, natural selection acts only by way of the effect on the whole. Thus, if we know the adaptive significance of the kinetic constants, the effect on the probability of survival and reproduction of anxiety or high blood sugar or slow fluctuations in these, we could identify the direction of its evolution. The significance may be a function of average levels, stability properties, or transient events after perturbation. There is no ground *a priori* to expect selection to favor any particular property. For example, dynamic stability may be either advantageous or harmful compared to low-frequency oscillations.

In ecological communities such as those described in cases 1 and 2, some of the variables are species undergoing selection in terms of their own population dynamics, whereas the kinetic constants of phosphate and nitrate are not under the control of these ions themselves. There will be selection within the planktonic species for those traits which increase the rates of survival and reproduction for their bearers. Whether, however, this increases the population level of the species, reduces it, or leaves it unchanged depends on the complementary subsystem. For instance, in case 1 a higher reproductive rate in the herbivore H_1 will be selected for within H_1, but since its complementary function is zero, the level of H_1 will remain unchanged. On the other hand, a reduced probability of being parasitized is a "kinetic constant" that depends on both the parasitoid and its host, a shared parameter. Since it appears as a negative change in the equation for the rate of growth of the parasitoid population, selection in the host will be represented as a direct positive input to H_1 and also as a negative input to the parasitoid. The path from the parasitoid to its host is also negative, and the complement of this path has ordinary negative feedback. Thus both genes for parasite avoidance and for higher reproductive rate will be selected for, but the former will increase the herbivore population while the latter leaves it unchanged.

Selection within the species is determined by the *relative* fitness of genotypes within the species. But the consequences of that selection for the species or for the whole community depends on the network structure as a whole. Traits that may have equal selective value within the species may appear in quite different ways in the network. There is no necessary relation between selection within a species and the consequences for the whole (or even for that species). In particular, there is no justification for the belief that selection results in greater efficiency or stability or in any other civic virtue.

The problem of environmental management posed in case 1 is a combination of organismic and community types of evolution. While each species is evolving according to its own population dynamics, the agronomist has control of several parameters (represented in the example by insecticide level) that are adjusted with the aim of maximizing some whole-system goal that, depending on the broader context, may be yield, profit, or insecticide sales. In any case, natural selection continues in even the most unnatural situations and may work either in the same direction as artificial selection and environmental management, or to thwart it. It must always be taken into account.

Conclusions

We have argued that complex systems may be studied qualitatively by an examination of network properties of the wholes. This procedure may be the only one available in partially specified systems. It is also advantageous for the understanding of what is taking place even when precise measurement is possible, and indeed helps determine what should be measured. Our emphasis on the holistic and qualitative is intended not as a complete program for science but as a corrective for the one-sided analytical quantitative approach that is still dominant.

REFERENCES

1. GANTMACHER, O. 1959. Matrix Theory. Chelsea Publishing Co. New York, N.Y.
2. GARDNER, M. R. & W. R. ASHBY. 1970. Connectance of large dynamic (cybernetic) systems: critical values for stability. Nature **228:** 784.
3. LEVINS, R. 1968. Evolution in Changing Environments. Princeton Univ. Press. Princeton, N.J.
4. LEVINS, R., M. L. PRESSICK & H. HEATWOLE. 1973. Patterns of coexistence in insular ants. Amer. Scientist. In press.
5. MASON, S. J. 1953. Feedback theory—some properties of signal flow graphs. Proc. Inst. Radio Eng. **41:** 1144–1156.
6. MAY, R. M. 1971. Stability in multispecies community models. Math. Biosciences **12:** 59–79.
7. VANDERMEER, J. 1972. On the covariance of the community matrix. Ecol. **53:** 187–189.
8. WILSON, H. R. & J. D. COWAN. 1972. A Mathematical Theory of the Functional Dynamics of Cortical and Thalamic Nervous Tissue. (Mimeograph)

AUTHOR CITATION INDEX

SUBJECT INDEX

About the Editors

HERMAN HENRY SHUGART received his undergraduate and masters degrees at the University of Arkansas and his doctoral degree at the University of Georgia. His training was in ornithology, vertebrate ecology, and systems ecology. He came to Oak Ridge National Laboratory in 1972 where he has published over fifty reports, articles, and book chapters, mostly concerned with mathematical modeling of ecological systems. The major objectives underlying his broad research interest concern community dynamics and mechanistic modeling of complex ecological systems.

ROBERT V. O'NEILL is a senior researcher in the Environmental Sciences Division at Oak Ridge National Laboratory. He has a broad educational background with an undergraduate degree in philosophy and classical languages, a doctoral degree in physiological ecology from the University of Illinois, and postdoctoral training in systems ecology at Oak Ridge. Since 1968, his modeling experience has ranged from individual invertebrate populations to decomposition and consumer processes to total ecosystem models for forests, lakes, and streams.